Permanent Magnet Synchronous Machines

Permanent Magnet Synchronous Machines

Special Issue Editor

Sandra Eriksson

MDPI • Basel • Beijing • Wuhan • Barcelona • Belgrade

MDPI

Special Issue Editor
Sandra Eriksson
Uppsala University
Sweden

Editorial Office
MDPI
St. Alban-Anlage 66
4052 Basel, Switzerland

This is a reprint of articles from the Special Issue published online in the open access journal *Energies* (ISSN 1996-1073) from 2018 to 2019 (available at: https://www.mdpi.com/journal/energies/special_issues/Permanent_Magnet_Synchronous_Machines).

For citation purposes, cite each article independently as indicated on the article page online and as indicated below:

LastName, A.A.; LastName, B.B.; LastName, C.C. Article Title. *Journal Name* **Year**, *Article Number*, Page Range.

ISBN 978-3-03921-350-4 (Pbk)
ISBN 978-3-03921-351-1 (PDF)

Contents

About the Special Issue Editor vii

Sandra Eriksson
Permanent Magnet Synchronous Machines
Reprinted from: *Energies* **2019**, *12*, 2830, doi:10.3390/en12142830 1

Zih-Cing You and Sheng-Ming Yang
A Restarting Strategy for Back-EMF-Based Sensorless Permanent Magnet Synchronous
Machine Drive
Reprinted from: *Energies* **2019**, *12*, 1818, doi:10.3390/en12091818 6

Camila Paes Salomon, Claudio Ferreira, Wilson Cesar Sant'Ana, Germano Lambert-Torres, Luiz Eduardo Borges da Silva, Erik Leandro Bonaldi, Levy Ely de Lacerda de Oliveira and Bruno Silva Torres
A Study of Fault Diagnosis Based on Electrical Signature Analysis for Synchronous Generators
Predictive Maintenance in Bulk Electric Systems
Reprinted from: *Energies* **2019**, *12*, 2830, doi:10.3390/en12081506 22

Petter Eklund and Sandra Eriksson
The Influence of Permanent Magnet Material Properties on Generator Rotor Design
Reprinted from: *Energies* **2019**, *12*, 1314, doi:10.3390/en12071314 38

Sandra Eriksson
Design of Permanent-Magnet Linear Generators with Constant-Torque-Angle Control for Wave
Power
Reprinted from: *Energies* **2019**, *12*, 1312, doi:10.3390/en12071312 57

Klemen Drobnič, Lovrenc Gašparin, and Rastko Fišer
Fast and Accurate Model of Interior Permanent-Magnet Machine for Dynamic Characterization
Reprinted from: *Energies* **2019**, *12*, 783, doi:10.3390/en12050783 76

Mingcheng Lyu, Gongping Wu, Derong Luo, Fei Rong and Shoudao Huang
Robust Nonlinear Predictive Current Control Techniques for PMSM
Reprinted from: *Energies* **2019**, *12*, 443, doi:10.3390/en12030443 96

Xuliang Yao, Jicheng Zhao, Guangxu Lu, Hao Lin and Jingfang Wang
Commutation Error Compensation Strategy for Sensorless Brushless DC Motors
Reprinted from: *Energies* **2019**, *12*, 203, doi:10.3390/en12020203 115

MeiLing Tang and Shengxian Zhuang
On Speed Control of a Permanent Magnet Synchronous Motor with Current Predictive
Compensation
Reprinted from: *Energies* **2019**, *12*, 65, doi:10.3390/en12010065 137

Alexandra C. Barmpatza and Joya C. Kappatou
Finite Element Method Investigation and Loss Estimation of a Permanent Magnet Synchronous
Generator Feeding a Non-Linear Load
Reprinted from: *Energies* **2018**, *11*, 3404, doi:10.3390/en11123404 152

**Miguel García-Gracia, Ángel Jiménez Romero, Jorge Herrero Ciudad and
Susana Martín Arroyo**
Cogging Torque Reduction Based on a New Pre-Slot Technique for a Small Wind Generator
Reprinted from: *Energies* **2018**, *11*, 3219, doi:10.3390/en11113219 . . . . **171**

Mohamed Dahbi, Said Doubabi and Ahmed Rachid
Current Spikes Minimization Method for Three-Phase Permanent Magnet Brushless DC Motor
with Real-Time Implementation
Reprinted from: *Energies* **2018**, *11*, 3206, doi:10.3390/en11113206 . . . . **186**

Ozgur Ustun, Omer Cihan Kivanc, Seray Senol and Bekir Fincan
On Field Weakening Performance of a Brushless Direct Current Motor with Higher Winding
Inductance: Why Does Design Matter?
Reprinted from: *Energies* **2018**, *11*, 3119, doi:10.3390/en11113119 . . . . **200**

**Damian Caballero, Borja Prieto, Gurutz Artetxe, Ibon Elosegui and
Miguel Martinez-Iturralde**
Node Mapping Criterion for Highly Saturated Interior PMSMs Using Magnetic Reluctance
Network
Reprinted from: *Energies* **2018**, *11*, 2294, doi:10.3390/en11092294 . . . . **217**

Yuqing Yao, Chunhua Liu and Christopher H.T. Lee
Quantitative Comparisons of Six-Phase Outer-Rotor Permanent-Magnet Brushless Machines for
Electric Vehicles
Reprinted from: *Energies* **2018**, *11*, 2141, doi:10.3390/en11082141 . . . . **236**

Jae Suk Lee
Stability Analysis of Deadbeat-Direct Torque and Flux Control for Permanent Magnet
Synchronous Motor Drives with Respect to Parameter Variations
Reprinted from: *Energies* **2018**, *11*, 2027, doi:10.3390/en11082027 . . . . **254**

About the Special Issue Editor

Sandra Eriksson, Associate Professor, finished her MSc in engineering Physics in 2003 and her PhD in engineering science with a specialization in science of electricity in 2008, both at Uppsala University, Sweden. She currently holds a position as associate professor at the Division of Electricity, Uppsala University. Her main topics of interest are design of permanent magnet electrical machines, alternative permanent magnet materials, as well as control strategies and electrical systems for renewable energy systems.

energies

MDPI

Editorial

Permanent Magnet Synchronous Machines

Sandra Eriksson🆔

Department of Engineering Sciences, Uppsala University, Box 534, 751 21 Uppsala, Sweden;
sandra.eriksson@angstrom.uu.se

Received: 2 July 2019; Accepted: 22 July 2019; Published: 23 July 2019

Abstract: Interest in permanent magnet synchronous machines (PMSMs) is continuously increasing worldwide, especially with the increased use of renewable energy and electrification of transports. This special issue contains the successful invited submissions of fifteen papers to a Special Issue of *Energies* on the subject area of "Permanent Magnet Synchronous Machines". The focus is on permanent magnet synchronous machines and the electrical systems they are connected to. The presented work represents a wide range of areas. Studies of control systems, both for permanent magnet synchronous machines and for brushless DC motors, are presented and experimentally verified. Design studies of generators for wind power, wave power and hydro power are presented. Finite element method simulations and analytical design methods are used. The presented studies represent several of the different research fields on permanent magnet machines and electric drives.

Keywords: permanent magnet synchronous generator; permanent magnet synchronous motor; electric propulsion systems; renewable energy; energy conversion

---

1. Introduction

This special issue contains the successful invited submissions [1–15] to a Special Issue of *Energies* on the subject area of "Permanent Magnet Synchronous Machines". Interest in permanent magnet (PM) synchronous machines is continuously increasing worldwide. With a growing global energy demand and awareness of climate aspects, electrification is increasing in several areas. Permanent magnet synchronous generators are in demand for wind power, as well as for novel renewable energy technologies such as wave power and tidal power. Another emerging market for permanent magnet machines is as electric motors, mainly for cars but also for heavier road transport, as well as the electrification of ships and aircraft.

This special issue focuses on PM synchronous machines and the electrical systems they are connected to. PM synchronous machines are a multidisciplinary research topic involving research areas such as electromagnetism, mechanical design, thermal management and material issues, as well as economic and environmental aspects. Both theoretical and experimental work and especially the combination of these are important. Recently, an interest in reducing the use of rare earth metals has been raised, and therefore research exploring substitution and reduction of rare earth metals in PM machines is being performed. Topics of interest for research on PM synchronous machines include, but are not limited to:

- Permanent magnet synchronous machine design;
- Modeling of PM machines;
- Innovative designs of PM machines;
- Drive systems for PM motors;
- Electrical systems and control strategies for PM generators;
- Substitution or reduction of rare earth metals in PM machines;
- Demagnetization risk for PMs in synchronous machines;

- Thermal design and losses;
- Mechanical design;
- PM pilot exciters;
- PM assisted synchronous reluctance machines.

Studies of permanent magnet machines are of more value if they are experimentally verified. However, the effort of experimental verification should not be diminished. Building a prototype and performing measurements is often a time-consuming and difficult task. The majority of the papers presented here include experimental verification of their work.

2. Research Presented in This Special Issue

A subject that is currently of high interest is the electrification of the transport sector. This trend can be seen in this special issue as a majority of the papers deals with PM motors with an emphasis on motor control. Five of the fifteen contributions target renewable energy sources [2–4,9,10], whereas the rest are focused on electric motors, with most examining electric vehicles [1,5–8,11–15]. Three of the papers address brushless DC (BLDC) motors [7,11,12]. Six of the contributions present work on motor control, with the work of all six being experimentally verified [1,6–8,11,15]. One paper is focused on fault diagnosis for hydropower generators [2]. Almost all PM machines have rare earth metal based magnets made of NdFeB. However, alternatives exist and novel magnets have been heavily researched. Almost all the papers in this special issue are based on NdFeB-magnets, as stated by the authors in references [10–13] or as assumed from the contents in references [1,4–9,14,15]. However, paper [3] investigates a span of magnetic properties, which could fit any magnet material. The generators in reference [2] have electromagnets.

The research presented in this special issue has been divided into two parts presented below: contributions on motor control and contributions on machine design and modeling.

2.1. Review of the Contributions on Motor Control

Control of motors commonly requires a speed and position sensor, however, it would be beneficial to not be dependent on a sensor and therefore different types of sensorless control are subject to research. This issue is addressed in reference [1], where a restarting strategy is presented for a back EMF-based sensorless drive for a PM synchronous machine, and the motor voltage is estimated without using the rotor position and speed. The method is experimentally verified and it is concluded that the induced current can be suppressed for various conditions within four to five periods and therefore will not lead to overcurrent fault.

In reference [6] a robust nonlinear predictive current control is presented and compared to conventional predictive current control for a PM synchronous motor. The robust nonlinear predictive current control was experimentally evaluated and compared to the conventional predictive current control under inductance and flux linkage parameter mismatch, and showed superiority on control precision as well as disturbance rejection. Model predictive control is also addressed in reference [8] regarding speed control for a non-salient PM synchronous motor. The developed model predictive control method is compared to traditional control methods both by simulations and experiments. It is concluded that the model predictive control can improve the speed tracking performance of the motor.

Both reference [6] and reference [15] investigate the stability of control methods with regards to machine parameter variations. In reference [15], the stability of deadbeat-direct torque and flux control is investigated with respect to parameter variations for interior PM synchronous motors. The behavior of the controller when the values for inductance and permanent magnet flux linkage varied, was studied using eigenvalue migration as a stability evaluation method. The control method was evaluated both with simulations and experiments, and was compared to current vector control. It was concluded that both control systems show stable operation along a whole driving cycle.

The studies presented in reference [7] and reference [11] both address the problem with current ripple in BLDC motors. In reference [7] a novel commutation error compensation strategy based on the line voltage difference integral is presented for a sensorless BLDC. A PI controller is designed to compensate commutation errors. Experimental verification of the method shows that the current ripple is effectively reduced. In reference [11], the method of placing an R-C filter before the MOSFET transistor gates was suggested to reduce the current spikes in the start-up and during sudden set point changes. The method was evaluated analytically, using both simulations and experiments. The introduction of the R-C filter reduced current ripple at start-up but increased current ripple at steady state operation. Therefore, a relay was introduced so that the R-C filter only was activated if current spikes were detected. Experimental results showed a 13% decrease in the current spike amplitude.

2.2. Review of the Contributions on Machine Design and Modeling

PM machines can have many different machine topologies. Comparison between different topologies is performed through the finite element method (FEM) simulations in reference [3] and reference [14]. In reference [14], four different machine types for six-phase outer rotor machines are compared. The four topologies studied are: V-shaped interior PM machine, Surface mounted PM machine, PM flux-switching machine and Vernier PM machine. The study concludes that the Vernier PM machine has best operating performance and best fault tolerance capability according to the comparison performed in this study. The study in reference [3] focuses on evaluating a novel PM material by studying a span of values for material properties and their influence on rotor design. The three rotor topologies studied are: surface mounted PM rotor, interior spoke type PM rotor and interior PM rotor with radially magnetized magnets. Machines are optimized for torque production for a fixed PM magnetic energy and demagnetization of PMs is considered. It is concluded that the highest torque is reached for surface mounted PMs with high remanence, whereas the reinforcement obtained with the spoke type topology is needed for low remanence materials. The interior radially magnetized PM topology shows the least sensitivity to demagnetization.

Machine design is the subject of reference [4] and reference [5]. The parametrization of an interior PM machine for dynamic characterization is considered in reference [5]. Two nonlinear models for a two-axis representation of an interior PM machine is presented, one based on the current model and one based on the flux-linkage model. The two models are compared with simulations and experiments. A procedure for fast and reliable parameterization for the flux-linkage model was presented. It is concluded that with the suggested parameterization method, the flux-linkage method is preferable, as the execution time of the simulations was up to 20% shorter in comparison to the current model.

In reference [4], a simulation method for initial design of linear PM generators run with constant torque angle control was presented. By considering the control strategy already at the design stage, simulations can be simplified substantially. It was shown that the choice of rated current density and shortening of the end winding length were crucial for the design. The choice of rated current density becomes a trade-off between a compact and less expensive generator and a generator with high efficiency and high maximum damping force.

Design and modeling of PM generators for wind turbines is the subject of [9,10] and reference [3]. The reduction of cogging torque for a surface mounted PM generator is addressed in reference [10]. A pre-slot technique is presented, where a stator slot wedge consisting of both magnetic and non-magnetic material is inserted in the front of each slot. FEM simulations confirm that the cogging torque is reduced by up to 47.8%. In reference [9], performance investigation and loss estimation are performed using FEM simulations for a PM generator connected to a non-linear load. A generator with surface mounted magnets mounted in an asymmetrical way is compared to a conventional symmetric surface mounted PM generator. The machine with asymmetrically placed magnets has less harmonics and a smaller cogging torque. Iron loses are investigated and a loss separation technique is used. The rectifier is run with different duty cycles and for different rotational speeds. A close loop control is also tested with a varying duty cycle. A conclusion from the study is the significance of studying the machine with FEM simulations. In addition, the efficiency is shown to increase when the closed loop control is used. In reference [13], a simulation method using a novel magnetic reluctance network is proposed for time-efficient interior PM machine design. A node mapping criteria is suggested to accurately take magnetic saturation into account. The method allows for geometrical variations such as varying magnet depth and angle between the magnets. The method is validated by FEM simulations for several v-shaped PM machines and with experiments for a prototype machine, showing good correspondence.

Increasing the field weakening performance for a surface mounted permanent magnet BLDC motor by increasing winding inductance is the subject addressed in reference [12]. A motor design with sub-fractional slot-concentrated winding and unequal tooth geometry is presented with an alternate teeth-wound stator and unequal teeth width. A field weakening controller is presented and an experimental study is performed. The design changes increase the field weakening properties of the machine. However, the field weakening is still not satisfactory for use in electric cars and is lower than for traditional interior PM synchronous motors and induction motors.

A study regarding predictive maintenance and fault analysis in synchronous generators is presented in reference [2]. The method presented is based on electric signature analysis used for condition monitoring on generators in bulk electric systems. The proposed method can detect both mechanical and electrical faults. However, two-pole machines provide an extra challenge, as they have the same mechanical and electrical frequency. The method was proven to be useful by testing the method on an in-service hydropower generator connected to the power system. Two types of faults were detected: an early stage of stator short circuit and a mechanical misalignment. The generators in this study have electromagnets. However, the presented method could also be of interest for PM machines.

3. Conclusions

A review of the fifteen papers included in this special issue for "Permanent magnet synchronous machines" has been presented. The research area is of large interest as both renewable energy and the electrification of the transport sector are in focus. The papers have different focus and include studies on design, modeling and control of electrical machines as well as experimental verifications. The papers included in this special issue have contributed to moving the research field of electrical machines and drives forward. The special issue gives an overview of the ongoing research and indicates where the research focus is today within this area.

Acknowledgments: I would like to thank the publisher for inviting me to be the editor for this special issue. I found the edition and selections of papers for this special issue very inspiring and rewarding. I would like to thank the editorial staff and reviewers for their efforts and help during the process.

References

1. You, Z.-C.; Yang, S.-M. A Restarting Strategy for Back-EMF-Based Sensorless Permanent Magnet Synchronous Machine Drive. *Energies* **2019**, *12*, 1818. [CrossRef]
2. Salomon, C.P.; Ferreira, C.; Sant'Ana, W.C.; Lambert-Torres, G.; Borges da Silva, L.E.; Bonaldi, E.L.; de Oliveira, L.E.L.; Torres, B.S. A Study of Fault Diagnosis Based on Electrical Signature Analysis for Synchronous Generators Predictive Maintenance in Bulk Electric Systems. *Energies* **2019**, *12*, 1506. [CrossRef]
3. Eklund, P.; Eriksson, S. The Influence of Permanent Magnet Material Properties on Generator Rotor Design. *Energies* **2019**, *12*, 1314. [CrossRef]
4. Eriksson, S. Design of Permanent-Magnet Linear Generators with Constant-Torque-Angle Control for Wave Power. *Energies* **2019**, *12*, 1312. [CrossRef]
5. Drobnič, K.; Gašparin, L.; Fišer, R. Fast and Accurate Model of Interior Permanent-Magnet Machine for Dynamic Characterization. *Energies* **2019**, *12*, 783. [CrossRef]
6. Lyu, M.; Wu, G.; Luo, D.; Rong, F.; Huang, S. Robust Nonlinear Predictive Current Control Techniques for PMSM. *Energies* **2019**, *12*, 443. [CrossRef]

7. Yao, X.; Zhao, J.; Lu, G.; Lin, H.; Wang, J. Commutation Error Compensation Strategy for Sensorless Brushless DC Motors. *Energies* **2019**, *12*, 203. [CrossRef]

8. Tang, M.; Zhuang, S. On Speed Control of a Permanent Magnet Synchronous Motor with Current Predictive Compensation. *Energies* **2019**, *12*, 65. [CrossRef]

9. Barmpatza, A.C.; Kappatou, J.C. Finite Element Method Investigation and Loss Estimation of a Permanent Magnet Synchronous Generator Feeding a Non-Linear Load. *Energies* **2018**, *11*, 3404. [CrossRef]

10. García-Gracia, M.; Jiménez Romero, Á.; Herrero Ciudad, J.; Martín Arroyo, S. Cogging Torque Reduction Based on a New Pre-Slot Technique for a Small Wind Generator. *Energies* **2018**, *11*, 3219. [CrossRef]

11. Dahbi, M.; Doubabi, S.; Rachid, A. Current Spikes Minimization Method for Three-Phase Permanent Magnet Brushless DC Motor with Real-Time Implementation. *Energies* **2018**, *11*, 3206. [CrossRef]

12. Ustun, O.; Kivanc, O.C.; Senol, S.; Fincan, B. On Field Weakening Performance of a Brushless Direct Current Motor with Higher Winding Inductance: Why Does Design Matter? *Energies* **2018**, *11*, 3119. [CrossRef]

13. Caballero, D.; Prieto, B.; Artetxe, G.; Elosegui, I.; Martinez-Iturralde, M. Node Mapping Criterion for Highly Saturated Interior PMSMs Using Magnetic Reluctance Network. *Energies* **2018**, *11*, 2294. [CrossRef]

14. Yao, Y.; Liu, C.; Lee, C.H. Quantitative Comparisons of Six-Phase Outer-Rotor Permanent-Magnet Brushless Machines for Electric Vehicles. *Energies* **2018**, *11*, 2141. [CrossRef]

15. Lee, J.S. Stability Analysis of Deadbeat-Direct Torque and Flux Control for Permanent Magnet Synchronous Motor Drives with Respect to Parameter Variations. *Energies* **2018**, *11*, 2027. [CrossRef]

energies

MDPI

Article

A Restarting Strategy for Back-EMF-Based Sensorless Permanent Magnet Synchronous Machine Drive

Zih-Cing You [†] and Sheng-Ming Yang [*,†]

Department of Electrical Engineering, National Taipei University of Technology, Taipei 10608, Taiwan;
carefree60024@gmail.com
* Correspondence: smyang@ntut.edu.tw
† This paper is an extended version of our paper published in: "Zih-Cing You and Sheng-Ming Yang. A Control Strategy for Flying-Start of Shaft Sensorless Permanent Magnet Synchronous Machine Drive. International Power Electronics Conference, IPEC 2018 ECCE Asia, Niigata, Japan, May 21–24, pp. 651–656".

Received: 16 April 2019; Accepted: 10 May 2019; Published: 13 May 2019

Abstract: Safely starting a spinning position sensorless controlled permanent magnet synchronous machine is difficult because the current controller does not include information regarding the motor position and speed for suppressing the back-electromotive force (EMF)-induced current. This paper presents a restarting strategy for back-EMF-based sensorless drives. In the proposed strategy, the existing back-EMF and position estimator are used and no additional algorithm or specific voltage vector injection is required. During the restarting period, the current controller is set to a particular state so that the back-EMF estimator can rapidly estimate motor voltage without using rotor position and speed. Then, this voltage is used to decouple the back-EMF of the motor in the current controller in order to suppress the induced current. After the back-EMF is decoupled from the current controller, sensorless control can be restored with the estimated position and speed. The experimental results indicated that the induced current can be suppressed within four to five sampling periods regardless of the spinning conditions. Because of the considerably short time delay, the motor drive can restart safely from various speeds and positions without causing overcurrent fault.

Keywords: permanent magnet synchronous machine (PMSM); flying start; sensorless control

---

1. Introduction

A shaft position sensor is generally used to detect the rotor position for the implementation of vector control in permanent magnet synchronous machine (PMSM) drives. However, such a sensor increases the cost and decreases the reliability of the motor drive. The shaft position sensor can be eliminated by using the machine itself as the position sensor. This technique is commonly called sensorless control. Sensorless control strategies generally belong to two categories: (1) saliency-based strategies and (2) back-electromotive force (EMF)-based strategies. In saliency-based strategies, the position is estimated by demodulating the injection-induced current [1–5]. In back-EMF-based strategies, the position is estimated by tracking the back-EMF of the motor [6–8]. Because these two approaches have complementary speed range limitations, two different sensorless control algorithms are generally combined to achieve a full speed range operation [9–11]. Many studies have reported satisfactory motor drive performance with sensorless control [2,3,9,11,12].

For a sensorless controlled PMSM, a stable startup from zero speed can be achieved by using any practical saliency-based control algorithm. However, starting a spinning sensorless controlled PMSM (known as flying start) is difficult and risky due to the lack of position and speed feedback during the restarting period. Without these feedbacks, the back-EMF of the motor cannot be decoupled from the current controllers. Consequently, the regeneration current is induced as soon as the switches of inverter are turned on. The induced current causes both undesirable motor dynamics and the rapid rise of the DC-link voltage. Moreover, a large induced current may cause a drive overcurrent fault. Therefore, a restarting strategy that can effectively counteract the influences of the back-EMF is essential for the safety and reliability of sensorless PMSM drives.

Several restarting strategies have been reported for PMSMs in recent years. Most of these strategies involve applying zero-voltage vector pulses intermittently in order to identify the initial rotor position and speed, as well as to mitigate the regeneration current [13–18]. In Ref. [16], additional zero voltage vector pulses were applied to reduce the speed estimation error resulting from the limited time interval between two zero voltage vector pulses. To eliminate the influence of motor parameters and speed variations on the estimation performance, an adjustment procedure for the time duration of zero voltage vector pulses was developed in [17] according to the methods described in [14–16]. Although the aforementioned methods can be feasibly implemented on sensorless PMSM drives, the methods are generally complicated, sensitive to speed variations, and increase the computational burden on the controller.

Because motor restarting is generally practiced at medium and high speeds, a restarting strategy for back-EMF-based sensorless PMSM drives is proposed in this paper. The proposed strategy utilizes the existing back-EMF and position estimator. According to the analytical results, the back-EMF estimator can estimate back-EMF accurately during the restarting period even without the position and speed feedback. The estimated back-EMF is then added to the current control loop as the decoupling voltage to suppress the regeneration current. Simultaneously, the rotor position and speed are also estimated by tracking the estimated back-EMF. Consequently, no additional algorithm or specific voltage vector pulses are required to identify the initial rotor position and speed. The proposed strategy is based on the scheme in [19] but with extensive improvements made to the algorithm and the experimental verifications. A supplementary transient current suppression algorithm is developed to suppress the transient current within five sampling period. The experimental results for the motor restarting from various rotor positions and speeds are additionally conducted to verify the feasibility of the proposed restarting strategy.

2. Sensorless Control System

This paper presents a mixed saliency-based and back-EMF-based sensorless control algorithm for PMSM drives. Figure 1 displays the block diagram of the control system. The saliency-based sensorless algorithm estimates the rotor position at zero speed and low speeds through high-frequency (HF) square-wave voltage injection, whereas the back-EMF-based sensorless algorithm estimates the rotor position at intermediate and high speeds. A transition procedure merges the results to estimate the rotor position when the motor is operating in the transition speed region. These aforementioned algorithms are briefly explained in this section.

The stator voltage for the PMSM in the rotor reference frame can be expressed as follows:

$$
\begin{bmatrix} v_{qs}^r \\ v_{ds}^r \end{bmatrix} = \begin{bmatrix} r_s + sL_{qs} & \omega_r L_{ds} \\ -\omega_r L_{qs} & r_s + sL_{ds} \end{bmatrix} \begin{bmatrix} i_{qs}^r \\ i_{ds}^r \end{bmatrix} + \begin{bmatrix} \omega_r \lambda_m \\ 0 \end{bmatrix} \tag{1}
$$

where v_{qs}^r, v_{ds}^r, i_{qs}^r, and i_{ds}^r are the q- and d-axis voltages and currents, respectively; L_{qs} and L_{ds} are the q- and d-axis inductance, respectively; r_s, ω_r, and λ_m are the phase resistance, rotor speed, and magnet flux, respectively; and s is the differential operator.

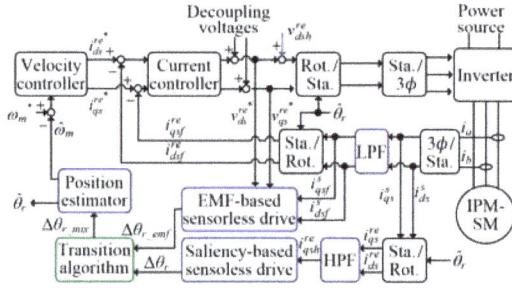

Figure 1. Overall control system for the full-speed region sensorless speed drive.

2.1. Saliency-Based Sensorless Control

As displayed in Figure 1, a square-wave voltage (v_{dsh}^{re}) is injected in the estimated *d*-axis and the saliency spatial signal is extracted from the induced *q*-axis current. The superscript *re* indicates that the quantity is in the estimated rotor reference frame, and v_{inj} denotes the magnitude of the injection voltage. The induced difference currents are given as follows:

$$\begin{bmatrix} \Delta i_{qsh}^{re} \\ \Delta i_{dsh}^{re} \end{bmatrix} = sign(\pm v_{inj}) \cdot \frac{\pm v_{inj} \cdot \Delta T}{(L_\Sigma^2 - L_\Delta^2)} \begin{bmatrix} L_\Delta sin(2\Delta\theta_r) \\ L_\Sigma + L_\Delta cos(2\Delta\theta_r) \end{bmatrix} \tag{2}$$

and

$$L_\Sigma = \left(L_{qs} + L_{ds}\right)/2, L_\Delta = \left(L_{qs} - L_{ds}\right)/2 \tag{3}$$

$$\Delta\theta_r = \theta_r - \hat{\theta}_r \tag{4}$$

where the subscript *h* denotes the HF quantities, θ_r is the rotor position, ΔT is the inverse of injection frequency, and $\hat{\theta}_r$ denotes the estimated rotor position. As indicated in (2), when the estimated rotor frame is not aligned with the actual one, a $2\Delta\theta_r$ position-dependent current signal is generated in both the *d*- and *q*-axis currents. The $\pm$ sign compensation is necessary due to the square-wave voltage injection. Moreover, a high-pass filter is implemented to remove the fundamental component for calculating the difference current. When the position error is sufficiently small, the current signal in the *q*-axis can be rewritten as

$$\Delta i_{qsh}^{re} \approx \frac{v_{inj} \cdot \Delta T \cdot L_\Delta}{(L_\Sigma^2 - L_\Delta^2)} \cdot 2\Delta\theta_r = k_{err} \cdot \Delta\theta_r \tag{5}$$

Thus,

$$\Delta\theta_r \approx \Delta i_{qsh}^{re}/k_{err} \tag{6}$$

The rotor position can be estimated from the measured *q*-axis difference current by using a closed-loop estimator. Note that the injection voltage and frequency is 60 V and 9 kHz, respectively.

2.2. Back-EMF-Based Sensorless Control

The rotor position can be estimated by tracking the extended back-EMF voltage [6]. Equation (1) can be rewritten as:

$$\begin{bmatrix} v_{qs}^s \\ v_{ds}^s \end{bmatrix} = \begin{bmatrix} r_s + L_{ds}s & P\omega_m L_\Delta \\ -P\omega_m L_\Delta & r_s + L_{ds}s \end{bmatrix} \begin{bmatrix} i_{qs}^s \\ i_{ds}^s \end{bmatrix} + \begin{bmatrix} e_{qs} \\ e_{ds} \end{bmatrix} \tag{7}$$

where $e_{qs} = E_b \cdot \cos(\theta_r)$ and $e_{ds} = -E_b \cdot \sin(\theta_r)$ represent the extended back-EMF along the q- and d-axes, respectively; $E_b = L_\Delta (pi^r_{qs} - \omega_r i^r_{ds}) + \omega_r \lambda_m$; and P represents the pole pairs of rotor poles. When the motor parameters are known, the extended back-EMF is calculated as follows:

$$
\begin{bmatrix} \hat{e}_{qs} \\ \hat{e}_{ds} \end{bmatrix} = \hat{E}_b \begin{bmatrix} \cos(\theta_{r\_emf}) \\ -\sin(\theta_{r\_emf}) \end{bmatrix} = \begin{bmatrix} v^{s*}_{qs} \\ v^{s*}_{ds} \end{bmatrix} - \begin{bmatrix} \hat{r}_s + \hat{L}_{ds}s & P\hat{\omega}_m \hat{L}_\Delta \\ -P\hat{\omega}_m \hat{L}_\Delta & \hat{r}_s + \hat{L}_{ds}s \end{bmatrix} \begin{bmatrix} i^s_{qsf} \\ i^s_{dsf} \end{bmatrix}
\tag{8}
$$

where the subscript f is the fundamental frequency components, "*" denotes the command value, "^" denotes the estimated value, ω_m is the mechanical speed., and θ_{r_emf} is the rotor position estimated from the extended back-EMF. A low-pass filter with cutoff frequency of 3 kHz is used to remove the HF current components to avoid HF noise. A position error dependent signal ($\Delta\theta_{emf}$) is then extracted from the following vector product:

$$
\Delta\theta_{emf} = \frac{(-\hat{e}_{qs}\sin\hat{\theta}_r - \hat{e}_{ds}\cos\hat{\theta}_r)\text{sign}(\hat{\omega}_m)}{|\hat{E}_b|} = \text{sign}(\hat{\omega}_m) \cdot \frac{E_b}{|\hat{E}_b|} \cdot \sin(\theta_{r\_emf} - \hat{\theta}_r)
\tag{9}
$$

The estimated back-EMF voltage is normalized with its magnitude. Figure 2 presents the formulas for calculating $\hat{e}_{qs}$, $\hat{e}_{ds}$, and $\Delta\theta_{emf}$. When the position error is sufficiently small, $\Delta\theta_{emf}$ can be approximated as follows:

$$
\Delta\theta_{emf} \approx \text{sign}(\hat{\omega}_m) \cdot (\theta_{r\_emf} - \hat{\theta}_r)
\tag{10}
$$

Thus, the rotor position can be estimated by controlling $\Delta\theta_{emf}$ to zero with a closed-loop estimator.

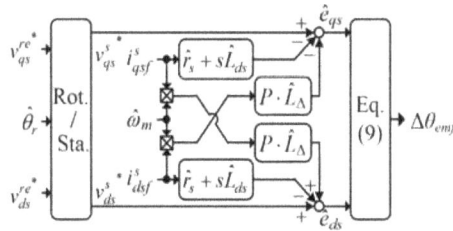

Figure 2. Extended back-EMF estimator.

2.3. Transition Period Control

In the transition speed region, a speed-dependent weighting function combines the position errors generated in (6) and (10) as follows:

$$
\Delta\theta_{r\_mix} \approx (1 - G_\omega) \cdot \Delta\theta_r + G_\omega \cdot \Delta\theta_{r\_emf}
\tag{11}
$$

where G_ω is a linear weighting function with a maximum value of 1 and minimum value of 0. The position estimator illustrated in Figure 3 is used to estimate the rotor position and speed by converging $\Delta\theta_{r\_mix}$ to zero. J denotes the combined rotor and load inertia, B denotes the frictional torque coefficient, and T_e is the motor torque command. The estimator gains k_{ir}, k_{pr}, and k_{dr} are tuned using the pole-placement method to track the actual rotor position and speed with the desired dynamic response.

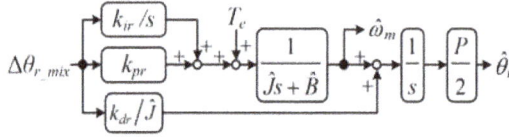

Figure 3. Rotor position and speed estimator.

3. Restarting A Spinning PMSM without Position and Speed Feedback

As displayed in Figure 1, the current controllers require the motor position and speed feedback for vector control and calculating the decoupling voltages. However, when starting a spinning PMSM without position or speed feedback, the current controllers become a stationary frame controller accordingly. Therefore, the current controllers are regulating the AC quantities and their performance is degraded due to the absence of the decoupling voltages and position feedback, as displayed in Figure 4. The voltage commands become stationary frame quantities, and the back-EMF becomes a disturbance to the current controllers. Because the cross-coupling voltages (i.e., $P\omega_m L_\Delta i^s_{qs}$ and $P\omega_m L_\Delta i^s_{ds}$) are generally much smaller than the back-EMF, these voltages are neglected in the following analysis. The current commands are set to zero to obtain zero torque expectantly during restarting period. From Figure 4, the transfer functions between the current and back-EMF can be approximated as

$$\frac{i^s_{qs}}{e_{qs}} \approx -\frac{s}{L_{ds}s^2 + \left(r_s + k_{pq}\right)s + k_{iq}} \tag{12}$$

$$\frac{i^s_{ds}}{e_{ds}} \approx -\frac{s}{L_{ds}s^2 + \left(r_s + k_{pd}\right)s + k_{id}} \tag{13}$$

where k_{pq}, k_{iq} and k_{pd}, k_{id} are the proportional and integral gain for the q- and d-axis current controller, respectively.

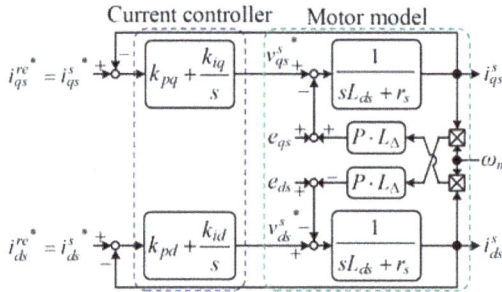

Figure 4. Equivalent stationary frame current control loop.

Figure 5 shows the stator current responses at a set current controller bandwidth (BW) of 500 Hz and 1 kHz. The PMSM parameters used for the calculations are presented in Appendix A. The amplitude of the induced current is highly dependent on the rotor speed and current controller BW. A higher rotor speed and lower controller BW yield a larger current. Moreover, the amplitude of the induced current at the rated speed is approximately 2.5 times the rated current at the current controller BW of 1 kHz. Consequently, the induced current may cause overcurrent fault and bring the motor back to the coasting state.

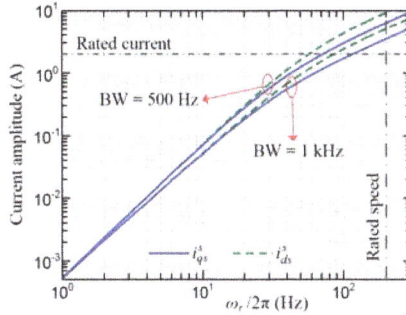

Figure 5. Frequency response of the induced current.

Solving (12) and (13) through the inverse Laplace transform can yield the steady-state induced currents as follows:

$$i_{qs}^s(t) = 2(E_b/Z_{eq}) \cdot \cos(\omega_r t + \varphi_q) \tag{14}$$

$$i_{ds}^s(t) = -2(E_b/Z_{ed}) \cdot \sin(\omega_r t + \varphi_d) \tag{15}$$

where Z_{eq} and Z_{ed} are the equivalent impedances and φ_q and φ_d are the equivalent phases. The equivalent impedances and phases are given as follows:

$$\begin{bmatrix} Z_{eq} \\ Z_{ed} \end{bmatrix} = \begin{bmatrix} \sqrt{4\omega_r^2(r_s + k_{pq})^2 + 4(k_{iq} - L_{ds}\omega_r^2)^2} / \omega_r \\ \sqrt{4(L_{ds}\omega_r^2 - k_{id})^2 + 4\omega_r^2(r_s + k_{pd})^2} / \omega_r \end{bmatrix} \tag{16}$$

$$\begin{bmatrix} \varphi_q \\ \varphi_d \end{bmatrix} = \begin{bmatrix} \sin^{-1}\left(-2(k_{iq} - L_{ds}\omega_r) \cdot Z_{eq}\right) \\ -90° - \sin^{-1}\left(-2\omega_r(r_s + k_{pd}) \cdot Z_{ed}\right) \end{bmatrix} \tag{17}$$

The braking torque (T_{eb}) produced by the induced current can be calculated by substituting (14) and (15) into the following expression:

$$T_{eb} = \frac{3}{4}P\lambda_m \cdot \left(\cos\theta_r \cdot i_{qs}^s - \sin\theta_r \cdot i_{ds}^s\right) - \frac{3}{2}PL_\Delta \cdot \left(\sin\theta_r \cdot i_{qs}^s + \cos\theta_r \cdot i_{ds}^s\right)\left(\cos\theta_r \cdot i_{qs}^s - \sin\theta_r \cdot i_{ds}^s\right) \tag{18}$$

Figure 6 illustrates the values of T_{eb} for various speeds. Both average and pulsating braking torques exist when the motor is restarted at high speeds. A large average braking torque can cause the motor to brake unexpectedly. This torque also represents the regenerative power generated by the motor. A large regenerative power may cause a rapid rise in the DC-link voltage and potential damage to the front-end power supply.

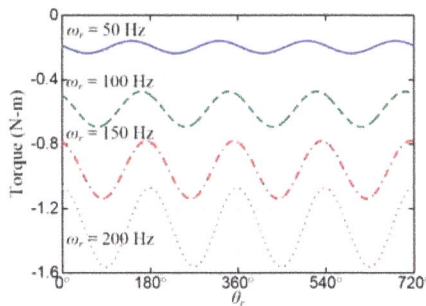

Figure 6. Braking torque produced by the induced current at different rotor speed.

4. Proposed Restarting Strategy

As the potential danger mentioned in previous section, the back-EMF-induced current must be suppressed to safely start a spinning PMSM. This can be achieved through the accurate and prompt decoupling of the back EMF as soon as the current controller is activated. The strategy presented in the following text uses the estimators displayed in Figures 2 and 3 extensively.

4.1. Back-EMF Estimation without Speed Feedback

The relationship between the estimated and actual back EMF can be derived by substituting (7) into (8), which yields the following equations:

$$\hat{e}_{qs} = G_1 \cdot e_{qs} + v_{qs}^{s*} - G_1 \cdot v_{qs}^s + P\left(G_1 \omega_m L_\Delta - \hat{\omega}_m \hat{L}_\Delta\right) i_{ds}^s \tag{19}$$

$$\hat{e}_{ds} = G_1 \cdot e_{ds} + v_{ds}^{s*} - G_1 \cdot v_{ds}^s + P\left(\hat{\omega}_m \hat{L}_\Delta - G_1 \omega_m L_\Delta\right) i_{qs}^s \tag{20}$$

where $G_1 = (s\hat{L}_{ds} + \hat{r}_s)/(sL_{ds} + r_s)$. All the currents are fundamental components, and the subscript f is neglected for convenience. The mismatch in parameters such as L_{ds} and r_s causes amplitude and phase errors between the estimated and actual back-EMF. Moreover, inverter nonlinearity, such as the dead-time effect, also results in errors in the estimated back-EMF. To mitigate these errors, these motor parameters are measured with reasonable accuracy and the dead-time effect is compensated [20–22]. Consequently, $G_1 \approx 1$, $v_{qs}^{s*} - G_1 v_{qs}^s \approx 0$, and $v_{ds}^{s*} - G_1 v_{ds}^s \approx 0$. Thus, (19) and (20) are simplified to the following equations:

$$\hat{e}_{qs} \approx e_{qs} + P\left(\omega_m L_\Delta - \hat{\omega}_m \hat{L}_\Delta\right) i_{ds}^s \tag{21}$$

$$\hat{e}_{ds} \approx e_{ds} + P\left(\hat{\omega}_m \hat{L}_\Delta - \omega_m L_\Delta\right) i_{qs}^s \tag{22}$$

Because the rotor speed is not yet identified, the estimated speed is set to zero. Therefore, (21) and (22) are transformed into the following equations:

$$\hat{e}_{qs} \approx e_{qs} + P\omega_m L_\Delta i_{ds}^s \tag{23}$$

$$\hat{e}_{ds} \approx e_{ds} + P\omega_m L_\Delta i_{qs}^s \tag{24}$$

The stator currents in the aforementioned equations can be eliminated by combining (12) and (13) and (23) and (24). Moreover, note that $e_{qs} = -s \cdot e_{ds}/\omega_r$ and $e_{ds} = s \cdot e_{qs}/\omega_r$. Then, the estimated back EMF can be expressed as

$$\frac{\hat{e}_{qs}}{e_{qs}} = \frac{\left(2 - L_{qs}/L_{ds}\right) s^2 + \left(r_s + k_{pq}\right) s/L_{ds} + k_{iq}/L_{ds}}{s^2 + \left(r_s + k_{pq}\right) s/L_{ds} + k_{iq}/L_{ds}} \tag{25}$$

$$\frac{\hat{e}_{ds}}{e_{ds}} = \frac{\left(2 - L_{qs}/L_{ds}\right) s^2 + \left(r_s + k_{pd}\right) s/L_{ds} + k_{id}/L_{ds}}{s^2 + \left(r_s + k_{pd}\right) s/L_{ds} + k_{id}/L_{ds}} \tag{26}$$

Figure 7 depicts the frequency responses of the aforementioned functions at various saliency ratios. The current controller BW is set to 1 kHz. The estimated back-EMF approaches the actual back EMF at low speeds. However, significant errors exist at high speeds for machines with a high saliency ratio. The saliency ratio of most of the PMSMs available on the market is less than 2. In addition, the BW of the current controller usually is set as high as possible. Therefore, the amplitude and phase errors for PMSMs are acceptable at speeds lower than the BW of the current controller.

Most importantly, the above analysis indicates that the back EMF at the restarting period can be estimated with reasonable accuracy without position or speed feedback by using the estimator presented in Figure 2.

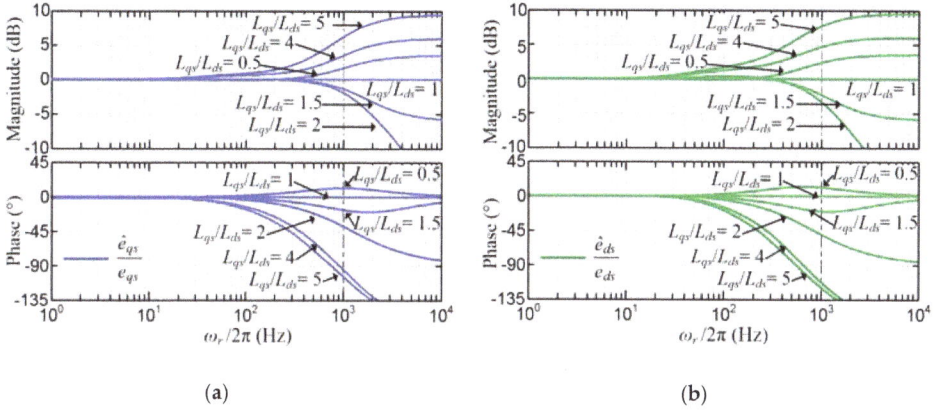

Figure 7. Frequency response of (a) $\hat{e}_{qs}/e_{qs}$; (b) $\hat{e}_{ds}/e_{ds}$.

4.2. Back-EMF Decoupling

Figure 8 illustrates the current controller and back-EMF estimator when the restarting procedure is implemented. Note that the estimated speed is set to zero because it is not identified yet. The back EMF of the motor is estimated as soon as the inverter and current controller are activated, and the estimated back-EMF is applied to decouple the actual back-EMF. Because the back-EMF estimator is effectively in the stationary frame during the restarting period, and most importantly, contains no integrator, the settling time of the back-EMF estimator is inherently zero. Therefore, the back-EMF can be estimated within one sampling period. Then, the inverter outputs the estimated back-EMF to the motor within two sampling periods after the drive is activated. Because the back-EMF can be decoupled within a very short time, the induced current can be suppressed promptly.

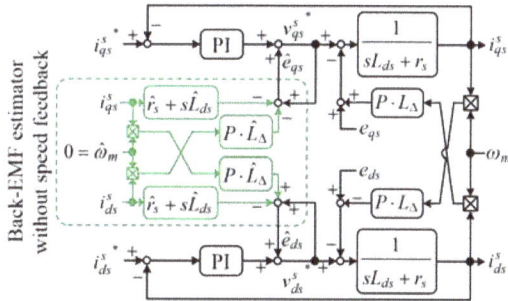

Figure 8. Current controller with the decoupling voltage from the back-EMF estimator during restarting period.

4.3. Transient Current Suppression

Because the current controller initiates closed-loop control as soon as the restarting procedure begins, the initial values in the integrators should be set accordingly. Otherwise, the stator current may require a relatively long time to reach the steady state even if the back-EMF is decoupled. From Equation (7), both the resistive voltage drop and cross-coupling voltage are neglected because they have minor contributions to the current response. The initial motor current can be approximated as follows:

$$L_{ds}s \begin{bmatrix} i_{qs}^s \\ i_{ds}^s \end{bmatrix} \approx - \begin{bmatrix} e_{qs} \\ e_{ds} \end{bmatrix} \qquad (27)$$

Because the back-EMF varies slower than the sampling period, the currents can be reasonably assumed to increase linearly during the restarting period. Moreover, because transient current suppression occurs in a very short time, the initial controller voltage is analyzed in the discrete domain.

The sampling period is T_s, and the drive activates at t_1. The controller voltages are calculated at $t_1 + T_s$, and output to the motor at $t_1 + 2T_s$ due to the pulse width modulation delay. The subsequent motor currents are sampled at $t_1 + 3T_s$. To force the motor current sampled at $t_1 + 3T_s$ to be zero, the initial controller voltages (i.e., v_{qsc} and v_{dsc}) should be set as follows:

$$v_{qsc}(t_1 + 3T_s) = L_{ds}\frac{i_{qs}^s(t_1 + 3T_s) - i_{qs}^s(t_1)}{(t_1 + 3T_s) - (t_1 + 2T_s)} \tag{28}$$

$$v_{dsc}(t_1 + 3T_s) = L_{ds}\frac{i_{ds}^s(t_1 + 3T_s) - i_{ds}^s(t_1)}{(t_1 + 3T_s) - (t_1 + 2T_s)} \tag{29}$$

Equations (28) and (29) cannot be evaluated because future currents are used. However, the currents sampled at $t_1 + 3T_s$ can be predicted according to the constant back-EMF assumption as

$$i_{qs}^s(t_1 + 3T_s) = 3 \cdot i_{qs}^s(t_1 + T_s) - 2 \cdot i_{qs}^s(t_1) \tag{30}$$

$$i_{ds}^s(t_1 + 3T_s) = 3 \cdot i_{ds}^s(t_1 + T_s) - 2 \cdot i_{ds}^s(t_1) \tag{31}$$

When (30) and (31) are substituted into (28) and (29), the initial controller voltages can be calculated as

$$v_{qsc}^s(t_1 + T_s) = 3L_{ds} \cdot \frac{i_{qs}^s(t_1 + T_s) - i_{qs}^s(t_1)}{T_s} \tag{32}$$

$$v_{dsc}^s(t_1 + T_s) = 3L_{ds} \cdot \frac{i_{ds}^s(t_1 + T_s) - i_{ds}^s(t_1)}{T_s} \tag{33}$$

Figure 9 illustrates the activation procedure of the restarting strategy. The predicted initial controller voltages are applied as soon as the drive is activated. These voltages force the currents to decrease to zero within three sampling periods.

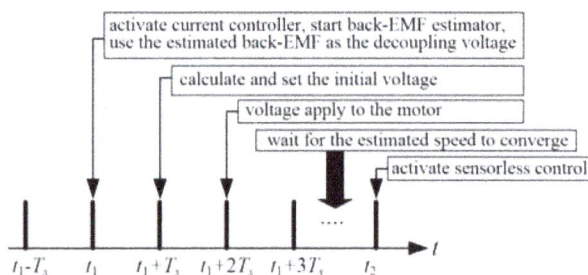

Figure 9. Activation of the restarting strategy.

5. Experimental Results

A 400 W, a four-pole PMSM was used for experimental verifications. The parameters of the PMSM are provided in Appendix A. Figure 10 shows the experimental system. The sensorless control and restarting control algorithms were implemented using a Texas Instruments TMS320F28335 digital signal processor. The sampling frequencies for current and velocity control were 18 kHz (T_s = 56 μs) and 2.2 kHz, respectively. The bandwidths of the current and velocity controller were tuned to 1000 and 25 Hz, respectively. The DC-link voltage was 300 V. The transition speed for the saliency-based and back-EMF-based sensorless control algorithm was 600–900 rpm. A load motor provided external load to the test motor. The actual rotor position and speed were monitored by an encoder with a resolution of 2500 pulse/rev.

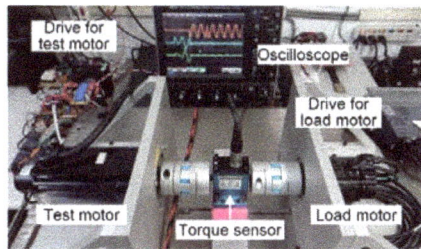

Figure 10. Experimental system.

Figure 11 depicts the measured steady-state currents and estimated back-EMF when the current controllers were activated without speed or position feedback. No decoupling voltage was applied. Both the current commands were set to zero, and the motor speed was regulated at 3000 rpm by the load motor. Significant induced currents existed due to the undecoupled back-EMF. The motor currents at various speeds were measured and compared with the calculated values obtained using (14) and (15). The calculated and measured results are summarized in Table 1. As indicated in Table 1, the measurements highly agreed with the calculated values. The results list in Table 1 also demonstrate the validity of the assumption in (12)–(13) that the cross-coupling voltages are small enough to be ignored. Figure 12a,b illustrates a comparison of the measured and calculated braking torques at 1080 and 1500 rpm, respectively. The braking torque was measured using a torque sensor. The frequency of the torque ripple was twice the rotor speed. Moreover, when HF components were ignored, the measured torque was highly consistent with the values calculated using (18).

Figure 13 presents the current responses when the back-EMF was decoupled and the initial controller voltage was set to zero. The motor speed was regulated at 3000 rpm by the load motor, and the current commands were set to zero. The drive was activated with the proposed restarting strategy at t_1. The back EMF was estimated at $t_1 + T_s$ and applied to the motor at $t_1 + 2T_s$. However, motor currents require approximately eight sampling periods to settle down after the drive is activated. Figure 14 displays the results when the initial controller voltages were calculated using (32) and (33) under experimental conditions similar to those in Figure 13. The motor currents settled down within only four sampling periods. This indicates that the motor currents increase linearly in such short time, and the assumption used for the initial controller voltages calculation in Section 4.3 is reasonable. Table 2 provides a summary of the maximum amplitude and settling time of the measured currents when the motor started from various rotor positions at 3000 rpm. All the maximum currents were less than half the rated current, and the settling times were 4–5 times T_s. These results indicate that the back-EMF-induced current can be effectively suppressed with the proposed restarting strategy.

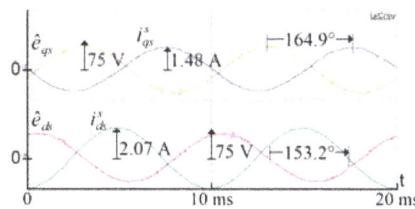

Figure 11. Measured steady-state currents and estimated back-EMF when the current controllers were activated without speed and position feedback and with no decoupling voltage.

Table 1. Comparison of the measured and calculated currents at various speeds.

ω_m (rpm)	Current	Meas.	Calc.	Meas.	Calc.
		Amplitude (A)		Phase (°)	
1500	i_{qs}^s	0.67	0.61	−149.7	−147.8
	i_{ds}^s	0.84	0.78	−133.7	−137.4
3000	i_{qs}^s	1.48	1.4	−164.9	−165.1
	i_{ds}^s	2.07	1.96	−153.2	−158.8
4500	i_{qs}^s	2.43	2.15	−170.3	−173
	i_{ds}^s	3.36	3.1	−162.2	−169.9

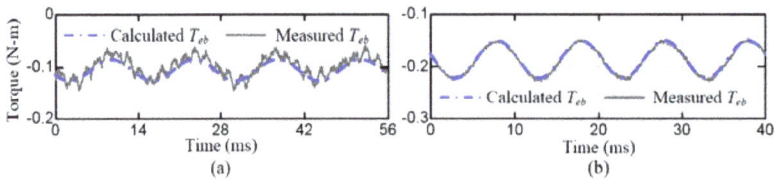

Figure 12. Comparison of the measured and calculated braking torque at (**a**) 1800 rpm and (**b**) 1500 rpm.

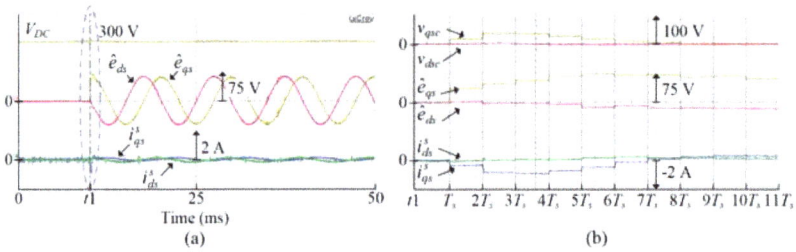

Figure 13. (**a**) Current responses when the back-EMF was decoupled but the initial controller voltage was set to zero (the motor speed was regulated at 3000 rpm by the load motor); (**b**) Amplified waveforms around t_1.

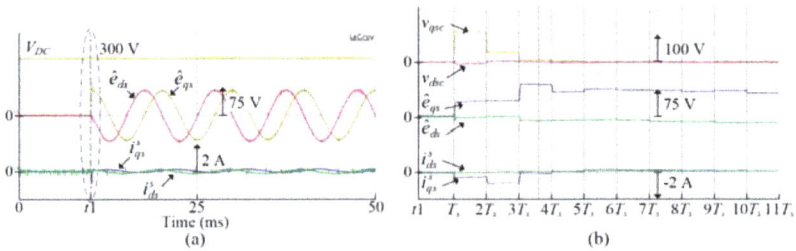

Figure 14. (**a**) Current responses when the back-EMF was decoupled and the initial controller voltage was set to the value calculated by (32) and (33) (the motor speed was regulated at 3000 rpm by the load motor); (**b**) Amplified waveforms around t_1.

Table 2. Measured currents at various positions when the restarting strategy is implemented.

θ_r (°)	Maximum Amplitude (A)		Settling Time
	i_{qs}^s	i_{ds}^s	
0	−0.86	0.17	$4T_s$
60	−0.41	0.77	$4T_s$
90	0.15	0.87	$5T_s$
180	0.91	−0.2	$4T_s$
240	0.48	−0.76	$4T_s$
270	0.23	−0.84	$5T_s$

Figures 15 and 16 show the results obtained when the motor was running at 3000 rpm with the rated load and then restarted from an unexpected error occurrence at $\theta_r = 0°$ and 90°, respectively. Figures 17 and 18 illustrate the results obtained when the motor was running at −4500 rpm with the rated load and then restarted from an unexpected error at $\theta_r = 180°$ and 270°, respectively. The motor was initially controlled through sensorless control and subjected to the rated load. The motor drive was turned off at t_0 to emulate the occurrence of an error, such as sensor failure or temporary power disruption. Subsequently, the motor coasted down rapidly between t_0 and t_1 due to the large external load torque. The drive was then activated using the proposed restarting strategy at t_1. After the estimated speed and position reached the steady state at t_2, sensorless control was performed again with constant current commands to increase the motor speed. Finally, the system switched back to the regular sensorless speed control after the motor speed reached the preset command (3000 and −4500 rpm).

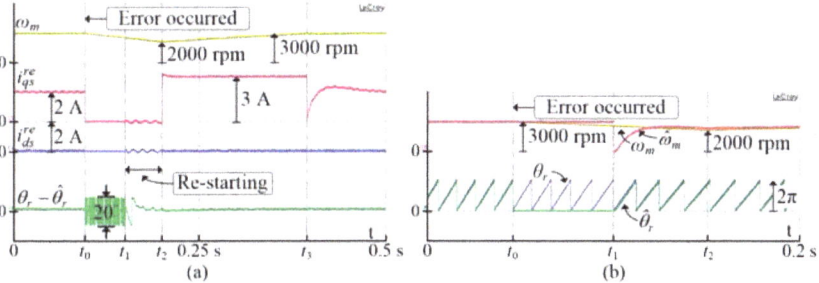

Figure 15. (a) Restarting of the motor from $\theta_r = 0°$ with a positive speed under the rated load; (b) Amplified waveforms.

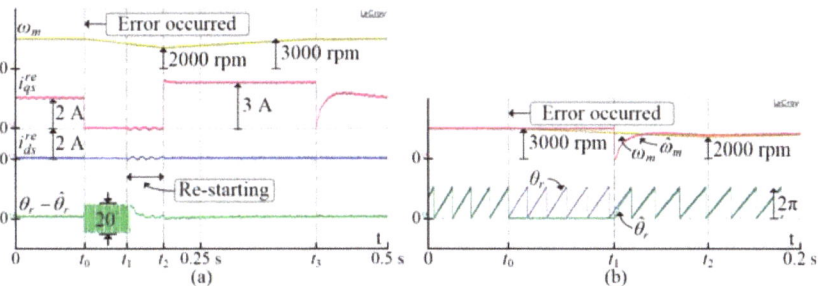

Figure 16. (a) Restarting of the motor from $\theta_r = 90°$ with a positive speed under the rated load; (b) Amplified waveforms.

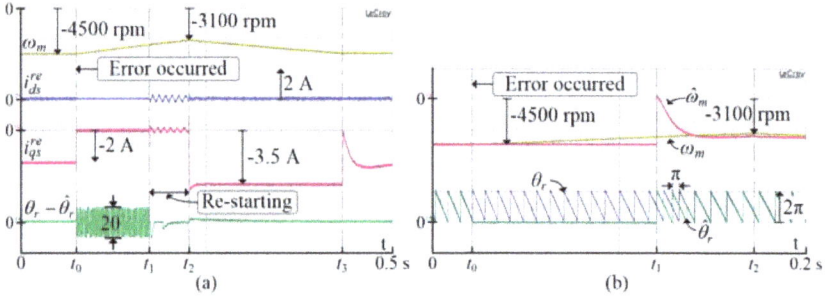

Figure 17. (**a**) Restarting of the motor from $\theta_r = 180°$ with a negative speed under the rated load; (**b**) Amplified waveforms.

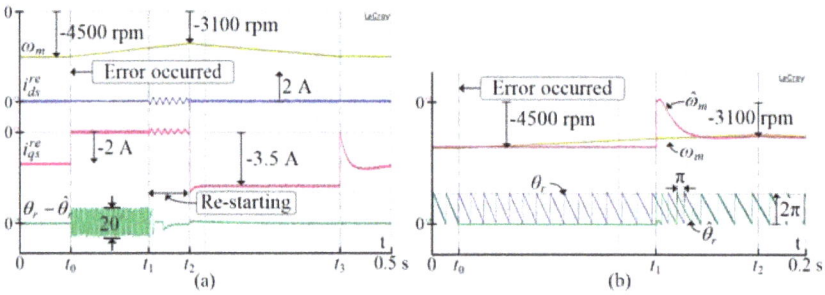

Figure 18. (**a**) Restarting of the motor from $\theta_r = 270°$ with a negative speed under the rated load; (**b**) Amplified waveforms.

According to the above results, all the transient currents at t_1 were very small, which indicates that the back-EMF-induced currents had been effectively suppressed irrespective of the starting speed or position. The amplified waveforms also indicate that the rotor position and speed could be estimated correctly irrespective of the rotation direction. Although the estimated position reached the actual position rapidly, the estimated speed required approximately 0.02 s to reach the actual speed. Therefore, the duration between t_1 and t_2 should be higher than 0.02 s for a smooth restart of the sensorless control. Most importantly, although the position estimator takes longer time to estimate the actual speed correctly, the induced current is still suppressed effectively because the back-EMF can be estimated accurately without the speed feedback. Actually, this is one of the key features of the proposed restarting strategy. Note also that for a negative speed, a 180° phase error appeared in the estimated position due to the normalization of $\Delta\theta_{emf}$. The correct rotor position was obtained after the rotation direction was identified and the phase error was compensated. Additionally, as shown in Figures 15–18, the position error $(\theta_r - \hat{\theta}_r)$ after t_2 approximates to 0° because the motor parameters are measured with reasonable accuracy and the nonlinearity of the inverter is well compensated.

Figure 19 displays the results of an experiment conducted under similar conditions to those for the experiment displayed in Figure 18 but with direct restarting of the motor. The back-EMF is not decoupled and the initial voltage is not applied when the motor is directly restarted. Although the motor speed and position could be still estimated with high accuracy, significant back-EMF-induced currents appeared between t_1 and t_2 because the back-EMF was not decoupled. Moreover, the motor coasted down faster in the experiment displayed in Figure 19 than in the experiment displayed in Figure 18 due to the large braking torque.

Figure 20 presents the currents and position errors when the motor speed was regulated by the load motor at 3000 rpm, current commands were set to zero, and BW of the current controller was set to 500 Hz. The current controller was activated at t_1, and the restarting strategy was implemented at t_1' to examine its effectiveness. It can be seen that the position error is obviously larger than that in Figures 15–18 because low BW of the current controllers causes significant phase delay on the estimated back-EMF, as mentioned in Section 4.1. In addition, a large transient current was induced between t_1 and t_1'. However, the induced current was suppressed as soon as the restarting procedure was implemented. This result also confirms that the proposed restarting strategy is effective even with a considerably low current controller bandwidth.

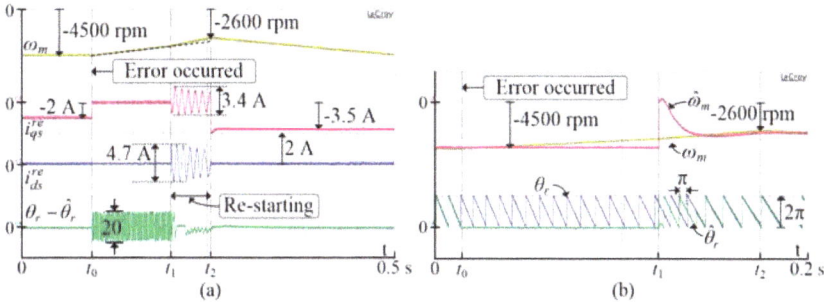

Figure 19. (**a**) Restarting the motor directly from $\theta_r = 270°$ with a negative speed under the rated load; (**b**) Amplified waveforms.

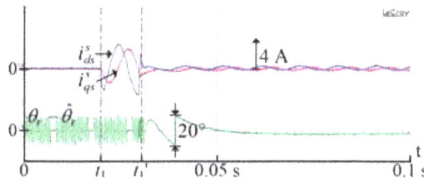

Figure 20. Currents and position errors during restarting period when the current controller BW was set to 500 Hz.

6. Conclusions

This paper proposes a restarting strategy for back-EMF-based sensorless controlled PMSMs when the rotor is spinning. The proposed restarting strategy presents the following features: (1) the restarting strategy is easy to implement because it is developed based on the existing back-EMF-based sensorless control algorithm; (2) the restarting procedure does not increase the computation burden because no voltage vector injection is required to identify the rotor position and speed; (3) the induced current can be suppressed effectively before the rotor position and speed are identified; (4) the rotor speed and position are still tracked accurately even when the motor is coasting down very quickly.

The analytical results indicate that the back-EMF of the motor can be estimated with good accuracy even without the rotor position or speed feedback provided that the bandwidth of the current controller is much higher than the rotor speed. According to this result, the back-EMF is estimated and used as the decoupling voltage for mitigating the back-EMF-induced currents when activates the current controller. Furthermore, the initial voltages in the integrators of current controllers are calculated properly and applied to greatly reduce the transient current. The experimental results indicate that the induced current can be suppressed within four to five sampling periods for various spinning conditions. Because of the considerably short time delay, the motor drive can restart safely from various speeds and positions without causing overcurrent fault.

Author Contributions: Conceptualization, Z.C.Y. and S.M.Y.; methodology, Z.C.Y.; software, Z.C.Y.; validation, Z.C.Y. and S.M.Y.; formal analysis, Z.C.Y.; investigation, Z.C.Y.; resources, S.M.Y.; data curation, Z.C.Y.; writing—original draft preparation, Z.C.Y.; writing—review and editing, S.M.Y.; visualization, Z.C.Y.; supervision, S.M.Y.; project administration, S.M.Y.; funding acquisition, S.M.Y.

Funding: This research received no external funding.

Conflicts of Interest: The authors declare no conflict of interest.

Appendix A

Table A1. Main motor parameters.

Parameter	Value	Unit
Rated speed/pole pairs	6000/2	rpm
Rated current	2	A
Magnet flux (λ_m)	0.106	Wb-turns
Stator resistance	1.53	Ω
d-axis inductance (L_{ds})	4.8	mH
q-axis inductance (L_{qs})	7.1	mH

References

1. Liu, J.M.; Zhu, Z.Q. Sensorless control strategy by square-waveform high-frequency pulsating signal injection into stationary reference frame. *IEEE J. Emerg. Sel. Topics Power Electron.* **2014**, *2*, 171–180. [CrossRef]
2. Yoon, Y.D.; Sul, S.K.; Morimoto, S.; Ide, K. High-Bandwidth sensorless algorithm for AC machines based on square-wave-type voltage injection. *IEEE Trans. Ind. Appl.* **2011**, *47*, 1361–1370. [CrossRef]
3. Murakami, S.; Shiota, T.; Ohto, M.; Ide, K.; Hisatsune, M. Encoderless servo drive with adequately designed IPMSM for pulse-voltage-injection-based position detection. *IEEE Trans. Ind. Appl.* **2012**, *48*, 1922–1930. [CrossRef]
4. Jung, S.; Ha, J.I. Analog filtering method for sensorless AC machine control with carrier-frequency signal injection. *IEEE Trans. Ind. Electron.* **2015**, *62*, 5348–5358. [CrossRef]
5. Yang, S.C.; Yang, S.M.; Hu, J.H. Design consideration on the square-wave voltage injection for sensorless drive of interior permanent-magnet machines. *IEEE Trans. Ind. Electron.* **2017**, *64*, 159–168. [CrossRef]
6. Chen, Z.; Tomita, M.; Doki, S.; Okuma, S. An extended electromotive force model for sensorless control of interior permanent-magnet synchronous motors. *IEEE Trans. Ind. Electron.* **2003**, *50*, 288–295. [CrossRef]
7. Morimoto, S.; Kawamoto, K.; Sanada, M.; Takeda, Y. Sensorless control strategy for salient-pole pmsm based on extended emf in rotating reference frame. *IEEE Trans. Ind. Appl.* **2002**, *38*, 1054–1061. [CrossRef]
8. Kim, H.; Harke, M.C.; Lorenz, R.D. Sensorless control of interior permanent-magnet machine drives with zero-phase lag position estimation. *IEEE Trans. Ind. Appl.* **2007**, *39*, 1726–1733.
9. Patel, N.; O'meara, T.; Nagashima, J.; Lorenz, R. Encoderless IPM traction drive for EV/HEV's. In Proceedings of the Conference Record of the 2001 IEEE Industry Applications Conference. 36th IAS Annual Meeting, Chicago, IL, USA, 30 September–4 October 2001; Volume 3, pp. 1703–1707.
10. Wang, G.; Yang, R.; Xu, D. DSP-Based control of sensorless IPMSM drives for wide-speed-range operation. *IEEE Trans. Ind. Electron.* **2013**, *60*, 720–727. [CrossRef]
11. Lara, J.; Chandra, A.; Xu, J. Integration of HFSI and extended-EMF based techniques for PMSM sensorless control in HEV/EV applications. In Proceedings of the IECON 2012—38th Annual Conference on IEEE Industrial Electronics Society, Montreal, QC, Canada, 25–28 October 2012; pp. 3688–3693.
12. Ide, K.; Takaki, M.; Morimoto, S.; Kawazoe, Y.; Maemura, A.; Ohto, M. Saliency-Based sensorless drive of adequate designed IPM motor for robot vehicle application. In Proceedings of the 2007 Power Conversion Conference, Nagoya, Japan, 2–5 April 2007; pp. 1126–1133.
13. Yoo, H.; Kim, J.H.; Sul, S.K. Sensorless operation of a PWM rectifier for a distributed generation. *IEEE Trans. Power Electron.* **2007**, *22*, 1014–1018. [CrossRef]
14. Son, Y.C.; Jang, S.J.; Nasrabadi, R.D. Permanent Magnet AC Motor Systems and Control Algorithm Restart Methods. U.S. Patent 8054030B2, 8 November 2011.
15. Son, Y.C.; Jang, J.; Welchko, B.A.; Patel, N.R.; Schulz, S.E. Method and System for Initiating Operation of an Electric Motor. U.S. Patent 8319460B2, 27 November 2012.

16. Taniguchi, S.; Mochiduki, S.; Yamakawa, T.; Wakao, S.; Kondo, K.; Yoneyama, T. Starting procedure of rotational sensorless PMSM in the rotating condition. *IEEE Trans. Ind. Appl.* **2009**, *45*, 194–202. [CrossRef]
17. Lee, K.; Ahmed, S.; Lukic, S.M. Universal restart strategy for high-inertia scalar-controlled PMSM drives. *IEEE Trans. Ind. Appl.* **2016**, *52*, 4001–4009. [CrossRef]
18. Iura, H.; Ide, K.; Hanamoto, T.; Chen, Z. An estimation method of rotational direction and speed for free-running AC machines without speed and voltage sensor. *IEEE Trans. Ind. Appl.* **2016**, *47*, 153–160. [CrossRef]
19. You, Z.C.; Yang, S.M. A Control strategy for flying-start of shaft sensorless permanent magnet synchronous machine drive. In Proceedings of the 2018 International Power Electronics Conference (IPEC-Niigata 2018-ECCE Asia), Niigata, Japan, 20–24 May 2018; pp. 651–656.
20. Yang, S.M.; Lin, K.W. Automatic control loop tuning for permanent-magnet AC serve motor drives. *IEEE Trans. Ind. Electron.* **2016**, *63*, 1499–1506. [CrossRef]
21. Inoue, Y.; Yamada, K.; Morimoto, S.; Sanada, M. Effectiveness of voltage error compensation and parameter identification for model-based sensorless control of IPMSM. *IEEE Trans. Ind. Appl.* **2009**, *45*, 213–221. [CrossRef]
22. Hejny, R.W.; Lorenz, R.D. Evaluating the practical low-speed limits for back-EMF tracking-based sensorless speed control using drive stiffness as a key metric. *IEEE Trans. Ind. Appl.* **2011**, *47*, 1337–1343. [CrossRef]

energies

MDPI

Article

A Study of Fault Diagnosis Based on Electrical Signature Analysis for Synchronous Generators Predictive Maintenance in Bulk Electric Systems

Camila Paes Salomon [1], Claudio Ferreira [1], Wilson Cesar Sant'Ana [2],
Germano Lambert-Torres [2,*], Luiz Eduardo Borges da Silva [3], Erik Leandro Bonaldi [2],
Levy Ely de Lacerda de Oliveira [2] and Bruno Silva Torres [1]

1   Institute of Electric Systems and Energy, Itajuba Federal University, Itajuba 37500-903, Brazil;
    camilapsalomon@gmail.com (C.P.S.); claudiof@unifei.edu.br (C.F.);
    brunotorres@institutognarus.com.br (B.S.T.)
2   Gnarus Institute, Itajuba 37500-052, Brazil; wilson.cesar.santana@gmail.com (W.C.S.);
    erik@pssolucoes.com.br (E.L.B.); levy@pssolucoes.com.br (L.E.d.L.d.O.)
3   Institute of System Engineering and Information Technology, Itajuba Federal University, Itajuba 37500-903,
    Brazil; leborgess@gmail.com
*   Correspondence: germanoltorres@gmail.com; Tel.: +55-35-99-986-0378

Received: 6 March 2019; Accepted: 18 April 2019; Published: 21 April 2019

Abstract: The condition of synchronous generators (SGs) is a matter of great attention, because they can be seen as equipment and also as fundamental elements of power systems. Thus, there is a growing interest in new technologies to improve SG protection and maintenance schemes. In this context, electrical signature analysis (ESA) is a non-invasive technique that has been increasingly applied to the predictive maintenance of rotating electrical machines. However, in general, the works applying ESA to SGs are focused on isolated machines. Thus, this paper presents a study on the condition monitoring of SGs in bulk electric systems by using ESA. The main contribution of this work is the practical results of ESA for fault detection in in-service SGs interconnected to a power system. Two types of faults were detected in an SG at a Brazilian hydroelectric power plant by using ESA, including stator electrical unbalance and mechanical misalignment. This paper also addresses peculiarities in the ESA of wound rotor SGs, including recommendations for signal analysis, how to discriminate rotor faults on fault patterns, and the particularities of two-pole SGs.

Keywords: bulk electric system; condition monitoring; electrical signature analysis; fault diagnosis; predictive maintenance; synchronous generator

1. Introduction

Electrical power systems consist of a large number of interconnected synchronous generators (SGs) operating in parallel, connected to transmission and distribution networks to supply large load centers. These machines are fundamental elements of power systems, and their condition affects network reliability and stability [1]. The parallel operation of SGs presents several advantages such as the increase of supply reliability, the improvement of efficiency, and lower cost. However, it increases the complexity of the stability control of the SGs when a fault occurs [2]. Thus, there is a growing interest in new technologies to improve SG protection and maintenance schemes [3–5].

Among the maintenance philosophies that have been applied to SGs, condition-based maintenance (CBM) is highlighted. This type of maintenance is based on the continuous monitoring of a condition parameter in a machine (vibration, temperature, electrical signals, etc.) [6]. CBM consists of the analysis of the monitored parameters to evaluate if certain indicators present signs of decreasing performance or incipient fault. Thus, the actual condition of the asset is evaluated, and it is possible to decide what and when the maintenance action must be done [7].

In the context of CBM, electrical signature analysis (ESA) has been increasingly applied to rotating electric machines fault diagnostics. In ESA, the monitored parameters are the electrical signals of the machine in the frequency domain. This is performed by using a Fast Fourier Transform (FFT) algorithm and other signal processing procedures. The faults are detected and identified by analyzing fault patterns on voltage and current spectra, which are frequency components whose magnitudes vary when a fault happens. Moreover, faults can be detected at an early stage and the frequency components' magnitudes are generally related to the fault severity. The main advantages of ESA are no intrusiveness, dependence on only electrical quantities, technical and economic viability, and possibility of obtaining fault patterns applicable for all types of wound rotor SGs [6,8–10].

There are several published works about ESA-based methodologies for fault detection in synchronous machines. ESA has been used mainly for detection of stator winding inter-turn short circuit [11–16], rotor winding inter-turn short circuit [11,13,17–20], air-gap eccentricity [21–25], and rotating diode failure [26,27]. There are also works approaching various electrical and mechanical faults in SGs [28].

Despite the increasing use of ESA in CBM, it is important to highlight that some particularities arise when this technique is applied to SGs. Firstly, the choice of signals to be analyzed must be considered, because both current and voltage are outputs of these machines. There is also an issue related to faults that coincide in the same fault pattern. For instance, the rotor rotation frequency fault pattern is indicative of rotor winding inter-turn short circuit and rotor mechanical faults [9,11,13,19,29]. Thus, the analysis of only the rotation frequency components allows the detection of rotor problems. However, this is not sufficient to identify the type of fault (electrical or mechanical). There is also an issue related to two-pole type SGs. In these machines, the rotor rotation frequency matches the power line frequency. Therefore, the fault patterns and the power system harmonics match the same components. Thus, the ESA-based fault diagnostic can be obscured because of the SGs intrinsic harmonics or harmonics related to non-linear loads [30–34]. Finally, in general, the works found in literature present ESA application to isolated SGs, both in a laboratory environment and in SGs onboard ships [13,15,24]. However, large SGs are usually interconnected to power systems, so ESA should also be applied to SGs in this condition.

Given this scenario, this paper presents a study on fault diagnosis of SGs in bulk electric systems using ESA. Firstly, a methodology for SG fault detection using ESA is proposed, addressing ESA fault patterns and the mentioned peculiarities of this technique for fault detection in SGs. Then, as the main contribution of this work, case studies of fault detection using ESA in an in-service SG in a Brazilian hydroelectric power plant are presented. The results show the potential of ESA for condition monitoring of SGs interconnected to power systems and are valuable, since SGs in this context are concerned with monitoring and are subjected to diverse conditions of the bulk power system. As mentioned previously, other works usually present results in a laboratory environment with isolated SGs and controlled conditions, which do not fully depict the SGs in practical situations.

As a note, it is known that SGs conditions might include normal condition, oil-membrane oscillation, imbalance, no orderliness, short circuit, and so on. However, this work focuses on conditions covering healthy and faulty situations, including fault detection and identification. Moreover, it is worth noting that artificial intelligence-based methodologies have been increasingly applied to rotating machinery fault detection and condition monitoring nowadays [35–37]. These techniques can be used in future works as auxiliary tools to complement the fault diagnosis by using the proposed methodology, being not in the scope of the present work.

The rest of the paper is divided as follows. Section 2 presents the ESA background. Section 3 presents ESA fault patterns for SGs. Section 4 approaches the peculiarities of ESA application to SGs. Section 5 presents the proposed methodology for fault detection in SGs in interconnected power systems. Section 6 presents results of fault detection in SG in a power plant. Finally, Section 7 presents the conclusions.

2. Electrical Signature Analysis Background

ESA consists of the frequency-domain representation, processing, and analysis of electrical signals and has been usually applied in electric machinery condition monitoring. In general, ESA comprises of an FFT (Fast Fourier Transform) algorithm to represent the time-domain signals in the shape of spectra, which are referred in this work as electrical signatures. ESA is based on the assumption that a significant change in a machine condition results in the change of its electrical signature. Moreover, there are specific frequency components whose magnitudes change in the presence of faults. These frequency components are related to the type and location of fault, being dependent on the power line frequency and structural characteristics of the motor or generator. Thus, it is possible to obtain a set of ESA fault patterns for fault detection and identification in electrical machines [38]. Moreover, the fault patterns can be applicable to all types of wound rotor SGs.

The ESA techniques used in this work are current (CSA) and voltage signature analysis (VSA), and extended Park's vector approach (EPVA), which will be explained in the next sections.

2.1. Current and Voltage Signatures

CSA and VSA consist purely of the frequency-domain analysis of the current and voltage signals from the machine stator. The electrical signatures are obtained through the application of FFT to voltage and current signals. FFT is an algorithm with the purpose of computing the Discrete Fourier Transform (DFT) in a faster way. Considering a list of complex numbers, the DFT transforms that into a list of coefficients of a finite combination of complex sinusoids. Each DFT component is given by [39]:

$$X_m = \sum_{n=0}^{N-1} x_n \cdot e^{-j2\pi mn/N}, \ m \in Z \tag{1}$$

where m is the DFT index (harmonic order), n is the time-domain index, N is the number of samples, X_m is the mth DFT coefficient, and x_n is the time-domain list of equally-spaced complex samples.

FFT decomposition allows the determination of the magnitude and the phase of each frequency component of the electrical signal under analysis. As mentioned previously, these components can compose a set of fault patterns for machine diagnosis, and generally the magnitudes of these components increase according to the fault severity.

In practice, the spectral components' magnitudes are usually presented normalized in relation to the fundamental component magnitude (line frequency), due to the changes of current in function of load. Moreover, the use of a logarithm scale (in general, dB) is common, because of the big difference between the magnitudes and the exponential characteristic of the evolution presented for several known faults [7].

In general, the fault components are expressed in the frequency spectrum as:

$$f_e = f_1 \pm k \cdot f_c \tag{2}$$

where f_e is the fault frequency as it appears in the spectrum, considering the power line frequency modulation; f_1 is the power line frequency; k is a positive integer value, indicating the harmonic order; and f_c is the specific fault characteristic frequency.

2.2. Extended Park's Vector Approach

EPVA is a technique based on a quantitative analysis of the Park circle distortion, whose main characteristic is considering information of the three phases of electrical signals. It is specifically used as a fault indicator of stator electrical unbalance. A brief explanation of this technique is provided below [12,40].

The components of Park's vector (i_D and i_Q) for a set of three phase balanced currents are computed by applying the Clarke transformation and given by:

$$i_D = \left(\frac{\sqrt{2}}{\sqrt{3}}\right)i_A - \left(\frac{1}{\sqrt{6}}\right)i_B - \left(\frac{1}{\sqrt{6}}\right)i_C$$
$$i_Q = \left(\frac{1}{\sqrt{2}}\right)i_B - \left(\frac{1}{\sqrt{2}}\right)i_C \tag{3}$$

where i_D and i_Q are the current components of Park's vector in direct and quadrature axes, respectively; and i_A, i_B, and i_C are balanced line currents in phases *A*, *B* and *C*, respectively.

When the conditions are ideal, the resultant Park circle is a perfect circle, whose center locates at the origin of the coordinates. Considering the wave shape parameters of the balanced line currents *A*, *B* and *C*, it is obtained:

$$i_D = \left(\frac{\sqrt{6}}{2}\right)i_M \cos(\omega t - \theta)$$
$$i_Q = \left(\frac{\sqrt{6}}{2}\right)i_M \sin(\omega t - \theta) \tag{4}$$

where i_M is the peak value of the line current; ω is the angular frequency, in (rad/s); θ is the initial phase angle, in (rad); and *t* is the time variable.

When the conditions are not ideal, for instance, when an electrical unbalance is present, the Park circle presents some distortion. Thus, the currents contain direct and inverse sequence components, which can be represented as:

$$i_A = i_d \cos(\omega t - \theta_d) + i_i \cos(\omega t - \gamma_i)$$
$$i_B = i_d \cos\left(\omega t - \theta_d - \frac{2\pi}{3}\right) + i_i \cos\left(\omega t - \gamma_i + \frac{2\pi}{3}\right)$$
$$i_C = i_d \cos\left(\omega t - \theta_d + \frac{2\pi}{3}\right) + i_i \cos\left(\omega t - \gamma_i - \frac{2\pi}{3}\right) \tag{5}$$

where i_d is the maximum value of the direct sequence current; i_i is the maximum value of inverse sequence current; θ_d is the initial phase angle of the direct sequence current, in (rad); and γ_i is the initial phase angle of the inverse sequence current, in (rad). By substituting (5) in (3), it is obtained:

$$i_D = \left(\frac{\sqrt{3}}{\sqrt{2}}\right)(i_d \cos(\omega t - \theta_d) + i_i \cos(\omega t - \gamma_i))$$
$$i_Q = \left(\frac{\sqrt{3}}{\sqrt{2}}\right)(i_d \sin(\omega t - \theta_d) - i_i \sin(\omega t - \gamma_i)) \tag{6}$$

In order to obtain the resultant signal for analysis, the computation of the square of Park's vector module is executed:

$$i_D^2 + i_Q^2 = \left(\frac{3}{2}\right)(i_d^2 + i_i^2) + 3i_d i_i \cos(2\omega t - \theta_d - \gamma_i) \tag{7}$$

Finally, the FFT algorithm is applied to the square of the Park's vector module. In the presence of electrical unbalance, the resulting spectrum contains a DC level and a component located at twice the power line frequency. This component is defined as the EPVA electrical unbalance fault pattern.

As a final comment, the procedure has been explained for current signals, but the same can be used for voltage signals.

3. ESA Fault Patterns for Synchronous Generators

This section presents ESA fault patterns for SGs that will be used in the proposed methodology of this work. These patterns are valid for different types of SGs and have been presented in the literature and proved experimentally [41]. An observation is that both voltage (VSA or EPVA) and current (CSA or EPVA) signatures must be analyzed.

3.1. Rotor Winding Inter-Turn Short Circuit

SG rotor winding inter-turn short circuit fault presents the characteristics: increase of rotor current, an increase of winding temperature, distortion of voltage waveform, unusual vibration, and possibility of mechanical faults occurring.

The proposed ESA fault patterns are rotor rotation frequency pattern and even harmonics on the electrical signature.

The rotor rotation frequency is defined by the monitoring of line frequency with sidebands in integer multiples of rotor rotation frequency. The rotor rotates at synchronous speed. Thus, the rotor rotation frequency is given by:

$$f_r = \frac{f_1}{P} \tag{8}$$

where f_r is the rotor rotation frequency, f_1 is the line frequency, and P is the number of pole pairs. A note is that the expression "pole pairs" is used to refer to the "physical" number of pole pairs in the rotor of the SG and this is valid for all the times this expression appears in this paper. Thus, the rotation frequency pattern is given by [6,13,19,38]:

$$f_{pfr} = f_1 \pm k \cdot f_r \tag{9}$$

where f_{pfr} are the spectral components analyzed for SG rotor mechanical problems and k is a positive integer.

Figure 1a illustrates the rotor rotation frequency, which must be analyzed on voltage and current signatures.

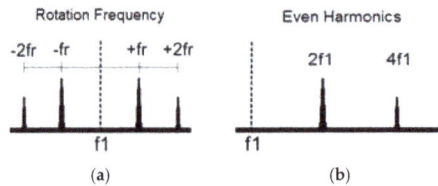

Figure 1. Fault patterns: (**a**) rotor rotation frequency (pattern for rotor winding inter-turn short circuit—electrical fault, and for rotor mechanical faults); (**b**) even harmonics (pattern for rotor winding inter-turn short circuit—electrical fault).

Another proposed fault pattern for this type of fault is the analysis of even harmonics (f_{php}) in the electrical signature, which are given by [13,19,33,34]:

$$f_{php} = 2 \cdot k \cdot f_1 \tag{10}$$

This pattern must be analyzed on voltage and current signatures and is illustrated in Figure 1b.

3.2. Stator Winding Inter-Turn Short Circuit

Another type of fault that SGs can suffer is stator winding inter-turn short circuit, whose characteristics are: the emergence of pulsating currents and generation of rotating fields in the opposite direction of the original one.

The proposed fault patterns for this fault are the zero sequence harmonics (mainly the third harmonic) in the electrical signature, and the EPVA electrical unbalance pattern [12,13].

The zero sequence harmonics fault pattern (f_{phsz}) is given by:

$$f_{phsz} = 3 \cdot k \cdot f_1 \tag{11}$$

This fault pattern must be analyzed in voltage and current signatures and is illustrated in Figure 2a.

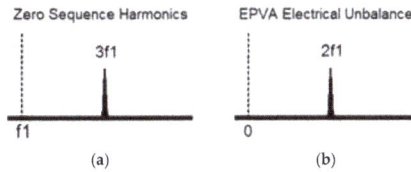

Figure 2. Fault patterns: (**a**) zero sequence harmonics; (**b**) EPVA electrical unbalance (patterns for stator electrical unbalance—electrical fault).

The EPVA electrical unbalance pattern (f_{epva}) is defined as the component of twice the line frequency in the EPVA spectrum of voltage and current, being represented as:

$$f_{epva} = 2 \cdot f_1 \tag{12}$$

Figure 2b illustrates this pattern, which must be analyzed on EPVA voltage and current signatures. As a note, these patterns are also valid for other faults that result in stator electrical unbalance (phase-to-neutral short circuit, phase-to-phase short circuit, open circuit fault).

3.3. Rotor Mechanical Faults

Rotor mechanical faults include mechanical misalignment, mechanical unbalance, static airgap eccentricity, and dynamic airgap eccentricity. Some effects of these faults include: an increase in vibration, higher electromagnetic stress, an increase of unbalanced magnetic pull (UMP), increase of bear wear, and rotor and stator rubbing.

The fault pattern proposed for these faults is the rotor rotation frequency on voltage and current signatures, as illustrated in Figure 1a, because these faults also cause an increase in the magnitude of this component [25,29,38].

4. Peculiarities of Electrical Signature Analysis of Synchronous Generators

The last section presented ESA fault patterns for SGs. However, it is important to contextualize that ESA has been largely applied to fault detection of induction motors in an industrial environment and the application to SGs is more recent and entails some peculiarities. Thus, this section compiles some important issues related to ESA application to SGs condition monitoring.

4.1. Signal Analysis

The first peculiarity of ESA application to fault detection of SGs is related to the choice of signals to be analyzed because both voltages and currents are outputs of these machines. In the literature, ESA experts recommend to analyze the stator voltage and current signals. In the case of isolated SGs, the analysis should focus on the voltage signature (VSA and EPVA) and relate to the current signature (CSA and EPVA). If the component magnitude under analysis is greater in voltage than in current (in dB), one may suppose that there is an incipient fault in the SG. Otherwise, one may suppose that the incipient fault is in the load [28].

These considerations are valid for isolated SGs. This work assumed that for SGs interconnected to power systems, the focus should be on current signature (CSA and EPVA). This because if an SG is connected to an infinite bus, then the system voltage would prevail over the SG generated voltage.

4.2. Discrimination of Mechanical and Electrical Rotor Faults on Fault Patterns

A second peculiarity is related to the discrimination of faults whose effects occur in the same fault pattern. As one may note, the rotor rotation frequency pattern is indicative of rotor mechanical problems, and rotor winding inter-turn short circuit [11,13,19,25,29]. Thus, by analyzing only this fault pattern, it is possible to detect rotor problems, but not to distinguish if it is an electrical or a mechanical problem. In order to get a reliable fault diagnosis for SGs based on ESA, it would be necessary to separate the effects of electrical and mechanical faults in the rotation frequency components.

A methodology based on the analysis of symmetrical components to distinguish SG rotor faults has been proposed in [9]. This approach focused on the first right sideband of the rotation frequency, presenting a theoretical analysis and experimental results. The cited work found that the rotor winding inter-turn short circuit fault caused the increase of positive (mainly), negative, and zero sequence magnitudes of the rotation frequency's first right sideband. On the other hand, the rotor mechanical unbalance caused the increase only of the positive sequence magnitude of the rotation frequency's first right sideband. Thus, these findings should be associated with the rotor rotation frequency fault pattern, to provide a better fault diagnosis using ESA.

4.3. Two-Pole Synchronous Generators

The SGs at hydroelectric power plants are usually built with a salient pole for two or more pole pairs (low-speed or hydrogenerators), and SGs at thermal and nuclear power plants are normally built with a non-salient pole for one or two pole pairs (high speed or turbogenerators) [42]. As the proposed ESA methodology is applicable for both types of SGs, a last peculiarity emerges for the specific case of two-pole SGs. In this case, the rotation frequency is equal to the line frequency (fundamental frequency). Thus, the fault patterns match the harmonics of the fundamental frequency. The harmonics in an SG include the intrinsic harmonics due to the structural characteristics of the SGs and the magnetomotive force waveform, harmonics due to non-linear loads fed by the SG (they reflect in the SGs signals because of the armature reaction effect), and harmonics due to possible internal faults of the SGs (for instance, the second harmonic as indicative of rotor faults). In practice, even healthy machines may present even harmonics, and this can confuse the fault diagnosis when using ESA [30–33]. Even harmonics are indicative of rotor winding short circuit and the second harmonic matches the rotation frequency pattern's first right sideband. For a correct fault diagnostic, it is necessary to do an in-depth study about the different harmonics in SG signals and the interactions among them.

A previous study [34] on this issue concluded that the SG intrinsic even harmonics (due to possible internal asymmetry in the machine or mechanical misalignment condition) did not confuse the diagnosis of rotor winding inter-turn short circuit fault by using ESA. The cited work was substantiated by experimental results. However, for future works, it is important to perform experimental tests with non-linear loads at the SG output to analyze the effect of harmonics in ESA fault patterns and their influence on fault detection.

Finally, it is not the scope of this work to approach in depth each cited peculiarity. The main objective was to point them out and show how they have been approached. Moreover, the influence of saturation and power factor in SG fault detection is not addressed, being proposed for future works.

5. Proposed Methodology for Fault Diagnosis of SGs in Bulk Electric Systems

This section describes the proposed ESA-based methodology for condition monitoring of SGs, considering the ESA fault patterns and the SGs peculiarities presented in the last sections.

A system has been developed for SGs CBM based on ESA. The system includes two PS TTD-01 differential AC voltage transducers, 2.5% peak-to-peak error, used to measure two phase-to-phase voltages. Two TT 50-SD differential AC current transducers, 2.0% peak-to-peak error are used to measure two line-currents. The transducers are installed at the secondary side of the SGs voltage and current transformers, and are accessed through the panels of the generating units (GUs).

The transducers are connected to an acquisition system by National Instruments composed by the Ethernet cDaq-9181 carrier and the NI-9239 module for data acquisition. This system is provided with four channels for simultaneous signal acquisition, and the acquisitions are accomplished with 131072 points each one and 16.67 kHz sample frequency. The analog-to-digital converter resolution is 24 bits.

The acquisition system communicates with a personal computer through a signal analysis software (SAS) developed in C# to command the acquisitions and perform the signal analysis and the machine diagnosis. The set of fault patterns presented in (9)–(12), including the symmetrical components approach described previously, is implemented in the SAS. The SAS is the heart of the proposed system, being the part of the system where the fault patterns and rules presented in Sections 3 and 4 are implemented to perform fault detection and identification, even for an SG connected to the grid.

The first step of the proposed methodology is the acquisition of the machine's electrical signals. Once acquired, the signals are processed, the FFT algorithm is applied to get the signal spectral components, other signal processing techniques or post-processing analysis are performed, the database is fed with this information, and the SAS performs the analysis. Thus, the electrical signatures can be analyzed in a qualitative and quantitative way, especially considering the predefined fault patterns. Figure 3 presents a schematic diagram of the described ESA-based methodology for SGs CBM.

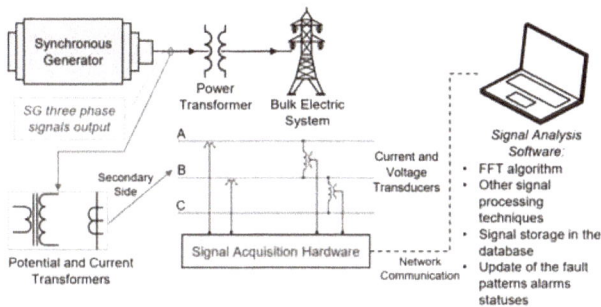

Figure 3. Electrical signature analysis (ESA) based methodology for condition-based maintenance of synchronous generators (SGs).

The described process is performed continuously over time. The system can be programmed to acquire a number of signals in predefined time intervals, each day. Then, a database is created and continuously fed. The analysis for fault detection is performed considering the database stored from the system installation time until the moment of analysis. This is performed by analyzing the trend curves of the fault patterns magnitudes. Thus, it is possible to note any abnormal increase in a component magnitude and its evolution over time.

The analysis is accomplished by setting alarms for the fault pattern's component's magnitudes. The alarm levels are obtained empirically, according to maintenance specialist expertise. For means of complementation, the analysis can also be comparative. When there are data of the machine in a healthy condition (baseline), the analysis is comparative with the baseline condition. However, when the baseline is not available, and there are equal machines (of the same model and nameplate data) being monitored, the analysis can be comparative between the machines. In both cases, the selected data for analysis must be in the same operation range, considering the machine voltage and load level.

The SAS presents features so that the user can access screens with the fault pattern's trend curves and spectrum of a selected electrical signal in order to perform an in-depth analysis. The SAS also provides a report containing the general condition of the SG, including a traffic light (red, yellow or green) indicating the condition regarding the monitored faults. This is automatically updated when the database receives a new signal acquisition, including the summary of the report, which can be accessed on the main screen of the SAS.

It is important to reinforce that the fault types are detected and identified considering the fault patterns and recommendations presented in the Sections 3 and 4, and these fault patterns are comprehensive and include faults normally able to be detected in a predictive maintenance system. The approach of other faults in these systems can be proposed for future works. Finally, with the support of the described system, the decision to stop a machine or not due to indications from the electrical signature will be performed by the maintenance staff and requires some experience and knowledge of the process and the machine's behavior history [7].

6. Results of Fault Detection in In-Service SGs in Bulk Electric Systems by Using ESA

This section presents results of fault detection in in-service SGs in a bulk electric system by using ESA.

The developed methodology of SG CBM based on ESA has been applied to in-service generating units (GUs) of a Brazilian hydroelectric power plant called Goiandira. This power plant is located at Brazilian state of Goias and has two GUs (GU#1 and GU#2) with the following ratings: power = 15 MW, stator voltage = 13,800 V, stator current = 627.6 A, frequency = 60 Hz, 26 poles (salient-pole rotor), excitation voltage = 190 V, excitation current = 480 A, rotor rotation speed = 276.9 rpm.

Two types of faults were detected in this power plant by using the proposed ESA based methodology, as described as follows.

6.1. Early Stage of Stator Phase-to-Phase Short Circuit

Firstly, the power plant personnel had reported that GU#1 presented a covered winding coil. This is because a short circuit had occurred between turns 179 of phase 1 and 184 of phase 2 (the level of severity had not been informed) and the solution adopted to mitigate the problem was the isolation of the stator coils, which caused stator electrical unbalance. Figure 4 presents the part of the GU#1 stator winding where the fault happened. Thus, the electrical signatures were analyzed to search for fault indicators for this type of fault. The fault pattern of EPVA electrical unbalance, as described in (12), has been analyzed.

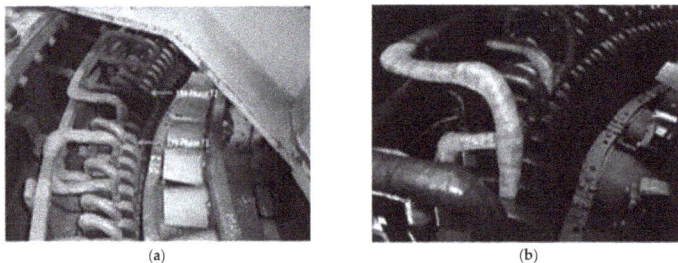

(a)   (b)

Figure 4. Fault indication and mitigation in the stator winding of generating unit (GU) #1: (a) fault indication; (b) fault mitigation.

Here it is important to point out that as a data history of GU#1 and GU#2 before the fault was not available, the analysis was accomplished as a comparison between the two GUs, considering the database stored from the installation time until the analysis moment. As explained in the last section, the data used in this analysis must be in the same operation range, considering the load condition. The data of the GUs were selected in a range between 89.2% and 92.4% of the rated load, considering the current as reference, and the levels of voltage were similar between the GUs. As the GUs were in a similar level of operation, considering the load condition, the selected data were appropriate for an ESA comparative analysis.

Considering the current and voltage signals in the database and the EPVA computed for them by the SAS, Figure 5 presents the trend curves of the EPVA electrical unbalance in voltage and current signatures of the GUs at Goiandira. For this analysis, the daily average of data has been considered, and this has also been performed for all the trend curves presented in this section. It can be noticed that, for both GUs, the voltage unbalance was about 0.4%. The current unbalance was between 0.5% and 1.5% for GU#2 and between 2.5% and 3.0% for GU#1. Thus, GU#1 presented a larger electrical unbalance at the EPVA current signature than GU#2. Thus, the fault indicator of the covered coil was detected in the EPVA electrical unbalance fault pattern.

In order to verify if the data sets of EPVA electrical unbalance of current signature of the GUs were statistically different and reinforce the ESA diagnosis effectiveness, a two-sample t-test has been performed [43] using Minitab software. The data used for the hypothesis testing were the percentage values of electrical unbalance of current signals in the period of analysis. Table 1 presents the results from the two-sample t-test for the fault pattern in analysis.

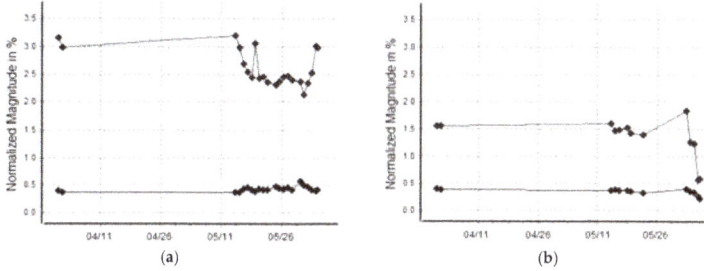

Figure 5. Trend of the Extended Park's vector approach (EPVA) electrical unbalance of voltage (lower curve) and current (upper curve) signatures for GUs at Goiandira power plant: (**a**) GU#1; (**b**) GU#2.

Table 1. Results of the two-sample t-test for the EPVA electrical unbalance at current signature.

GU	Number of Samples	Mean	Standard Deviation	Standard Error of the Mean
GU#1	117	2.535	0.329	0.030
GU#2	46	1.35	0.410	0.060

Estimated difference: 1.1842
Confidence interval of 95% for the difference: (1.0493; 1.3191)
t-Test of the difference = 0 (vs not =): p-value = 0.000

From Table 1, it can be concluded that the data set of the EPVA electrical unbalance at current signature of GU#1 is statistically different from the corresponding data set of GU#2, because p-value < 0.05. The estimated difference is that the mean of the component magnitudes of GU#1 is 1.1842% above the mean of the component magnitudes of GU#2. This result evidences the fault indicator of covered coil.

Figure 6 presents, for illustration purposes, examples of the EPVA electrical unbalance of current and voltage signatures of GUs #1 and #2 of Goiandira power plant. This was performed by choosing an arbitrary signal acquisition for each GU and considering the FFT computed by the SAS. The components are normalized in relation to the fundamental frequency magnitude. By comparing the spectra of Figure 6a,b, one can notice a significant difference between the component magnitudes of GUs #1 and #2, mainly in the EPVA current signature, which evidences the fault indicator of covered coil.

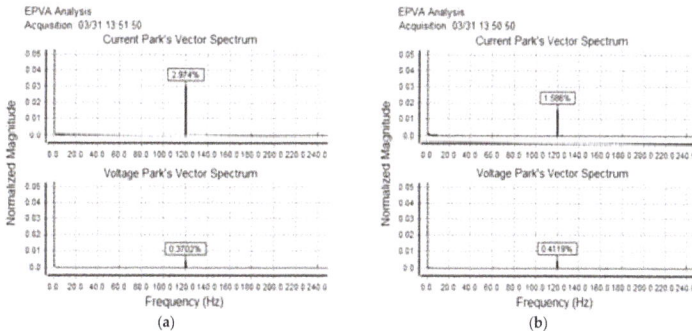

Figure 6. EPVA electrical unbalance of current and voltage signatures of the Goiandira GUs: (**a**) GU#1; (**b**) GU#2.

6.2. Mechanical Misalignment

During the performed analysis, the rotation frequency fault pattern, as presented in (9), was also noteworthy. This fault pattern is indicative of rotor mechanical problems. For the purpose of analysis, the same current and voltage data selected in the last section have been considered in the present section.

Considering the current and voltage signals in the database and the signatures computed by the SAS, Figure 7 presents the trend curves of the rotation frequency of the voltage and current signatures of GUs at Goiandira power plant, considering the average level of the components. It can be noticed that the component's magnitudes were about -90 dB for GU#1 and -95 dB for GU#2, for voltage signals. The component's magnitudes were about -70 dB for GU#1 and -82 dB for GU#2, for current signals.

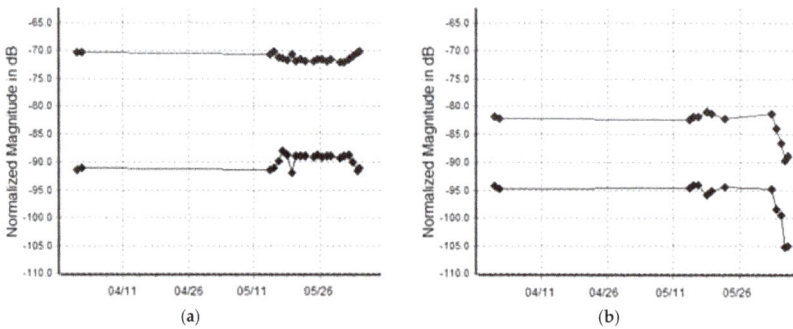

(a)   (b)

Figure 7. Trend of rotation frequency pattern of voltage (lower curve) and current (upper curve) signatures for GUs at Goiandira power plant: (**a**) GU#1; (**b**) GU#2.

There is a significant difference between the rotation frequency component's magnitudes of GU#1 and GU#2, mainly in current signature, and GU#1 presents higher magnitudes in relation to GU#2. This fault pattern is indicative of rotor winding inter-turn short circuit or rotor mechanical problems. After the analysis, the power plant personnel reported that GU#1 presented "dogleg" condition, which is an angular mechanical misalignment between the turbine shaft and the generator rotor shaft. Thus, the hypothesis raised by the analysis has been confirmed.

In order to verify if the data sets of rotation frequency of current signature of GUs were statistically different and reinforce the ESA diagnosis effectiveness, a two-sample t-test has been performed [43] using Minitab software. The data used for the hypothesis testing were the rotation frequency component's magnitudes in dB of current signals in the period of analysis. Table 2 presents the results for the two-sample t-test for the fault pattern in question.

Table 2. Results of two-sample t-test for rotation frequency of current signature.

GU	Number of Samples	Mean	Standard Deviation	Standard Error of the Mean
GU#1	117	−71.357	0.678	0.063
GU#2	46	−83.520	3.430	0.510

Estimated difference: 12.165

Confidence interval of 95% for the difference: (11.140; 13.190)

t-Test of the difference = 0 (vs not =): p-value = 0.000

It can be concluded that the data set of the rotation frequency component's magnitudes of GU#1 is statistically different from the corresponding data set of GU#2, because p-value < 0.05. The estimated difference is that the GU#1 component magnitudes mean is 12.165 dB above the mean of the component magnitudes of GU#2. This result provided evidence for the fault indicator of rotor mechanical misalignment.

Figure 8 presents, for illustration purposes, examples of rotation frequency components of the current and voltage signatures of GU#1 and GU#2 at Goiandira power plant. These were signal acquisitions chosen arbitrarily to show the signatures computed by the SAS. There is a significant difference between the components of GU#1 and GU#2, mainly in current signature, and the component's magnitudes of GU#1 are higher than GU#2. This is evidence of the fault indicator of rotor mechanical misalignment.

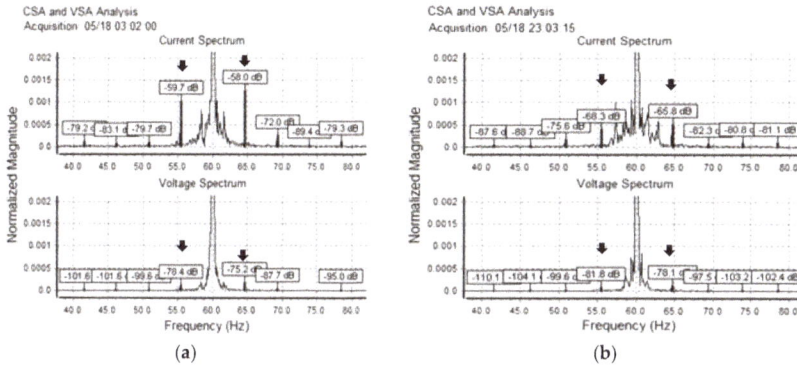

Figure 8. Rotation frequency of the current and voltage signatures of Goiandira GUs: (**a**) GU#1; (**b**) GU#2.

6.3. Final Remarks

It is important to reinforce that the proposed system can detect electrical and mechanical faults, including those presented in Section 3. The current section has focused only on two types of faults (stator electrical unbalance and mechanical misalignment) because these were the faults detected in practice in the in-service SG in the monitored hydroelectric power plant. When working in a laboratory environment, it is possible to simulate different types of faults in a machine and get results covering all of them [3]. However, when considering a power plant application, the studied types of faults will not always occur during the selected period of monitoring and analysis. Finally, the results obtained in a power plant are very valuable, because this environment is concerned with monitoring and is subjected to diverse conditions of bulk power system.

7. Conclusions

This paper has presented a study on ESA for condition monitoring of SGs in bulk electric systems. The paper has presented a methodology for SG predictive maintenance by using ESA and has also addressed peculiarities in the application of this technique to fault detection in wound rotor SGs. The main contribution of this work is the practical results of ESA for fault detection in an in-service SG connected to a power system.

The detected faults were an early stage of stator phase-to-phase short circuit, detected in the EPVA electrical unbalance pattern, and mechanical misalignment, detected in the rotation frequency pattern. The results show the potential of applying ESA to fault detection in SGs interconnected to a power system. Moreover, it has been shown that in this case, the emphasis of ESA must be on CSA. This is because the faults were more evident through the analysis of fault patterns in current signatures than in voltage signatures.

It may be proposed for future works to improve the automatisms and diagnostics of the system by using artificial intelligence techniques and expand the possible monitored conditions beyond fault detection and identification.

Author Contributions: C.P.S., C.F. and G.L.-T. conceived and designed the experiments; C.P.S., L.E.B.d.S., C.F., E.L.B., and L.E.d.L.d.O. performed the experiments; C.P.S., W.C.S., C.F., and G.L.-T. analyzed the data; L.E.B.d.S., E.L.B., L.E.d.L.d.O., and B.S.T. contributed analysis tools; and C.P.S., W.C.S. and G.L.-T. wrote the paper.

Acknowledgments: The authors would like to thank the National Council for Scientific and Technological Development (CNPq), the Coordination for the Improvement of Higher Education Personnel (CAPES), and the Brazilian Electricity Regulatory Agency Research and Development (ANEEL R&D) for supporting this project.

Conflicts of Interest: The authors declare no conflict of interest.

Nomenclature

Abbreviations

CBM	Condition-based maintenance
CSA	Current signature analysis
DFT	Discrete Fourier Transform
ESA	Electrical signature analysis
EPVA	Extended Park's vector approach
FFT	Fast Fourier Transform
GU#1	Generating unit #1
GU#2	Generating unit #2
GUs	Generating units
SAS	Signal analysis software
SGs	Synchronous generators
UMP	Unbalanced magnetic pull
VSA	Voltage signature analysis

Parameters

f_1	Power line frequency
f_c	Specific fault characteristic frequency
f_e	Fault frequency
f_{epva}	EPVA electrical unbalance pattern
f_{pfr}	Spectral components analyzed for SG rotor mechanical problems
f_{php}	Spectral components analyzed for SG even harmonics
f_{phsz}	Zero sequence harmonics fault pattern
f_r	Rotor rotation frequency
i_A	Current of phase A
i_B	Current of phase B
i_C	Current of phase C
i_D	Direct component of Park's vector
i_d	Maximum value of the direct sequence current
i_i	Maximum value of inverse sequence current
i_M	Peak value of the line current
i_Q	Quadrature component of Park's vector
k	Positive integer value
m	DFT index (harmonic order)
n	Time-domain index
N	Number of samples
P	Number of pole pairs
t	Time variable
X_m	m^{th} DFT coefficient
x_n	Time-domain list of equally-spaced complex samples
γ_i	Initial phase angle of the inverse sequence current, in (rad)
θ	Initial phase angle, in (rad)
θ_d	Initial phase angle of the direct sequence current, in (rad)
ω	Angular frequency, in (rad/s)

References

1. Monaro, R.M.; Vieira, J.C.M.; Coury, D.V.; Malik, O.P. A Novel Method Based on Fuzzy Logic and Data Mining for Synchronous Generator Digital Protection. *IEEE Trans. Power Deliv.* **2015**, *30*, 1487–1495. [CrossRef]
2. Gaona, C.A.P.; Blázquez, F.; Frias, P.; Redondo, M. A Novel Rotor Ground-Fault-Detection Technique for Synchronous Machines With Static Excitation. *IEEE Trans. Energy Convers.* **2010**, *25*, 965–973. [CrossRef]
3. Mendonça, P.L.; Bonaldi, E.L.; de Oliveira, L.E.L.; Lambert-Torres, G.; Borges da Silva, J.G.; Borges da Silva, L.E.; Salomon, C.P.; Santana, W.C.; Shinohara, A.H. Detection and modelling of incipient failures in internal combustion engine driven generators using Electrical Signature Analysis. *Electr. Power Syst. Res.* **2017**, *149*, 30–45. [CrossRef]
4. Chothani, N.G. Development and Testing of a New Modified Discrete Fourier Transform-based Algorithm for the Protection of Synchronous Generator. *Electr. Power Compos. Syst.* **2016**, *44*, 1564–1575. [CrossRef]
5. Safari-Shad, N.; Franklin, R.; Negahdari, A.; Toliyat, H.A. Adaptive 100% Injection-Based Generator Stator Ground Fault Protection With Real-Time Fault Location Capability. *IEEE Trans. Power Deliv.* **2018**, *33*, 2364–2372. [CrossRef]
6. Tavner, P.J. Review of condition monitoring of rotating electrical machines. *IET Electr. Power Appl.* **2008**, *2*, 215–247. [CrossRef]
7. Bonaldi, E.L.; de Oliveira, L.E.L.; Borges da Silva, J.G.; Lambert-Torres, G.; Borges da Silva, L.E. Predictive maintenance by electrical signature analysis to induction motors. In *Induction Motors—Modelling and Control*; Araujo, R., Ed.; InTech: Rijeka, Croatia, 2012; pp. 487–520. [CrossRef]
8. Merizalde, Y.; Hernández-Callejo, L.; Duque-Perez, O. State of the Art and Trends in the Monitoring, Detection and Diagnosis of Failures in Electric Induction Motors. *Energies* **2017**, *10*, 1056. [CrossRef]
9. Salomon, C.P.; Santana, W.C.; Borges da Silva, L.E.; Lambert-Torres, G.; Bonaldi, E.L.; de Oliveira, L.E.L.; Borges da Silva, J.G. Induction Motor Efficiency Evaluation using a New Concept of Stator Resistance. *IEEE Trans. Instrum. Meas.* **2015**, *64*, 2908–2917. [CrossRef]
10. Reljic, D.; Jerkan, D.; Marcetic, D.; Oros, D. Broken Bar Fault Detection in IM Operating Under No-Load Condition. *Adv. Electr. Comput. Eng.* **2016**, *16*, 63–70. [CrossRef]
11. Salomon, C.P.; Santana, W.C.; Borges da Silva, L.E.; Lambert-Torres, G.; Bonaldi, E.L.; de Oliveira, L.E.L. Comparison among Methods for Induction Motor Low-Intrusive Efficiency Evaluation Including a New AGT Approach with a Modified Stator Resistance. *Energies* **2017**, *11*, 691. [CrossRef]
12. Cruz, S.M.A.; Cardoso, A.J.M. Stator winding fault diagnosis in three-phase synchronous and asynchronous motors, by the extended park's vector approach. *IEEE Trans. Ind. Appl.* **2001**, *37*, 1227–1233. [CrossRef]
13. Sottile, J.; Trutt, F.C.; Leedy, A.W. Condition Monitoring of Brushless Three-Phase Synchronous Generators With Stator Winding or Rotor Circuit Deterioration. *IEEE Trans. Ind. Appl.* **2006**, *42*, 1209–1215. [CrossRef]
14. Fayazi, M.; Haghjoo, F. Turn to turn fault detection and classification in stator winding of synchronous generators based on terminal voltage waveform components. In Proceedings of the 9th Power Systems Protection & Control Conference (PSPC2015), Tehran, Iran, 14–15 January 2015; pp. 36–41. [CrossRef]
15. Nadarajan, S.; Panda, S.K.; Bhangu, B.; Gupta, A.K. Hybrid Model for Wound-Rotor Synchronous Generator to Detect and Diagnose Turn-to-Turn Short-Circuit Fault in Stator Windings. *IEEE Trans. Ind. Electron.* **2015**, *62*, 1888–1900. [CrossRef]
16. Nadarajan, S.; Panda, S.K.; Bhangu, B.; Gupta, A.K. Online Model-Based Condition Monitoring for Brushless Wound-Field Synchronous Generator to Detect and Diagnose Stator Windings Turn-to-Turn Shorts Using Extended Kalman Filter. *IEEE Trans. Ind. Electron.* **2016**, *63*, 3228–3241. [CrossRef]
17. Shuting, W.; Heming, L.; Yonggang, L.; Fanchao, M. Analysis of stator winding parallel-connected branches circulating current and its application in generator fault diagnosis. In Proceedings of the IEEE 40th Industry Applications Conference (IEEE IAS 2005), Kowloon, Hong Kong, China, 2–6 October 2005; pp. 42–45. [CrossRef]
18. Na, Y.; Yonggang, L.; Tianming, F.; Zhiqian, Y. A study of inter turn short circuit fault in turbogenerator rotor winding based on single-end fault information and wavelet analysis method. In Proceedings of the Eighth International Conference on Electrical Machines and Systems (EMS 2005), Nanjing, China, 27–29 September 2005; pp. 2211–2215. [CrossRef]

19. Yucai, W.; Yonggang, L.; Heming, L. Diagnosis of turbine generator typical faults by shaft voltage. In Proceedings of the 2012 IEEE Industry Applications Society Annual Meeting (IEEE IAS 2012), Las Vegas, NV, USA, 7–11 October 2012; pp. 1–6. [CrossRef]

20. Yucai, W.; Yonggang, L. Diagnosis of short circuit faults within turbogenerator excitation winding based on the expected electromotive force method. *IEEE Trans. Energy Convers.* **2016**, *31*, 706–713. [CrossRef]

21. Toliyat, H.A.; Al-Nuaim, N.A. Simulation and detection of dynamic air-gap eccentricity in salient-pole synchronous machines. *IEEE Trans. Ind. Appl.* **1999**, *35*, 86–93. [CrossRef]

22. Bruzzese, C.; Rossi, A.; Santini, E.; Benucci, V.; Millerani, A. Ship brushless generator shaft misalignment simulation by using a complete mesh-model for machine voltage signature analysis (MVSA). In Proceedings of the 2009 IEEE Electric Ship Technologies Symposium (ESTS 2009), Baltimore, MA, USA, 20–22 April 2009; pp. 113–118. [CrossRef]

23. Joksimovic, G.; Bruzzese, C.; Santini, E. Static eccentricity detection in synchronous generators by field current and stator voltage signature analysis—Part I: Theory. In Proceedings of the XIX International Conference on Electrical Machines (ICEM 2010), Rome, Italy, 6–8 September 2010; pp. 1–6. [CrossRef]

24. Bruzzese, C. Diagnosis of Eccentric Rotor in Synchronous Machines by Analysis of Split-Phase Currents—Part II: Experimental Analysis. *IEEE Trans. Ind. Electron.* **2014**, *61*, 4206–4216. [CrossRef]

25. Ilamparithi, T.; Nandi, S.; Subramanian, J. A disassembly-free offline detection and condition monitoring technique for eccentricity faults in salient-pole synchronous machines. *IEEE Trans. Ind. Appl.* **2015**, *51*, 1505–1515. [CrossRef]

26. Salah, M.; Bacha, K.; Chaari, A.; Benbouzid, M.E.H. Brushless three-phase synchronous generator under rotating diode failure conditions. *IEEE Trans. Energy Convers.* **2014**, *29*, 594–601. [CrossRef]

27. Cui, J.; Tang, J.; Shi, G.; Zhang, Z. Generator rotating rectifier fault detection method based on stacked auto-encoder. In Proceedings of the 2017 IEEE Workshop on Electrical Machines Design, Control and Diagnosis (WEMDCD 2017), Nottingham, UK, 20–21 April 2017; pp. 256–261. [CrossRef]

28. Penrose, H. *Electrical Motor Diagnostics*, 2nd ed.; Success by Design: Old Saybrook, CT, USA, 2008.

29. Sahraoui, M.; Ghoggal, A.; Zouzou, S.E.; Benbouzid, M.E. Dynamic eccentricity in squirrel cage induction motors—Simulation and analytical study of its spectral signatures on stator currents. *Simul. Modell. Pract. Theory* **2008**, *16*, 1503–15138. [CrossRef]

30. Leong, M.S.; Hee, L.M.; Kae, G.Y. Turbine Generator Synchronization—Two Case Studies. *Sound Vib.* **2012**, *46*, 8–11.

31. Jha, R.K.; Pande, A.S.; Singh, H. Design and Analysis of Synchronous Alternator for Reduction in Harmonics and Temperatureby Short Pitch Winding. *Int. J. Emerg. Technol. Adv. Eng.* **2013**, *3*, 651–656.

32. Hao, L.; Wu, J.; Zhou, Y. Theoretical Analysis and Calculation Model of the Electromagnetic Torque of Nonsalient-Pole Synchronous Machines With Interturn Short Circuit in Field Windings. *IEEE Trans. Energy Convers.* **2015**, *30*, 110–121. [CrossRef]

33. dos Santos, H.F.; Sadowski, N.; Batistela, N.J.; Bastos, J.P.A. Synchronous Generator Fault Investigation by Experimental and Finite Element Procedures. *IEEE Trans. Magn.* **2016**, *52*, 1–4. [CrossRef]

34. Salomon, C.P.; Santana, W.C.; Bonaldi, E.L.; de Oliveira, L.E.L.; Borges da Silva, L.E.; Borges da Silva, J.G.; Lambert-Torres, G.; Pellicel, A.; Lopes, M.A.A.; Figueiredo, G.C. A study of electrical signature analysis for two-pole synchronous generators. In Proceedings of the 2017 IEEE Int. Instrumentation and Measurement Technology Conf. (I2MTC 2017), Turin, Italy, 22–25 May 2017; pp. 1–6. [CrossRef]

35. Gopinath, R.; Santhosh Kumar, C.; Ramachandran, K.I.; Upendranath, V.; Sai Kiran, P.V.R. Intelligent fault diagnosis of synchronous generators. *Expert Syst. Appl.* **2016**, *45*, 142–149. [CrossRef]

36. Moosavi, S.S.; Djerdir, A.; Ait-Amirat, Y.; Khaburi, D.A. ANN based fault diagnosis of permanent magnet synchronous motor under stator winding shorted turn. *Electr. Power Syst. Res.* **2015**, *125*, 67–82. [CrossRef]

37. Zhang, X.; Chen, W.; Wang, B.; Chen, X. Intelligent fault diagnosis of rotating machinery using support vector machine with ant colony algorithm for synchronous feature selection and parameter optimization. *Neurocomputing* **2015**, *167*, 260–279. [CrossRef]

38. El Hachemi Benbouzid, M. A review of induction motors signature analysis as a medium for faults detection. *IEEE Trans. Ind. Electron.* **2000**, *47*, 984–993. [CrossRef]

39. Smith, S.W. *The Scientist and Engineer's Guide to Digital Signal Processing*, 2nd ed.; California Technical Publishing: San Diego, CA, USA, 1999.

40. Cruz, S.M.A.; Cardoso, A.J.M. Rotor Cage Fault Diagnosis in Three-Phase Induction Motors by Extended Park's Vector Approach. *Electr. Mach. Power Syst.* **2000**, *28*, 289–299. [CrossRef]
41. Salomon, C.P.; Santana, W.C.; Lambert-Torres, G.; Borges da Silva, L.E.; Bonaldi, E.L.; de Oliveira, L.E.L.; Borges da Silva, J.G.; Pellicel, A.; Figueiredo, G.C.; Lopes, M.A.A. Discrimination of Synchronous Machines Rotor Faults in Electrical Signature Analysis based on Symmetrical Components. *IEEE Trans. Ind. Appl.* **2017**, *53*, 3146–3155. [CrossRef]
42. Boldea, I. *Synchronous Generators—Electrical Generators Handbook*; CRC Press: Boca Raton, FL, USA, 2006.
43. Montgomery, D.C.; Runger, G.C. *Applied Statistics and Probability for Engineers*, 3rd ed.; John Wiley and Sons: New York, NY, USA, 2003.

energies

MDPI

Article

The Influence of Permanent Magnet Material Properties on Generator Rotor Design

Petter Eklund and Sandra Eriksson *

Division of Electricity, Uppsala University, SE-751 05 Uppsala, Sweden; Petter.Eklund@angstrom.uu.se
* Correspondence: Sandra.Eriksson@angstrom.uu.se

Received: 28 February 2019; Accepted: 29 March 2019; Published: 5 April 2019

Abstract: Due to the price and supply insecurities for rare earth metal-based permanent magnet (PM) materials, a search for new PM materials is ongoing. The properties of a new PM material are not known yet, but a span of likely parameters can be studied. This paper presents an investigation on how the remanence and recoil permeability of a PM material affect its usefulness in a low speed, multi-pole, and PM synchronous generator. Demagnetisation is also considered. The investigation is carried out by constrained optimisation of three different rotor topologies for maximum torque production for different PM material parameters and a fixed PM maximum energy. The rotor topologies used are surface mounted PM rotor, spoke type PM rotor and an interior PM rotor with radially magnetised PMs. The three different rotor topologies have their best performance for different kinds of materials. The spoke type PM rotor is the best at utilising low remanence materials as long as they are sufficiently resistant to demagnetisation. The surface mounted PM rotor works best with very demagnetisation resistant PM materials with a high remanence, while the radial interior PM rotor is preferable for high remanence materials with low demagnetisation resistance.

Keywords: permanent magnet synchronous generator; electrical machine design; permanent magnet material

---

1. Introduction

Low speed, high torque synchronous machines are primarily used in wind power, as direct driven generators. They can employ permanent magnet (PM) excitation to reduce complexity and increase efficiency. The most common group of PM materials used are based on neodymium-iron-boron (Nd-Fe-B) systems, which have very good performance [1]. In the last decade, there has been a volatility in price of these kinds of PMs, which has sparked the search for more economically stable alternatives [2]. One option is to use ferrites, but this requires a more mechanically complex and heavier rotor, such as a flux concentrating spoke type rotor [3–5], than the relatively simple surface mounted PM rotor that can be used with Nd-Fe-B [6,7]. Another approach is the development of new PM materials [8]. Suitable rare earth metal–free systems for use as PM materials are currently being investigated. Ref. [9] presents a theoretical study of Fe_xCo_{1-x} alloys. Ref. [10] is a theoretical study of magnetic properties of the alloys FeNi, CoNi, MnAl, and MnGa. Ref. [11] presents measurements for MnAl. These novel materials are not likely to outperform the Nd-Fe-B and Samarium-Cobalt (Sm-Co) PM materials but could provide a cost efficient rare-earth metal free alternative. In addition, research is performed on recycled rare-earth magnets, expected to have properties in between ferrites and commercial Nd-Fe-B [12]. It can be of interest to investigate how to best utilise possible new or recycled materials, even though the novel magnets are not developed and commercialised yet.

In this study, the suitability of different PM material properties for use with different PM synchronous generator rotor topologies is investigated. Different rotor topologies subject the PM to different permeance, which results in different load lines. In [1], it is discussed how different load lines interact with the demagnetisation magnetic flux density curve. Different rotor topologies have been studied previously. In [13], three topologies, similar to those studied here, are compared for use in 4-pole traction motors for trains. Similar topologies are compared for use as an aircraft starter-generator in [14]. Five different topologies, not including the spoke-type rotor, are compared in [15]. Ref. [16] also compares five different topologies, including the spoke-type rotor. Ref. [17] compares machines with Nd-Fe-B, Sm-Co and alnico for surface mounted 4-pole machines. Ref. [18] compares three machine configurations for interior v-shaped magnets for two different materials: the novel material $NdFe_{12}N_x$ and a conventional Nd-Fe-B magnet. A spoke-type rotor with ferrites is compared to a rotor with surface mounted Nd-Fe-B magnets for a wind power generator in [19,20]. A comparison of demagnetization risk for the same generator types is presented in [21].

The aim of this paper is to study how the magnetic properties of PM materials affect the machine design in low-speed radial flux PM generators. To our knowledge, there has not been any previous study combining material properties with rotor design and rotor topology choice. The results from this study could give hints on which property to improve when developing new PM materials as well as on the suitability of a material with certain magnetic properties for different generator topologies.

2. Scope and Limitations

The PM material is described by three parameters and three different PM rotor topologies are compared. The PM material parameters used are the remanent flux density B_r, the recoil permeability μ_{rec}, and a demagnetization parameter denoted $C_{B_{PM}^{min}}$. The second quadrant demagnetisation curve given by the parameters is shown in Figure 1. The demagnetisation parameter $C_{B_{PM}^{min}}$ gives a minimum value of magnetic flux density along the magnetisation allowed in the PM as fraction B_r. Considering demagnetisation is important since the different topologies can be expected to provide different levels of protection against demagnetisation [21].

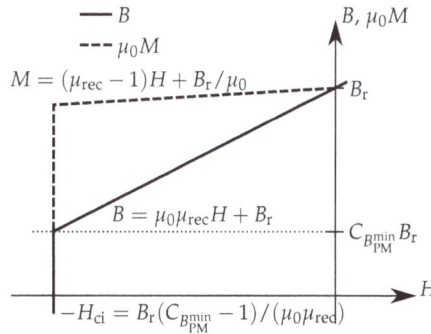

Figure 1. The second quadrant demagnetisation curve of the PM material in terms of the material parameters used: remanent flux density B_r, recoil permeability μ_{rec}, and the demagnetisation parameter $C_{B_{PM}^{min}}$. B is magnetic flux density, M is magnetisation, H the applied magnetic field, and H_{ci} the intrinsic coercivity.

The figure of merit used in comparison is torque per unit machine length and pole pair, with a fixed electrical loading of the stator and a fixed amount of magnetic energy in the rotor. The machine used in the study is a generator with 40 poles and a stator inner diameter of slightly over one meter; no length is set as all calculations are done per unit length. The generator would be suitable for use in a small scale direct drive wind turbine. A direct drive generator for a wind turbine is normally not operated at high loading at normal operation, so no saturation or flux weakening operation is expected. Typical torque curves for this type of generator can be seen in [19].

Magnetic properties for PMs are normally defined with temperate coefficients explaining how they change with temperature. The coefficients vary both in size and in sign and it is therefore not possible to include temperature variations of magnetic properties in this study. However, as the generator is a small scale, large-diameter generator which is considered sufficiently cooled (or heated if needed), it can be considered to have a rather small allowed temperature span. The material parameters in the results should be interpreted as those valid for the operating temperature.

The stator is made as generic as possible, without any specific winding scheme and will not be subject to optimisation. The end windings are not included in the model and all calculations are performed per unit of machine length, i.e., only two dimensions are considered. It is also kept as geometrically simple as possible to make meshing for the simulation easier. The rotor topologies are optimised only with respect to the rectangular shape of the PM, with the volume of PM material kept fixed for each point in the parameter space. The amount of iron is not part of the objective function but is changed as necessary to accommodate the PM. Structural integrity is not considered and it is assumed that the PM material can be shaped into blocks of the required sizes.

3. Method

The three different PM rotor topologies are optimised for torque production, for multiple points in a parameter space of PM material properties, and are then compared. The topologies compared are surface mounted PM rotor, spoke PM rotor and capped PM rotor, all of which will be presented in detail below. The stator is kept the same both geometrically and electrically. Each PM material point is defined by the values of the PM material parameters B_r, μ_{rec}, and $C_{B_{PM}^{min}}$. The values of B_r are varied between 0.3 T and 1.5 T, in steps of 0.1 T. The lower bound is chosen because a material with lower B_r is unlikely to be useful in an electrical machine, and the upper bound is chosen slightly above what is commercially available. The values of μ_{rec} are varied between 1 and 2, in eight logarithmically distributed steps. The lower bound is given by fundamental physics and the upper is judged to be high enough for the investigated space to contain most useful materials. Approximate PM material parameter values of commercially available PM material types are listed in Table 1 for reference.

Table 1. Properties of commercially available PM materials. Data compiled from data sheets of various PM manufacturers, for temperatures in the range of 20 °C to 30 °C. H_{ci} is the intrinsic coercivity of the PM, for alnico PMs, the normal coercivity is given instead (as the data sheets does not give H_{ci}). No $C_{B_{PM}^{min}}$ is given for alnico since that class of materials is usually used on a demagnetisation minor loop and $C_{B_{PM}^{min}}$ therefore depends on which minor loop is chosen. When μ_{rec} is not given, it is estimated by $\mu_{rec} = B_r^2/(4\mu_0|BH|_{max})$, i.e., by assuming that the energy maximum occurs in the linear part of the magnetisation curve, estimated μ_{rec} are marked with a "*".

Material Family	B_r [T]	H_{ci} [kA/m]	$|BH|_{max}$ [kJ/m³]	μ_{rec} [-]	$C_{B_{PM}^{min}}$ [-]
Alnico	0.55–1.37	38–151	10.7–83.6	1.3–6.2	-
Hard ferrite	0.20–0.46	140–405	6.4–41.8	1.05–1.2	−0.4–0.4
Nd-Fe-B	1.08–1.49	876–2710	220–430	1.0–1.1 *	−2.3–0.2
Sm-Co	0.87–1.19	1350–2400	143–251	1.0–1.1 *	−2.7−−0.62

The amount of PM material used in a given PM material point is chosen to give the same PM maximum energy, in order to allow fair comparison between materials. The PM maximum energy, assuming a linear second quadrant demagnetisation curve, is calculated as

$$E_{PM} = A_{PM}l|BH|_{max} = A_{PM}l\frac{B_r^2}{4\mu_{rec}\mu_0}, \tag{1}$$

where $|BH|_{max}$ is the maximum energy product of the PM material, A_{PM} is the cross section area of the PM and l the length of the machine, set to unity. The resulting A_{PM} for $E_{PM}/l = 164$ J/m varies from 3.67×10^{-4} m² to 1.84×10^{-2} m². The $E_{PM}/l = 164$ J/m is chosen to match that of a surface mounted PM covering $\frac{3}{4}$ of the pole pitch, with a height of twice the air gap length, $B_r = 1.3$ T, and $\mu_{rec} = 1.0$

(assuming $C_{B_{PM}^{min}} < 0.5$). This can be considered a typical generator design for a rotor with Nd-Fe-B magnets and thereby gives a representative E_{PM}/l-value.

A parameterised geometry is created for each of the rotor topologies and the stator. The geometry only represents a two-dimensional cross section, as common when simulating radial flux machines, to save computation time. This can be done since, for the most of the machine length, there is very little change of geometry in the axial direction and since end effects are expected to be small as long as the axial length is of sufficient length. The equations governing magnetostatics, in vector potential formulation, are combined with constitutive equations, boundary conditions and source terms, and solved using the finite element (FE) method on the parametrised geometry. The torque is calculated from the obtained field by integrating the Maxwell stress in the air gap. Constrained optimisation is done of the ratio between height (along magnetisation) and width of the PM, on a closed interval bounded by the requirement to keep the geometry consistent.

3.1. Geometries and Rotor Topologies

Both the stator geometry and all of the rotor geometries have their sizes defined in terms of the pole pitch, τ_p, number of pole pairs, P, PM height, h_{PM}, and PM width, w_{PM}. How the remaining geometrical parameters depend on these are listed in Table 2. The parameters are chosen to give a reasonable generator geometry for a small scale direct drive wind turbine. The choice of air gap length in a PM generator is mainly made from mechanical considerations. In literature, it is given that, for large machines, the air gap length needs to be $\delta \geq D_{si}/1000$ due to manufacturing tolerances [22], p. 306. To be on the safe side, an air gap length of $\delta = D_{si}/250$ is chosen.

Table 2. The relationships between the geometrical parameters and their default values.

Quantity	Symbol	Expression	Value
Pole pitch	τ_p	-	80 mm
Number of pole pairs	P	-	20
Number of slots per pole	q	-	15/2
Stator inner diameter	D_{si}	$2P\tau_p/\pi$	1.02 m
Air gap length (mechanical)	δ	$D_{si}/250$	4.07 mm
Slot pitch	τ_s	τ_p/q	10.7 mm
Slot depth	d_{slot}	$3\tau_p/4$	60 mm
PM height, along magnetisation	h_{PM}	Set by optimiser	
PM width, across magnetisation	w_{PM}	Set by optimiser	

The stator geometry is simple, with a minimum of details in order to make it easier to mesh for the simulations—see Figure 2. The slots and teeth are of the same width at the inner periphery, and the slots are rectangular. A nonintegral number of slots per pole, $\frac{15}{2}$, is chosen to reduce the cogging. The depth of the slots is chosen to give a typical current density in the conductor, while at the same time achieving a typical linear current density along the stator periphery, reasonable values are taken from [22], p. 298. The yoke is set large enough not to limit the magnetic flux, since the amount of iron used is not part of the optimisation.

The rotor topologies used are the surface mounted PM rotor, capped PM rotor, and spoke PM. While there are many other topologies to choose from and variations on each of them, see e.g., [13–16,18,19,23,24], the three chosen topologies are all relatively simple. All three topologies allow the PM to be represented as a rectangular block with a well defined height and width, along and across magnetisation, respectively. All rotors are designed to give the same air gap length, δ but the width of the pole face on the rotor is dependent on topology and PM size.

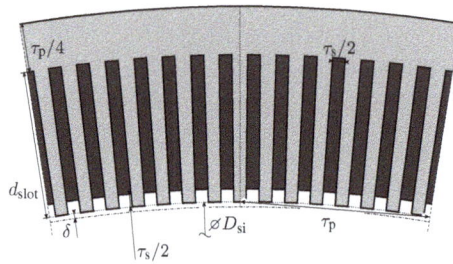

Figure 2. The geometry of the stator, only showing the two poles needed for symmetry. The grey shaded area is steel. Black shaded areas are the armature winding. The dot-dashed lower boundary represents the rotor surface. The white parts of the geometry are occupied by air.

The surface mounted PM rotor consists of an iron back ring and the PMs, mounted directly on the ring. The geometry is shown in Figure 3. The PMs are curved to give a constant mechanical air gap length in front of the PMs. The curvature is the same on both curved sides, and the straight sides are parallel, such that the cross sectional area still is $h_{PM} \times w_{PM}$. Wedges holding the PMs are not included in the geometry to make it simple to mesh for the FE solver.

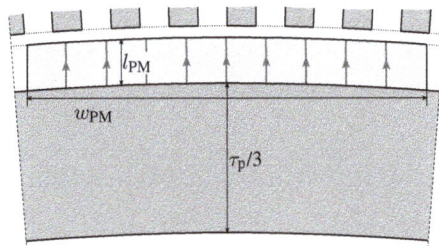

Figure 3. One pole of the surface mounted PM rotor. The dashed lines on the sides are the inter pole boundary. The dotted lines are simulation geometry boundaries. The grey shaded areas are the iron of the rotor back ring and parts of the stator. The light grey areas hatched with arrows are the PMs, with the arrows indicating the direction of magnetisation for a north pole. The white parts of the geometry are occupied by air.

The capped PM rotor is an interior PM rotor with radially magnetised PMs see Figure 4. Its design is similar to the surface mounted rotor, with the addition of a magnetically soft pole shoe on top of the PM. This evens out the magnetic flux density in the PM and air gap slightly and helps protect the PM from demagnetisation. This kind of design is described as suitable for use with alnico PMs in [23], and a similar interior PM rotor design, with few poles and used for motor applications, is compared to a few other interior PM rotor topologies in [15]. Height of the pole shoe is fixed and not subject to optimisation. The width is changed to match the PM.

The spoke type PM rotor, also called tangential or circumferential interior PM rotor, consists of a ring formed by PMs of alternating, tangential magnetisation, separated by magnetically soft pole pieces that guide the magnetic flux into the air gap. The design of one rotor of this type is shown in [23] and is described as suitable for use with ferrite PMs. It is also used in [3] in a rotor with ferrite PMs intended to be interchangeable with a surface mounted rotor with Nd-Fe-B PMs. Pole pieces are sized to fill up the space between the PMs, starting at a slightly larger radius than the inward face of the PM and extending to the rotor periphery—see Figure 5. To avoid leakage flux between the poles at the rotor periphery, there is a slot in the rotor surface with the PM at the bottom.

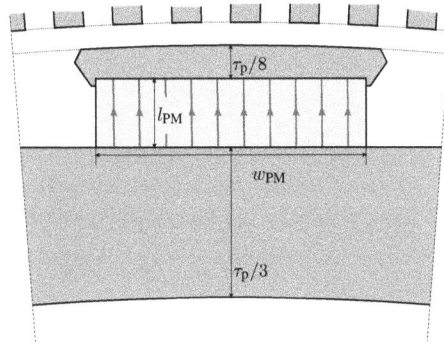

Figure 4. One pole of the capped PM rotor geometry. The dashed lines on the sides are the inter pole boundary. The dotted lines are simulation geometry boundaries. The grey shaded areas are the iron of the rotor back ring, the pole shoe and steel parts of the stator. The light grey area hatched with arrows is the PM, with the arrows indicating the direction of magnetisation for a north pole. The white parts of the geometry are occupied by air.

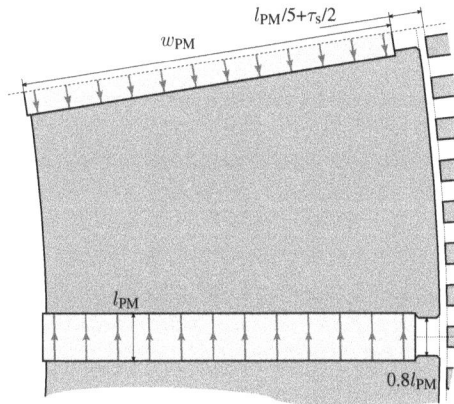

Figure 5. One pole of the spoke PM rotor geometry. The dashed line at the top is the inter pole boundary. The dotted lines are simulation geometry boundaries. The grey shaded areas are the iron of the pole pieces and parts of the stator. The light grey areas hatched with arrows are the PMs, with the arrows indicating the direction of magnetisation for the magnets surrounding a north pole. The white parts of the geometry are occupied by air.

3.2. Finite Element Modelling

The finite element software used is COMSOL Multiphysics® v. 5.2a (COMSOL AB, Stockholm, Sweden) together with the MATLAB® LiveLink™ package (MATLAB R2017a, Mathworks, Inc., Natick, MA, USA) to interface with the optimiser. The field equation solved is the one of magnetostatics, i.e., stationary simulations, with magnetic vector potential and all out of plane derivatives set to zero. This ensures that only the out of plane component of the vector potential needs to be calculated, since the in plane components are constant zero.

On the boundaries on the outside of the stator, a magnetic insulation boundary condition, i.e., zero normal component of the magnetic field, is used. For the boundaries between the pole pairs, periodic boundary conditions are used, in order to make use of the symmetry in the geometry and reduce the size of the computational domain. Between the stator and rotor mesh, a sliding boundary is used, to avoid re-meshing when rotating the rotor. The boundary conditions are shown in Figure 6.

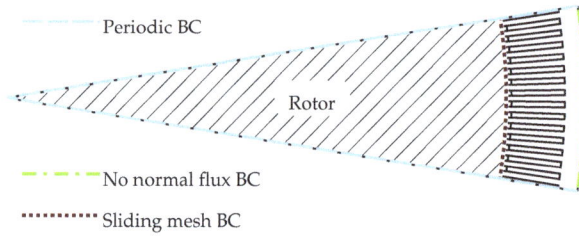

Figure 6. The boundary conditions (BC) used, the rotor part of the geometry has been hatcheted. Simulations are performed for two poles.

The constitutive equations used to relate $\vec{B}$ and $\vec{H}$ in the materials are

$$\vec{B} = \mu_0 \vec{H}, \tag{2}$$

where μ_0 is the permeability of free space, for the air gap and stator slots. In the PMs, the material is represented by

$$\vec{B} = \vec{B}_r + \mu_0 \mu_{rec} \vec{H} \tag{3}$$

and, for the soft iron parts, the equation is

$$\vec{B} = \frac{f(\|\vec{H}\|)\vec{H}}{\|\vec{H}\|}, \tag{4}$$

where f is a monotonically increasing function representing the *B-H* curve of the iron. It is computed by table lookup with linear interpolation with data supplied as the generic "Soft Iron" material from the material library of COMSOL Multiphysics®. The *B-H* curve of the soft iron can be seen in Figure 7. The soft iron no conductivity and no hysteresis, i.e., iron losses are not included in the simulations.

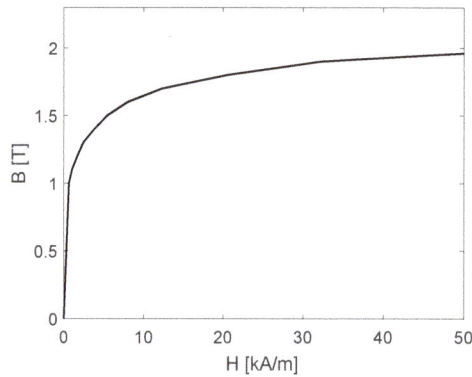

Figure 7. The *B-H* curve of the soft iron used for stator steel and all iron parts in the simulations.

The stator currents are introduced as regions of out-of-plane current density located in the slots. The distribution is sinusoidal, with an amplitude chosen to give a linear current density of $A_s = 45$ kA/m (RMS) aligned to give resistive load. The current density is 2.5 A/mm^2

The meshing is done using the built in algorithms of the FE software, with the element size in the air gap set to be less than a third of the air gap length. This is done to ensure the solution in the air gap will be of sufficient quality to allow accurate calculation of the torque. The torque generated by the generator is calculated using the Arkkios method [25]. To reduce the influence of cogging on the results, the rotor is rotated one slot pitch, in ten steps, and the mean torque is calculated. The mesh for an example geometry is shown in Figure 8. Note that the mesh in the rotor will vary for each iteration, i.e., it varies both with PM shape and with rotor type.

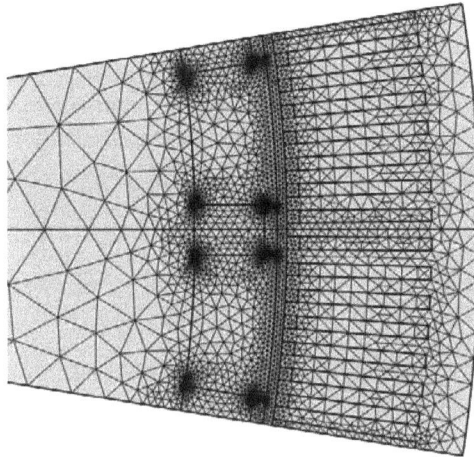

Figure 8. Example of a mesh from the simulations, here shown for a generator with spoke type rotor.

3.3. Optimiser

The variable optimised is the shape of the PM given by its height along magnetisation, h_{PM}, and width, across magnetisation, w_{PM}, with the objective to find the maximum torque. The cross sectional area of the PM, $A_{PM} = h_{PM}w_{PM}$, is kept constant. For the constrained optimisation, the MATLAB standard library function *fminbnd* is used. The function uses a combination of golden ratio search and successive parabolic interpolation to find the minimum of a unimodal function. An initial rough search is performed, to find a better interval for *fminbnd* to work on. The search starts by computing the torque in the lowest possible value of h_{PM}, storing the value, and then computing the torque for $3/2$ times the previous value of h_{PM} until either the maximum possible value of h_{PM} is reached, or the torque for the last h_{PM} is of smaller magnitude than that in previous steps.

3.4. Stability Analysis

To test the stability of the results against changes in the parameters, a stability analysis is performed by changing one parameter at the time. The parameters A_s, E_{PM}, δ, and τ_p are tested at $\pm 25\%$; additionally, δ is tested at 0.25 and two times the default value. The number of pole pairs is tested at ± 5. The depth of the stator slots is tested at $2/3$ of their default value. The stability tests are run on a sparser grid of B_r and μ_{rec} with only four values each ($\mu_{rec} = 1, 1.22, 1.66, 2.0$ and $B_r = 0.3, 0.7, 1.1, 1.5$). All values of $C_{B_{PM}^{min}}$ are still used.

3.5. Demagnetisation Prevention

To determine if the PM is risking demagnetisation, the component of the magnetic flux density parallel to the remanent magnetisation, $B_{\|B_r}$, is compared to a threshold value B_{PM}^{min}, and the volume fraction of the PM where $B_{\|B_r} < B_{PM}^{min}$, denoted Q_{demag}, is computed. The computation of Q_{demag} is done for two cases, with no current in the stator windings, i.e., no load (NL), and for a symmetrical short circuit (SC). The short circuit case is a symmetrical short circuit at constant rotational speed. Two tolerances are used to determine if the demagnetisation is acceptable. The first is that Q_{demag} at NL, Q_{demag}^{NL}, should be less than $Q_{demag}^{NL, tol} = 2\%$. The second is that Q_{demag} at SC, Q_{demag}^{SC}, should not be more than $Q_{demag}^{SC, tol} = 2\%$ greater than Q_{demag}^{NL}, i.e., $Q_{demag}^{SC} - Q_{demag}^{NL} < Q_{demag}^{SC, tol}$.

The search for a PM size that is resistant to demagnetisation starts by computing $Q_{\text{demag}}^{\text{NL}}$ and $Q_{\text{demag}}^{\text{SC}}$ for a h_{PM} larger than the h_{PM} giving the optimum torque when not constrained by demagnetisation. The new h_{PM} is computed by taking the minimum of twice the original h_{PM}, and half the sum of the maximum h_{PM} allowed by the geometry and the original h_{PM}.

If the demagnetisation is below tolerance, linear interpolation is used to estimate the h_{PM} for where the tolerance is met and both cases of Q_{demag} is calculated for that new h_{PM}. This is repeated until the interval of h_{PM}, in which the tolerance is met, lies is less than a tolerance (set to 0.2 mm). Bisection of the interval is used instead of linear interpolation when the change in $Q_{\text{demag}}^{\text{SC}}$ over the interval is less than $1/50$ of $Q_{\text{demag}}^{\text{SC, tol}}$, or the interval is too large (larger than the initial h_{PM} or five times the last interval) to speed up convergence.

If the new h_{PM} does not meet the demagnetisation tolerance, a larger value of h_{PM} is tried until either a h_{PM} which meets the tolerance is found, or maximum allowed h_{PM} is tried and the search aborted if the demagnetisation tolerance is still not met. Should minimum $Q_{\text{demag}}^{\text{SC}} - Q_{\text{demag}}^{\text{NL}}$ occur inside the span of the tested h_{PM}, but fail to meet the tolerance, a second order polynomial is used to estimate where the minima lies and a new point is calculated. The polynomial is obtained by curve fitting using the five data-points surrounding the minima when ordered by h_{PM}. Should 10 attempts to either expand or refine the set of tested points fail to locate a point where the demagnetisation tolerance is met, the search is aborted.

The value of $B_{\text{PM}}^{\text{min}}$ for each material is given by

$$B_{\text{PM}}^{\text{min}} = C_{B_{\text{PM}}^{\text{min}}} B_{\text{r}}, \tag{5}$$

where $C_{B_{\text{PM}}^{\text{min}}}$ is a proportionality constant, the values $-\infty$, -0.2, 0 and 0.2 are used. Using $C_{B_{\text{PM}}^{\text{min}}} = -\infty$ corresponds to disregarding demagnetisation. For finite $C_{B_{\text{PM}}^{\text{min}}}$, the intrinsic coercivity of the PM material can be computed as

$$H_{\text{ci}} = \frac{1 - C_{B_{\text{PM}}^{\text{min}}}}{\mu_{\text{rec}} \mu_0} B_{\text{r}} \tag{6}$$

if a sharp and vertical knee of the demagnetisation curve is assumed. Should the knee be rounded, the knee starts at the value given by Equation (6) and the intrinsic coercivity is larger, depending on the sharpness of the knee.

3.6. Stator Current Prediction

Since there is no defined winding scheme, the armature currents are modelled as a sinusoidal (in azimutal space coordinate) current density distribution with the period of one pole pair. This current density distribution can be decomposed into two sinousoids corresponding to the direct axis (d-axis), and quadrature axis (q-axis), respectively. These in turn give rise to a d-axis magneto-motive force (MMF), $\mathcal{F}_d$, and a q-axis MMF, $\mathcal{F}_q$. To each of these MMFs, there is also a magnetic flux linking to it, denoted Φ_d for the d-axis flux, and Φ_q for the q-axis flux.

Following the method in [26], the MMFs are computed from an arbitrary current distribution as

$$\mathcal{F}_d = \int_S J_z \cos\theta_{\text{el}} \, dS \tag{7}$$

and

$$\mathcal{F}_q = - \int_S J_z \sin\theta_{\text{el}} \, dS, \tag{8}$$

where J_z is the out of plane component of the current density, θ_{el} is the electrical angle in a rotor reference frame with zero at the q-axis after a north pole when turning counter-clockwise; and S is the area occupied by the windings. To introduce a certain MMF on either axis, the current distribution can be written as

$$J_z = \mathcal{F}_d / A_{\text{eff}} \cos\theta_{\text{el}} \tag{9}$$

or as

$$J_z = \mathcal{F}_q / A_{\text{eff}} \sin\theta_{\text{el}} \tag{10}$$

for the d-axis and the q-axis, respectively, where

$$A_{\text{eff}} = \int_S \cos^2\theta_{\text{el}} \, dS \tag{11}$$

is the effective winding area. The magnetic fluxes for the d-axis and the q-axis, respectively, can be computed as

$$\Phi_d = \frac{1}{A_{\text{eff}}} \int_S A_z \cos\theta_{\text{el}} \, dS, \tag{12}$$

$$\Phi_q = -\frac{1}{A_{\text{eff}}} \int_S A_z \sin\theta_{\text{el}} \, dS, \tag{13}$$

where A_z is the out of plane component of the magnetic vector potential.

The voltage per turn and unit length, V_d and V_q for d-axis and q-axis, respectively, is given by

$$V_d = \varrho_s \mathcal{F}_d + \frac{d\Phi_d}{dt} - \omega_{\text{el}}\Phi_q, \tag{14}$$

$$V_q = \varrho_s \mathcal{F}_q + \frac{d\Phi_q}{dt} + \omega_{\text{el}}\Phi_d, \tag{15}$$

where t is time, ω_{el} electrical angular frequency, and $\varrho_s = 1/(A_{\text{eff}}\sigma)$ is the resistance of the stator winding as seen by the magnetic circuit, $\sigma = 2.97 \times 10^7 \, \text{S}\,\text{m}^{-1}$ is the effective conductivity of the winding. The value is chosen to match copper at 70 °C [27] and 60% of the area occupied by the winding is filled by conductor material.

For resistive load and steady state, this gives

$$\mathcal{F}_d = \omega_{\text{el}}/(\varrho_l + \varrho_s) \, \Phi_q, \tag{16}$$

$$\mathcal{F}_q = -\omega_{\text{el}}/(\varrho_l + \varrho_s) \, \Phi_d, \tag{17}$$

where ϱ_l is a load resistance chosen such that $\omega_{\text{el}}/(\varrho_l + \varrho_s) = \sqrt{2}A_s\tau_p P/\sqrt{\Phi_d^2 + \Phi_q^2}$. This system of algebraic equations is then coupled to the field equations and solved together with these by the non-linear solver of the FE software.

For a symmetrical short circuit, $V_d = V_q = 0$. The system is linearised by setting $\Phi_d = \Lambda_d\mathcal{F}_d + \Phi_{\text{PM}}$ and $\Phi_q = \Lambda_q\mathcal{F}_q$, where Λ_d and Λ_q are the permeances for the magnetic axes, and $\Phi_{\text{PM}} = \Phi_d$ at $\mathcal{F}_d = \mathcal{F}_q = 0$ is the magnetic flux from the PM. If ω_{el} is constant, there exists a closed form solution to Equations (14) and (15), and the parameters can be estimated from two FE solutions. One FE solution should have both MMFs set to zero, and the other should have large currents—ten times nominal current is used.

The MMFs to be used for the demagnetisation calculation are taken at the time where $\mathcal{F}_d$ has its minimum value. This is found by computing the MMFs at a hundred instants during the first oscillation of the solution, and then performing a numerical search to refine the minima found.

4. Results and Discussion

The maximum achieved torque for each material point, and the shape of the PM are calculated. The torque given is per pole-pair and unit of machine length. There are three different reasons for missing material points. First, for low energy density materials, the volume of material required for the given energy does not fit into the geometry. Second, the PM material cannot be configured in such a way that the operating condition of $45 \, \text{kA m}^{-1}$ stator current loading with resistive load can be met. Third, the PM cannot be made sufficiently high, along the direction of magnetisation, to resist demagnetisation.

The torque for $C_{B_{PM}^{\min}} = 0.2$, the most demagnetisation sensitive case, and $C_{B_{PM}^{\min}} = -0.2$, the least demagnetisation sensitive case, are shown in Figure 9. The spoke type rotor shows similar shape of the torque over the B_r and μ_{rec} surface for both values of $C_{B_{PM}^{\min}}$, with a maxima on the $\mu_{rec} = 1$ line and a plateau of nearly constant torque extending to the upper-right, and lower values toward the top left and bottom right corners. Adapting the shape of the PM to resist demagnetisation for a PM material with lower demagnetisation resistance lowers the maximum torque, so the location of the maximum torque shifts to a higher value of B_r. For the lower demagnetisation resistance (left column in Figure 9), there are also a lot of points where the PM cannot be shaped to resist demagnetisation at all. The spoke type rotor imposes geometrical limits on PM height (along the PM magnetisation), which both gives the PM a high effective demagnetising factor along the magnetisation, and limits the achievable reluctance of the PM. The first of these introduces a risk of self-demagnetisation. The second makes the PM more sensitive to demagnetisation caused by the armature currents, and increases the reactance. For the capped PM rotor, the maximum torque is on the $B_r = 1.5 \, \text{T}$ line for both demagnetisation sensitivities.

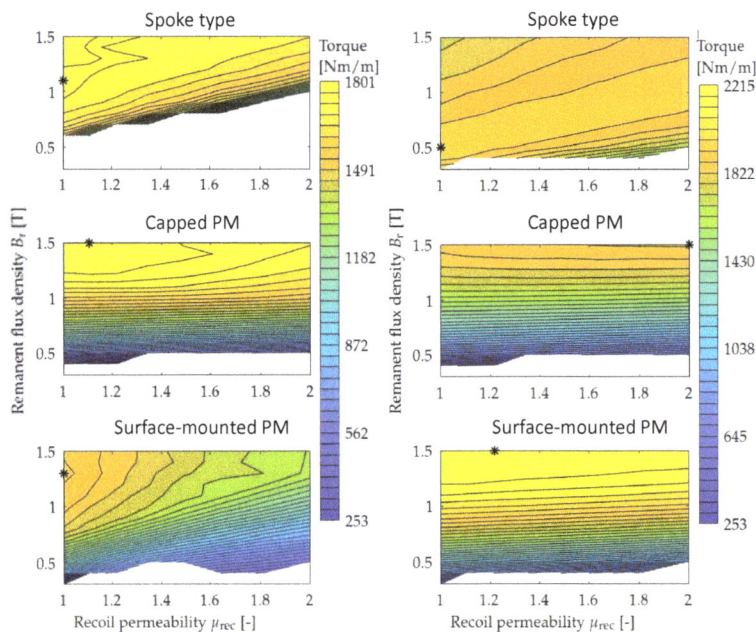

Figure 9. Contour plot of torque per pole pair and unit length as a function of remanence B_r and recoil permeability μ_{rec}. The left column is for $C_{B_{PM}^{\min}} = 0.2$ and the right column is for $C_{B_{PM}^{\min}} = -0.2$. The colour scale is shared between plots in the same column. The differences between two contour lines are $38.7 \, \text{Nm/m}$ and $49.1 \, \text{Nm/m}$ for the left and right columns, respectively. The asterisk indicates the point of highest torque for each plot.

For B_r below about 1 T, the protection against demagnetisation offered by the rotor is sufficient, i.e., there is no conflict between torque production and demagnetisation. In this region, the torque is, however, lower than that of the spoke type rotor.

The surface mounted PM rotor gives higher torque than the two other topologies for more demagnetisation resistant materials. With a more demagnetisation sensitive PM material, the produced torque is decreased as the PM has to be made higher (along the magnetisation) to avoid demagnetisation. This in turn reduces the flux, since the width of the PM is reduced to fulfil the requirement that A_{PM} should be kept constant for a given pair of B_r and μ_{rec} values.

The impact of changing the design to protect the PM from demagnetisation is further shown in Figure 10. The maximum torque obtained for a given PM material is plotted over remanence for two values of recoil permeability and all four values of $C_{B_{PM}^{min}}$. The capped PM rotor is most resistant to demagnetisation, making it useful for high B_r materials with low demagnetisation resistance. For this kind of PM material, the capped PM outperforms the other two topologies. The spoke type and surface mounted PM rotor on the other hand show larger sensitivity to demagnetisation. The surface mounted PM rotor has the best performance for high B_r materials that have good resistance against demagnetisation, but, for more sensitive materials, this performance diminishes as the design is adapted to protect the PM. The surface mounted PM rotor is also the only topology that needs to be modified to resist demagnetisation for $C_{B_{PM}^{min}} = -0.2$. For low B_r, below about 0.8 T, the spoke type rotor gives the highest torque of the three studied topologies. Adaptation to protect the PM from demagnetisation causes the torque that can be produced to drop, especially where it is the highest otherwise, similar to the surface mounted rotor. There are also greater constraints on how high the PM can be made for the spoke type rotor, which prevents the PM shape from being adapted to withstand demagnetisation.

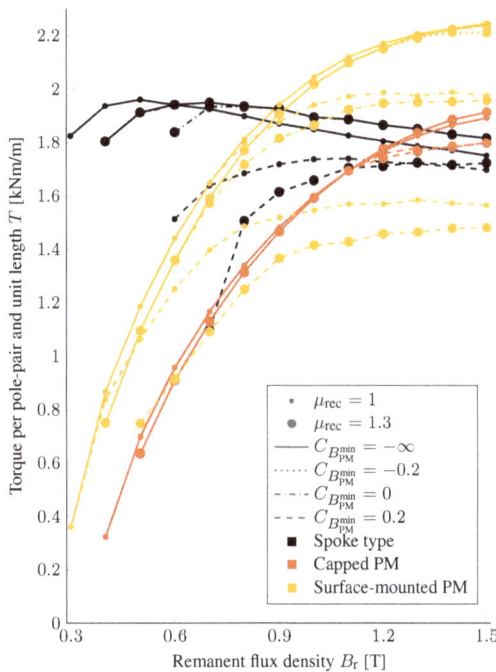

Figure 10. Torque per pole pair and unit length over remanent flux density for two different recoil permeabilities (indicated by marker size), four different demagnetisation sensitivities (indicated by line style), and the three topologies (indicated by line colour).

In Figure 11, the shape and size of the PM in the topology with the highest torque, in each material point for $C_{B_{PM}^{min}} = 0.2$ and $C_{B_{PM}^{min}} = -0.2$, are shown. It can be seen that higher demagnetisation resistance favours PMs with smaller height and greater width. In the low-B_r points, where the surface mounted PM rotor is the best for $C_{B_{PM}^{min}} = 0.2$, the spoke type rotor cannot accommodate a PM high enough to resist the demagnetisation. The capped PM rotor never gives the highest torque for $C_{B_{PM}^{min}} = -0.2$ (right in Figure 11).

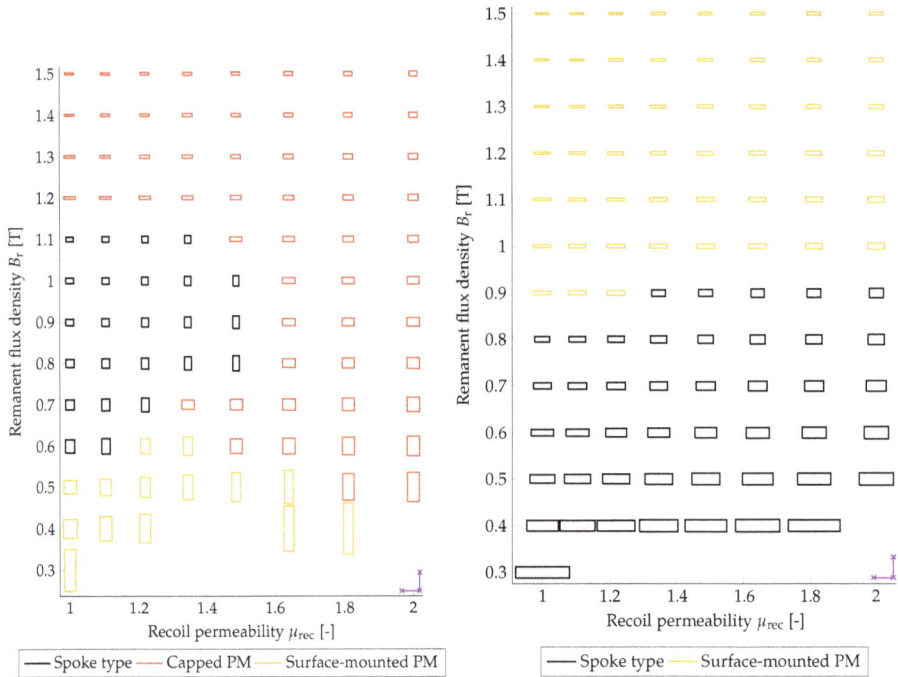

Figure 11. Shape of the PMs in the rotor topology that gives the highest torque for each pair of B_r and μ_{rec}. Vertical length corresponds to PM height (along magnetisation) h_{PM} and width to PM width w_{PM}. Colour of box indicates which topology gives the highest torque. The demagnetisation resistances shown are $C_{B_{PM}^{min}} = 0.2$ to the left and $C_{B_{PM}^{min}} = -0.2$ to the right. The purple angle in the lower right corner indicates the length of the pole pitch $\tau_p = 80$ mm.

In Figure 12, the magnetic field in the spoke type rotor optimised for $C_{B_{PM}^{min}} = -0.2$, $B_r = 0.5$ T, and $\mu_{rec} = 1$ is shown. Features of interest are the leakage flux in the region between the PM and the stator; and the radius where the radial component of the magnetic flux in teh pole piece becomes zero, and turns into a leakage flux for smaller radius. Both of these leakage fluxes depend both on the shape of the PM, which has been optimised, but also on the shape of the pole piece, which has not been optimised in this study. The latter is a limitation on the method for the spoke type and capped PM rotors, compared to the surface mounted PM rotor, as the pole shape will have a small impact on the performance that can be obtained. The surface mounted PM rotor does not have this limitation, as it does not have any PM size dependent iron in the rotor, except the rotor back ring which just needs to be thick enough to avoid saturation. A study on performance of generators with different shapes of poles can be found in [28].

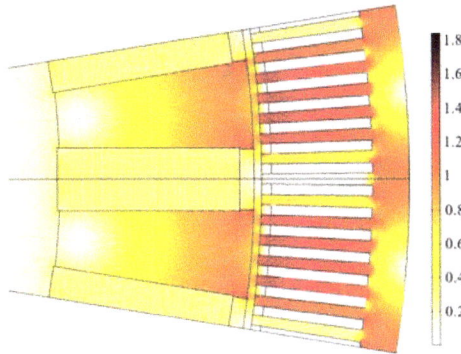

Figure 12. The magnetic flux density in the optimised spoke type rotor for a material point with $B_r = 0.5\,\text{T}$, $\mu_{rec} = 1$ and $C_{B_{PM}^{min}} = -0.2$ at no load. Field lines and magnitude in [T] are shown.

The result of the stability analysis is shown in Figure 13. The impact of the parameter changes is given as the arithmetic mean value and standard deviation of point-wise normalised optimised torque in the different cases—both for all points and for each topology on its own. Of most interest in Figure 13 are the standard deviations and the per topology means. The standard deviations show how much the shape, but not scaling of magnitude of the distribution, differs between the cases. The mean of each topology show how that topology has been affected by the change in the parameter. If all the means are the same, or the standard deviation for all the points is small, the value of the parameter changed in that case has small impact on how the different topologies compare to each other.

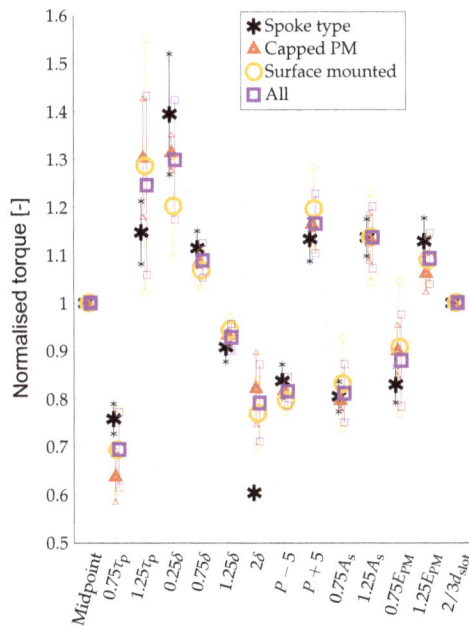

Figure 13. Optimised torque, normalised point-wise by the Midpoint case, of the different cases of the stability analysis. The large marker indicates the arithmetic mean value for each topology, or all of them together. The lines and small markers show mean±standard deviation. Cases are denoted by the parameter changed and how it was changed. Points where the optimisation failed have been removed.

The number of material points where the optimisation could not be completed, for the reasons mentioned earlier, are not shown in Figure 13. For the case *Midpoint*, this is 44, out of a total of 192 points. For the other cases, it varies between 38, and 94 in the case 2δ; the second highest number of missing points is 53 for the case $0.75\tau_p$. In the case 2δ, all but one of the material points are missing for the spoke type rotor, i.e., the geometry does not allow for enough PM material in this case.

The change in τ_p propagates to all dimensions of the geometry, except the PM volume, which is given by E_{PM}, B_r and μ_{rec}. The capped PM rotor shows the greatest change in mean value, in response to changed τ_p, and the surface mounted has the greatest variance. When it comes to changes in δ, the spoke type is the most sensitive; the mean value for the spoke type changes more than that of the other topologies and, for the largest δ, the topology is mostly unusable. Changing the number of pole pairs or stator current loading, both give a change in delivered torque that is similar between the different rotor topologies. Change of E_{PM} changes the torque of the spoke type rotor slightly more than it changes the torque obtained from the other two topologies. The slot depth show no change on obtained torque. It could have a rather significant impact on the thermal properties of the machine, but these are not explicitly modelled here.

The stator current loading, A_s, and slot depth d_{slot}, are both stator parameters. Changing either has a similar effect on all three topologies. This indicates that the design of the stator has a limited impact on the relative performance, as measured here in torque production, of the different rotor topologies. Thus, the choice of values of these parameters is of less importance to the results in the study. The number of pole pairs, P, can, by the same line of reasoning, be considered to be of lesser importance to the results (for constant τ_p). With the low impact on the outcome of the choice of values of P, A_s and d_{slot}, the choice of the values matter less to generality of the results obtained.

One problem for the generality of the result is the dependence on the pole pitch which varies greatly depending on the size of the machine. The air gap length δ also has this problem, but to a lesser extent since manufacturing tolerances imposes a lower limit on it, and the requirement to save PM material gives an incentive not to make it much larger than this. These limits are in relation to the stator bore diameter, see [22] (p. 306), which means that, while the absolute value of δ can vary a lot, the length relative to the rest of the geometry will not vary that much. In addition, the value of δ has, in the stability analysis, been tested from the smallest value mechanically feasible, to eight times that value. In this interval, it can be seen that the spoke type rotor is more sensitive to the change in air gap length than the other two rotor topologies. Decreasing δ allows more torque to be generated than with the other rotor topologies, while, for larger δ, it cannot reach the required operating point. Despite this drawback, the spoke type rotor outperforms the other rotor topologies for low B_r PMs, at least for air gap lengths up to 1.25 of the default value. The pole pieces in the spoke type rotor and the capped PM rotor are made from soft magnetic material. Making the pole pieces from laminated magnetically soft iron reduces the problems caused by slot-harmonic induced eddy currents in the pole face, which can occur with small air gap lengths. The poor performance of the spoke type rotor with large air gap lengths should therefore not be a problem, as there is no need for a large air gap.

The changes in optimised torque in response to a change in τ_p are problematic for the generality of the results. It should, however, be considered that, in most cases, making the machine larger, by increasing τ_p, would also mean increasing the volume of PM material used. Per unit length, this means that E_{PM} should increase as τ_p^2 to keep the machines geometrically similar, if the same material properties are assumed. The results presented in Figure 13 roughly indicate that the different trends in the altered E_{PM} and τ_p cases should cancel each other if E_{PM} is scaled as τ_p^2. There is still the matter of choosing a suitable value for E_{PM} for a given τ_p, which also has some effect on the results but less than the impact of changing τ_p with constant E_{PM}.

Generally it can be noted that the spoke type rotor is more dependent on the amount of PM material than the other two topologies, which are instead more dependent on B_r of the material. This is supported both by the stability analysis, where the spoke type rotor is the topology most impacted by changes that alters the amount of PM material in relation to the size of the rest of the geometry; and by the main result where the spoke type rotor has a much flatter torque over the B_r and μ_{rec} surface than the other two topologies.

5. Conclusions

Three different rotor topologies are optimised for torque production assuming identical stators, resistive load of a given current amplitude, and an equal amount of PM maximum energy for different values of PM material parameters. The parameters are the remanent flux density B_r, recoil permeability μ_{rec}, and relative minimum flux density $C_{B_{PM}^{min}}$, to model demagnetisation. Optimum torque for each material point and topology is calculated and used to compare the topologies.

It is found that the spoke type rotor achieves the highest torque for low B_r materials. It does, however, have issues with limits on how high the PM can be made, which sometimes makes it impossible to avoid demagnetisation or reach the desired working point. For high B_r material, the surface mounted PM rotor can give the highest torque if the PM material has good resistance against demagnetisation, low $C_{B_{PM}^{min}}$; otherwise, the capped PM, which offers more protection against demagnetisation, performs better. From this, it can be concluded that demagnetisation resistance is more important than remanence in a PM material for use in electrical machines, and that an inexpensive, low B_r material with reasonable demagnetisation resistance can be highly useful in this application—in particular, if it is used with a flux concentrating rotor, such as the spoke type rotor.

A stability analysis was carried out by changing design input parameters, one at a time. This shows that the results are reasonably stable under perturbations of most parameters. Changes in the pole pitch, τ_p, air gap length, δ, and PM maximum energy, E_{PM}, did show potential to cause qualitative changes of the outcome. The air gap length, relative to the rest of the size of the machine, should stay sufficiently close to the used value for the results to have some generality. The value of τ_p and E_{PM} should be changed together, as the size of the machine dictates how much PM material can be used. If E_{PM} is adjusted by the square of the relative change in τ_p, to preserve similarity, the impact of the change in E_{PM} and that of the change in τ_p cancel each other out. The range of values of E_{PM} that can fit into the geometry for a given τ_p is rather limited, while the values of τ_p can be varied over several orders of magnitude. Overall, the conclusions drawn should have some generality.

Author Contributions: Conceptualization, S.E. and P.E.; methodology, P.E.; software, P.E.; formal analysis, P.E.; investigation, P.E. and S.E.; writing—original draft preparation, P.E.; writing—review and editing, P.E. and S.E.; visualization, P.E.; supervision, S.E.; funding acquisition, S.E.

Funding: This study was carried out with funding from the Swedish Research Council, Grant No. 2012-4706; and from the Carl Trygger Foundation. This work was conducted within the StandUP for Energy strategic research framework.

Nomenclature

A_{eff}	m^2	Effective cross section area of winding
A_{PM}	m^2	PM area
A_s	A/m	Linear current density
A_z	Tm	Magnetic vector potential, axial component
B	T	Magnetic flux density
B_r	T	Remanent magnetic flux density
$B_{\|\|B_r}$	T	Magnetic flux density parallel to B_r
B_{PM}^{min}	T	Magnetic flux density, treshold
$\|BH\|_{max}$	J/m^3	Maximum energy product
$C_{B_{PM}^{min}}$		Demagnetization parameter
d_{slot}	m	Slot depth

D_{si}	m	Stator inner diameter
E_{PM}	J	PM maxmimum energy
$\mathcal{F}_d$	A	Direct axis MMF
$\mathcal{F}_q$	A	Quadrature axis MMF
H	A/m	Magnetising field
H_c	A/m	Coercivity
H_{ci}	A/m	Intrinsic coercivity
h_{PM}	m	PM height along magnetization
J_z	A/m^2	Current density
l	m	Machine length
M	A/m	Magnetisation
P		Number of pole pairs
t	s	Time
T	Nm/m	Torque per pole pair and unit length
w_{PM}	m	PM width across magnetization
q		Number of slots per pole and phase
Q_{demag}		Volume fraction
$Q_{\mathrm{demag}}^{\mathrm{SC}}$		Volume fraction at SC
$Q_{\mathrm{demag}}^{\mathrm{NL}}$		Volume fraction at NL
$Q_{\mathrm{demag}}^{\mathrm{SC,\,tol}}$		Tolerance for volume fraction at SC
$Q_{\mathrm{demag}}^{\mathrm{NL,\,tol}}$		Tolerance for volume fraction at NL
δ	m	Air gap length
θ_{el}	rad	Electrical angle, rotor reference frame
Λ_d	H/m	Permeance per unit length, d-axis
Λ_q	H/m	Permeance per unit length, q-axis
μ_0	Wb/Am	Permeability of vacuum
μ_{r}		Relative permeability
μ_{rec}		Relative recoil permeability
$\mathcal{V}_d$	V/m	Voltage per turn and unit length, d-axis
$\mathcal{V}_q$	V/m	Voltage per turn and unit length, q-axis
ϱ_{l}	Ω/m	Magnetic load resistance
ϱ_{s}	Ω/m	Magnetic winding resistance
τ_{p}	m	Pole pitch
τ_{s}	m	Slot pitch
Φ_d	Wb/m	Magnetic flux per unit length, d-axis
Φ_{PM}	Wb/m	Magnetic flux per unit length from PM
Φ_q	Wb/m	Magnetic flux per unit length, q-axis
ω_{el}	rad/s	Electrical angular frequency

References

1. Gutfleisch, O.; Willard, M.A.; Brück, E.; Chen, C.H.; Sankar, S.G.; Liu, J.P. Magnetic Materials and Devices for the 21st Century: Stronger, Lighter, and More Energy Efficient. *Adv. Mater.* **2011**, *23*, 821–842. doi:10.1002/adma.201002180. [CrossRef] [PubMed]
2. Lacal-Arántegui, R. Materials use in electricity generators in wind turbines—State-of-the-art and future specifications. *J. Clean. Prod.* **2015**, *87*, 275–283. doi:10.1016/j.jclepro.2014.09.047. [CrossRef]
3. Eklund, P.; Sjökvist, S.; Eriksson, S.; Leijon, M. A Complete Design of a Rare Earth Metal-Free Permanent Magnet Generator. *Machines* **2014**, *2*, 120. doi:10.3390/machines2020120. [CrossRef]
4. Kim, K.C.; Lee, J. The dynamic analysis of a spoke-type permanent magnet generator with large overhang. *IEEE Trans. Magn.* **2005**, *41*, 3805–3807. doi:10.1109/TMAG.2005.854934. [CrossRef]
5. Chen, Z.; Spooner, E. A modular, permanent-magnet generator for variable speed wind turbines. In Proceedings of the Seventh International Conference on Electrical Machines and Drives, Durham, UK, 11–13 September 1995; pp. 453–457. doi:10.1049/cp:19950913. [CrossRef]

6. Spooner, E.; Williamson, A. Direct coupled, permanent magnet generators for wind turbine applications. *IEE Proc. Electr. Power Appl.* **1996**, *143*, 1–8. doi:10.1049/ip-epa:19960099. [CrossRef]

7. Binder, A.; Schneider, T. Permanent magnet synchronous generators for regenerative energy conversion—A survey. In Proceedings of the 2005 European Conference on Power Electronics and Applications, Dresden, Germany, 11–14 September 2005, p. 10. doi:10.1109/EPE.2005.219668. [CrossRef]

8. Coey, J. Permanent magnets: Plugging the gap. *Scr. Mater.* **2012**, *67*, 524–529. doi:10.1016/j.scriptamat.2012.04.036. [CrossRef]

9. Delczeg-Czirjak, E.K.; Edström, A.; Werwiński, M.; Rusz, J.; Skorodumova, N.V.; Vitos, L.; Eriksson, O. Stabilization of the tetragonal distortion of Fe_xCo_{1-x} alloys by C impurities: A potential new permanent magnet. *Phys. Rev. B* **2014**, *89*, 144403. doi:10.1103/PhysRevB.89.144403. [CrossRef]

10. Edström, A.; Chico, J.; Jakobsson, A.; Bergman, A.; Rusz, J. Electronic structure and magnetic properties of $L1_0$ binary alloys. *Phys. Rev. B* **2014**, *90*, 014402. doi:10.1103/PhysRevB.90.014402. [CrossRef]

11. Fang, H.; Kontos, S.; Ångström, J.A.; Cedervall, J.; Svedlindh, P.; Gunnarsson, K.; Sahlberg, M. Directly obtained τ-phase MnAl, a high performance magnetic material for permanent magnets. *J. Solid State Chem.* **2016**, *237*, 300–306. doi:10.1016/j.jssc.2016.02.031. [CrossRef]

12. Kimiabeigi, M.; Sheridan, R.S.; Widmer, J.D.; Walton, A.; Farr, M.; Scholes, B.; Harris, I.R. Production and Application of HPMS Recycled Bonded Permanent Magnets for a Traction Motor Application. *IEEE Trans. Ind. Electron.* **2018**, *65*, 3795–3804. doi:10.1109/TIE.2017.2762625. [CrossRef]

13. Torrent, M.; Perat, J.; Jiménez, J. Permanent Magnet Synchronous Motor with Different Rotor Structures for Traction Motor in High Speed Trains. *Energies* **2018**, *11*, 1549. doi:10.3390/en11061549. [CrossRef]

14. Arumugam, P.; Dusek, J.; Aigbomian, A.; Vakil, G.; Bozhko, S.; Hamiti, T.; Gerada, C.; Fernando, W. Comparative design analysis of Permanent Magnet rotor topologies for an aircraft starter-generator. In Proceedings of the IEEE International Conference on Intelligent Energy and Power Systems (IEPS), Kiev, Ukraine, 2–6 June 2014; pp. 273–278. doi:10.1109/IEPS.2014.6874194. [CrossRef]

15. Wang, A.; Jia, Y.; Soong, W.L. Comparison of Five Topologies for an Interior Permanent-Magnet Machine for a Hybrid Electric Vehicle. *IEEE Trans. Magn.* **2011**, *47*, 3606–3609. doi:10.1109/TMAG.2011.2157097. [CrossRef]

16. Kim, H.J.; Moon, J.W. Improved rotor structures for increasing flux per pole of permanent magnet synchronous motor. *IET Electr. Power Appl.* **2018**, *12*, 415–422. doi:10.1049/iet-epa.2017.0432. [CrossRef]

17. Gundogdu, T.; Komurgoz, G. The Impact of the Selection of Permanent Magnets on the Design of Permanent Magnet Machines—A Case Study: Permanent Magnet Synchronous Machine Design with High Efficiency. *Przeglad Elektrotech.* **2013**, *89*, 103–108.

18. Shimizu, Y.; Morimoto, S.; Sanada, M.; Inoue, Y. Influence of permanent magnet properties and arrangement on performance of IPMSMs for automotive applications. In Proceedings of the 2016 19th International Conference on Electrical Machines and Systems (ICEMS), Chiba, Japan, 13–16 November 2016; pp. 1–6.

19. Eriksson, S.; Bernhoff, H. Inherent Difference in Saliency for Generators with Different PM Materials. *J. Renew. Energy* **2014**, *2014*, 567896. doi:10.1155/2014/567896. [CrossRef]

20. Eriksson, S.; Bernhoff, H. Rotor design for PM generators reflecting the unstable neodymium price. In Proceedings of the XXth International Conference on Electrical Machines, Marseille, France, 2–5 September 2012; pp. 1419–1423. doi:10.1109/ICElMach.2012.6350064. [CrossRef]

21. Sjökvist, S.; Eklund, P.; Eriksson, S. Determining demagnetisation risk for two PM wind power generators with different PM material and identical stators. *IET Electr. Power Appl.* **2016**, *10*, 593–597. [CrossRef]

22. Pyrhönen, J.; Jokinen, T.; Hrabovocová, V. *Design of Rotating Electrical Machines*, 2nd ed.; John Wiley and Sons: Hoboken, NJ, USA, 2014; pp. 298–308.

23. Binns, K.; Kurdali, A. Permanent-magnet a.c. generators. *Inst. Electr. Eng.* **1979**, *126*, 690–696. doi:10.1049/piee.1979.0154. [CrossRef]

24. Jun, C.S.; Kwon, B.I. Performance comparison of a spoke-type PM motor with different permanent magnet shapes and the same magnet volume. *IET Electr. Power Appl.* **2017**, *11*, 1196–1204. doi:10.1049/iet-epa.2016.0763. [CrossRef]

25. Arkkio, A. Analysis of Induction Motors Based on the Numerical Solution of the Magnetic Field and Circuit Equations. Ph.D. Thesis, Helsinki University of Technology, Espoo, Finland, 1987.

26. Eklund, P.; Eriksson, S. Winding Scheme Independent Method for Prediction of Short Circuit Current Distribution for a PMSM. In Proceedings of the XXIIIrd International Conference on Electrical Machines, Alexandroupoli, Greece, 3–6 September 2018.
27. Rumble, J.R. (Ed.) *CRC Handbook of Chemistry and Physics*; Chapter Electrical Resistivity of Pure Metals, (Internet Version 2018); CRC Press/Taylor & Francis: Boca Raton, FL, USA, 2018.
28. Ranlöf, M.; Lundin, U. Form Factors and Harmonic Imprint of Salient Pole Shoes in Large Synchronous Machines. *Electr. Power Compon. Syst.* **2011**, *39*, 900–916. [CrossRef]

energies

MDPI

Article

Design of Permanent-Magnet Linear Generators with Constant-Torque-Angle Control for Wave Power

Sandra Eriksson [ORCID]

Division of Electricity, Uppsala University, SE-751 21 Uppsala, Sweden; sandra.eriksson@angstrom.uu.se;
Tel.: +46-18-471-5823

Received: 28 February 2019; Accepted: 1 April 2019; Published: 5 April 2019

Abstract: This paper presents a simulation method for direct-drive permanent-magnet linear generators designed for wave power. Analytical derivations of power and maximum damping force are performed based on Faraday's law of induction and circuit equations for constant-torque-angle control. Knowledge of the machine reactance or the load angle is not needed. An aim of the simulation method is to simplify comparison of the maximum damping force, losses, and cost between different generator designs at an early design stage. A parameter study in MATLAB based on the derived equations is performed and the effect of changing different generator parameters is studied. The analytical calculations are verified with finite element method (FEM) simulations and experiments. An important conclusion is that the copper losses and the maximum damping force are mainly dependent on the rated current density and end winding length. The copper losses are inherently large in a slow-moving machine so special consideration should be taken to decrease the end winding length. It is concluded that the design of the generator becomes a trade-off between material cost versus high efficiency and high maximum damping force.

Keywords: coils; design tools; energy efficiency; linear generator; power control; stator; wave power

1. Introduction

Wave power has been studied for many years but there are still only a few realized commercial projects [1]. Many different types of wave energy converters exist and one of the most common is a point absorber. Alternative power take-off systems exist but a linear direct-drive permanent-magnet (PM) generator is a common solution and has previously been studied by different research teams, such as in [2–5]. At Uppsala University, a concept with a point-absorbing buoy and a direct-drive PM generator has been studied for 15 years [4]. Linear PM generators most commonly have magnets made of rare-earth metals [6], but generators with ferrite magnets have also been studied [7] and tested experimentally [8]. Advantages with PM generators are that you do not need to magnetize the rotor externally, through the magnetizing current in the stator or by directly magnetizing the rotor and therefore do not have any copper losses associated with the rotor. A disadvantage is that the magnetization is fixed and cannot be controlled. In this paper, the design of the moving part, commonly called translator, is not considered and it is only represented by a magnetic flux density amplitude in the generator airgap.

Different generations of linear generators for wave power have been built by Uppsala University and a generator that has been designed, installed and tested is used as a reference design in this work [9,10]. There are a limited number of pervious papers on linear generator design optimization and design choices. A comparison of different design aspects was presented in 2006 [4] for the same generator concept as in this study. That study was performed with FEM simulations for diode rectification and therefore focused on different aspects than this study. A design optimization including a cost estimation of a slotless tubular linear generator with longitudinal flux can be found in [11]. Reference [12] presented an optimization study of a generator design for no-load operation but did not motivate how different design choices depend on each other. In [13], the performance of linear PM generators with different pole pitch is compared. In [14], constant-torque-angle (CTA) control is implemented for a wave energy converter with a rotating 248-pole generator. To the author's knowledge, no previous study has been performed on how to optimize the generator design for CTA control including a thorough investigation on how different parameters affect the performance.

CTA control is a common choice for generators, and it is commonly implemented with PI-controllers and dq-transformation. CTA control has previously been implemented in wave energy converters. In [14] it is shown how CTA control can be implemented with PI regulators for a wave energy converter with rotating generator. In [15], a control system for an active rectifier with CTA control for a linear generator is simulated and verified experimentally. In [16], a different control system with active control of a PM linear generator is presented.

Designing a generator is a task with many degrees of freedom and is therefore very complex and normally time-consuming. For a complete generator design, a full-physics simulation is always needed, usually performed by solving Maxwell's equations in a two-dimensional finite element method (FEM) simulation and taking important three-dimensional effects into account, as presented in [17]. However, to understand how different parameters affect each other a simpler approach may be used, as in this paper, where combining Faraday's law of induction and CTA control simplifies equations and makes it possible to draw some interesting conclusions regarding linear generator design without computationally heavy simulations. The main simplification in this comparative study is that no saturation is taken into account. This will especially influence results at high loading and lead to an overestimation of the maximum damping force. However, the approximation of the damping force can be used to compare different generator designs.

The aim is of this study is, first, to present a fast computational method for comparing generator designs, where the generator design is adapted for a certain control strategy and, second, to gain increased understanding of how different design parameters influence the generator performance and to compare different generator designs. The evaluation is mainly performed based on three aspects; efficiency, maximum damping force, and relative material cost. Analytical derivations on generator performance and losses during CTA control, are presented. The equations have been implemented in MATLAB and the simulation method is verified through FEM simulations and experiments. A parameter study on linear generator design is performed and different generator designs are compared from different aspects.

2. Linear Generators

PM direct-drive linear generators for wave power commonly consist of a stator with steel and conductors and a PM covered translator directly connected to the point absorber. A large generator is needed due to the low speed. How large it needs to be is an economical trade-off between generator size and cost and on if the energy in the large, less common waves is worthwhile to capture. A linear generator operates at continuously varying speed as the generator changes direction twice per wave period. In addition, the speed will vary with the wave climate. When comparing different generator designs, a fixed speed is commonly used but variable speed operation needs to be considered.

One of the main differences between a conventional rotating generator and a direct-drive linear generator for wave power is that the stator height and translator height of the linear generator depends on the wave climate at the particular site, whereas the diameter of a rotating generator usually is a free design parameter. The translator is usually longer than the stator but depending on the wave height, the translator might for short periods of time partly leave the stator giving a partial stator-translator overlap [8]. The free stroke length of the translator, which is a function of the length difference between the translator and the stator, should be adapted to the wave climate at the particular site.

A sketch of the wave energy converter concept investigated at Uppsala University is shown in Figure 1 [18], together with an example geometry of a linear generator with geometrical parameters for the stator defined.

The electromagnetic force in tangential direction in the airgap of the generator is in wave power usually referred to as the damping force. In certain hydrodynamic control strategies for wave energy converters, a high damping force is crucial.

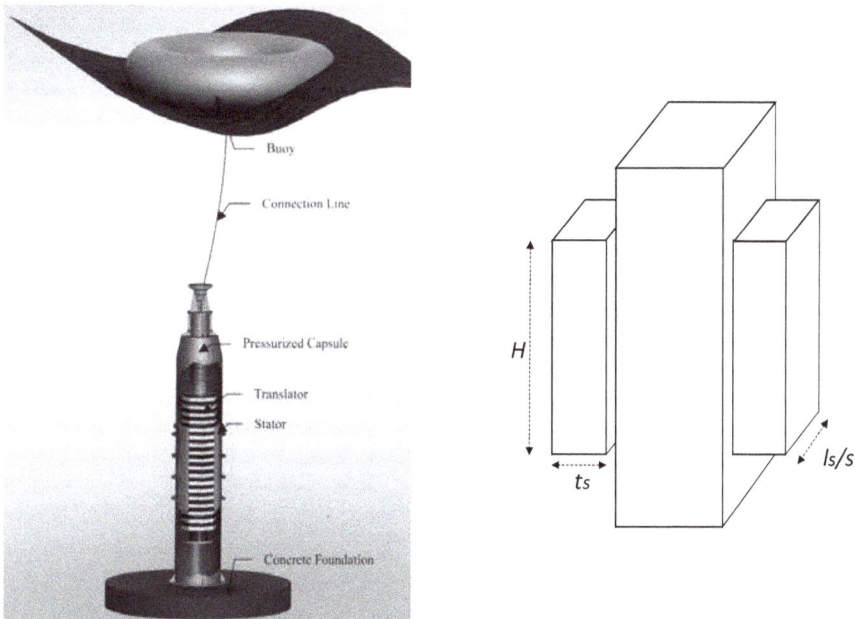

Figure 1. Left: Sketch of the wave energy converter concept from [18] licenced under CC BY 3.0 https://creativecommons.org/licenses/by/3.0. **Right**: Sketch of linear generator geometry. The figure shows two out of four sides of the generator stator and the geometrical parameters of the stator. The translator is in the middle and moves up and down. H denotes stator height, l_s the total stator length, t_s the stator thickness and s the number of stator sides.

Constant-Torque-Angle Control

There are several different ways to control a wave generator to obtain maximum power absorption from the waves. Most operational strategies involve controlling the damping force of the generator. This can electrically be done in different ways and the generator can either be connected to a passive or an active rectifier [19]. For an active rectifier the generator can be boosted with reactive power when necessary. This decreases the voltage drop in the generator since the angle between the electromotive force (emf) and the current is decreased. In this paper, CTA control is implemented since it decreases copper losses in the generator [20] which usually are quite large in a direct-drive linear generator for wave power.

With constant-torque-angle control, a complete phase angle compensation is achieved and the load angle in the machine, δ, equals the phase angle, θ. In the dq-frame, the CTA control is equivalent to zero d-axis current. The CTA control is implemented by dq-transformation and implementation of a control system with several PI-controllers, see for instance [14,15]. In [15] experimental verification of a CTA-based control system for a linear machine is presented. The control system uses two PI regulators per phase for the pulse-width modulation-controlled rectifier and a linear encoder gives the position and the phase.

The phasor diagram for the phase equivalent for a non-salient synchronous generator with CTA control can be seen in Figure 2 where X is the machine reactance. The output phase voltage, U_f, for a non-salient synchronous generator then becomes

$$U_f = (E_f - R_i I)/\cos\delta \tag{1}$$

where E_f is the emf per phase, R_i is the inner resistance and I is the current. The output power per phase, when the generator is run with CTA control, i.e., if $\theta = \delta$ becomes

$$P_{out} = U_f I \cos(\theta) = E_f I - R_i I^2 = P_f - P_{Cu} \tag{2}$$

where the first part is the generator power, P_f and the second part constitutes the resistive losses, P_{Cu}. Please note that E_f and I above are amplitudes and not phasors as for the CTA control the angle between them is zero. For a generator with surface-mounted magnets, i.e., a non-salient machine, the CTA control is very similar to the Maximum Torque per Ampere (MTPA) control [20]. As the name of the latter indicates, the CTA control will give lower current for a certain torque/force and therefore minimize the copper losses. However, at high loading, when the stator steel becomes saturated, the CTA control will differ a little from the maximum torque (or force) per ampere control [21].

By assuming CTA control, calculations for output power and damping force can be simplified, as will be shown in Section 4. In addition, knowledge of the reactance of the generator or the load angle is not needed to analyze the operation of the generator. This is a nice feature since the reactance is more difficult to estimate for a generator than the resistance.

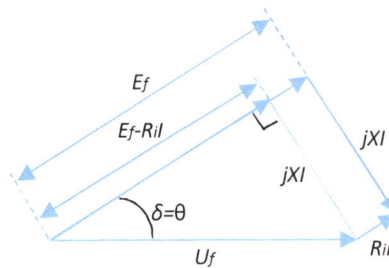

Figure 2. Phasor diagram for a non-salient synchronous generator with CTA control.

3. Generator Modelling

In this section, the different equations and simplifications used in the analytical derivations in Section 4 and in the simulations in Section 5, are motivated and introduced.

3.1. Induced Voltage and Current

Faraday's law of induction can be used to derive the induced voltage in a generator. The derivation results in the following formula, sometimes referred to as the "Generator equation", here adapted to a linear generator

$$E_f = \sqrt{2} N_{pp} B l_s p v \tag{3}$$

where B is the amplitude of the magnetic flux density in the airgap, l_s is the total stator length (equivalent to the axial length of a rotating generator), see Figure 1, p is the number of poles on the stator, v is the generator speed, and the number of effective turns per pole and phase, N_{pp}, is

$$N_{pp} = \frac{k_w q n_s}{2c} \tag{4}$$

where k_w is the winding factor, q is the number of slots per pole and phase, n_s is the number of conductors per slot and c is the number of parallel current paths per phase.

The pole pitch τ_p for a linear generator is

$$\tau_p = H/p \tag{5}$$

where H is the stator height. The translator is long enough to always be covered by the stator, i.e., there is no partial stator-translator overlap. The number of poles, p, here refers to the number of electric poles on the stator, as the translator is longer than the stator. Figure 3 shows a cross-section of approximately 3.5 pole pitches of a generator with $n_s = 8$ and $q = 5/4$.

For a three-phase machine the slot pitch is $\tau_p/3q$. If the copper takes up 1/3 of the slot pitch and the rectangular conductor is twice the size in the radial direction as in the tangential, the area of the conductor, A_{cond}, becomes

$$A_{cond} = 2 \left(\frac{1}{3} \frac{\tau_p}{3q} \right)^2 \tag{6}$$

For a current density of I_{dens} given in A/m^2 the current, I is

$$I = I_{dens} A_{cond} = 2 I_{dens} \left(\frac{1}{3} \frac{\tau_p}{3q} \right)^2 \tag{7}$$

The total conductor length per phase in the generator, L_{tot}, is

$$L_{tot} = 2 N_{pp} p (l_s + l_{end}) \tag{8}$$

where l_{end} is the length of the end windings per half-turn. The copper losses per phase are

$$P_{Cu} = R_i I^2 = 1.68 \cdot 10^{-8} L_{tot} I_{dens}^2 A_{cond} \tag{9}$$

where R_i is the inner resistance of the generator per phase and $1.68 \cdot 10^{-8}$ is the resistivity of copper at room temperature. The resistivity is strongly temperature dependent which will not be considered here. However, a large generator, with relatively low current density which is placed in sea water is not expected to reach very high temperatures.

3.2. Iron Losses and Efficiency

The iron losses consist mainly of eddy current losses, hysteresis losses and excess losses and depend on the magnetic flux density and frequency of the magnetic field in the stator steel as well as the stator steel volume [22]. To determine the stator steel volume, the thickness of steel, t_s, in the radial direction (magnetic flux direction) must be known. Here, the stator yoke is assumed to have the length of the pole pitch divided by 4. The tooth length is dependent of the number of conductors per slot, n_s, multiplied by the size of each conductor, and multiplied by 1.33 compensating for the fill factor in the radial direction. The tooth width is here set to half the slot pitch, so to get the average thickness of steel, $t_{s,av}$, the tooth length is multiplied with 0.5.

$$t_{s,av} = \tau_p/4 + n_s \cdot 2 \cdot \tau_p/(9q) \cdot 1.33 \cdot 0.5 \tag{10}$$

The stator steel volume, V_s, can then be calculated by multiplying the three dimensions of the assumed cubical stator and using the average thickness.

$$V_s = Hl_s t_{s,av} \tag{11}$$

The iron losses can be calculated assuming typical steel used in generators of type M270-50A, with losses of 2.7 W/kg for a magnetic flux density of 1.5 T (www.cogent-power.com, retrieved 19 September 2018). An average magnetic flux density of 1.5 T in the whole stator is assumed. The density is 7600 kg/m^3. A frequency dependency for iron losses of $f_{el}^{1.3}$ has been extrapolated from results for low frequencies in [22]. In addition, the iron losses are multiplied with a loss correction factor of 1.5 when extrapolating from 50 Hz, as motivated in [22]. The iron losses then become

$$P_{Fe} = (f_{el}/50)^{1.3} \cdot 1.5 \cdot 2.7 \cdot 7600 V_s \tag{12}$$

The electromagnetic efficiency, η, of the generator is

$$\eta = \frac{3P_{out}}{3P_{out} + 3P_{Cu} + P_{Fe}} \tag{13}$$

It should be noted that P_{Fe} usually is much smaller than $3P_{Cu}$. Here, mechanical losses are omitted as they are difficult to estimate and should be equal in different compared designs.

3.3. Cost Comparison

Generator cost is difficult to estimate as it depends on many different variables. Here, a simplified approach is used to calculate a relative cost for comparing different designs. The translator height is here assumed to vary with changing stator height, with the free stroke length kept constant at 1.998 m. Therefore, the variation of translator cost will both be dependent on l_s and on $(H + 1.998)$. The number of poles and number of conductors per slot will also slightly alter the magnet weight needed, but that effect is omitted here. For a design with ferrite magnets the magnet cost and weight is not so crucial, whereas it has a large impact on expensive magnets made of NdFeB. The translator body in a generator with low electrical frequency can be made of solid steel. The total generator cost is presented as a relative cost, normalized against the original design, see Section 5.1. Here, a conductor (copper) cost per kg three times higher than the cost per kg for stator steel is used as well as a translator cost of half of the stator cost for the original design. The density of copper is 8960 kg/m^3 and the density of stator steel is 7600 kg/m^3. The relative cost is then calculated as

$$Cost = C_n(3 \cdot 8960 L_{tot} A_{cond} + 7600 V_s + C_{tr} l_s (H + 1.998)) \tag{14}$$

where C_n is a normalization constant relating the generator cost to the original generator cost and C_{tr} is a normalization constant relating the translator cost to the cost of the stator for the original design.

4. Analytical Calculations

To find the output power of a generator where the induced voltage and current are known, the machine reactance or load angle is usually required. This complicates generator modelling and to get the exact value of the reactance a complete FEM simulation is required. Therefore, comparing many different generator designs becomes very time-demanding. In this section, the advantage of using CTA control is presented and expressions of the output power and the maximum damping force, independent of reactance and load angle, are derived.

4.1. Output Power for CTA Control

For CTA control the generator power per phase can be derived, according to (2)–(5) and (7) as

$$P_f = E_f I = \frac{\sqrt{2} k_w}{81 cq} B v I_{dens} n_s l_s H \tau_p \tag{15}$$

This expression shows that the output power from a generator run with CTA control can be found from only knowing the winding parameters and geometrical parameters of the generator and that the phase angle, load angle, or reactance can be unknown. This simplifies generator design and make design comparisons much more time-efficient.

For an operation scheme where q, c and k_w are fixed design choices, B is a fixed value expected to be reached from the magnetic circuit and the rated speed v is fixed for a certain site, the expression can be simplified to

$$P_f \propto I_{dens} n_s l_s H \tau_p \tag{16}$$

This expression shows which design parameters that can be varied to compare different generator designs and their output power. In the parameter study, in Section 6, the current density, the number of cables per slot and the geometrical variables l_s, H and τ_p are varied.

4.2. Copper Loss Fraction

The relation between the copper losses and the power in the generator is a measure of the efficiency of the generator (if only copper losses are considered). It becomes

$$\frac{P_{Cu}}{P_f} = \frac{\sqrt{2} \cdot 1.68 \cdot 10^{-8}}{B v} I_{dens} \frac{l_s + l_{end}}{l_s} \tag{17}$$

If the end windings are assumed related to the total stator length so that $l_{end} = \beta l_s$, the expression becomes

$$\frac{P_{Cu}}{P_f} = \frac{\sqrt{2} \cdot 1.68 \cdot 10^{-8}}{B v} I_{dens} (1 + \beta) \tag{18}$$

By assuming $l_{end} = \beta l_s$, the relation between copper losses and power mainly depends on the rated current density, which often is a design setting. A direct-drive slow-moving linear generator has inherently a large amount of copper losses and this derivation shows that it is not possible to optimize the design to decrease them. However, care should be taken to decrease the variable β, i.e., the end winding length, as much as possible, since this is the part of the conductor which does not contribute to induced voltage. Please note that these relationships only are valid during CTA control. Other control strategies will not compensate for the voltage drop in the generator as efficiently and therefore the current will be higher for the same amount of power and the copper losses will constitute a larger part of the power.

If the end windings are supposed independent of the stator length l_s and instead are assumed to be constant with a varying l_s the result is different. For correct modelling, the length of the end windings should depend on the conductor area since the size of the conductors will determine the end winding length as well as the coil pitch length.

4.3. Maximum Damping Force

From the power equation for CTA control, (2), it can be found that the maximum output power from the generator per phase, P_{out}^{max} is

$$P_{out}^{max} = E_f^2 / (4R_i) \tag{19}$$

The maximum output power is reached when the output power from the generator equals the power lost as copper losses, i.e., this is not a beneficial operating mode for the generator. The maximum power should be held below the maximum value to ensure stable operation of the generator. For the CTA control the maximum damping force for the three-phase generator, reached for the maximum output power, can be approximated as

$$F_{max} = 6P_{out}^{max}/v = 3E_f^2 / (2R_i v) \tag{20}$$

or given as a per unit value based on the rated damping force as

$$F_{max}^{pu} = E_f^2 / (2R_i P_f) = E_f / (2R_i I) = P_f / (2P_{Cu}) \tag{21}$$

which is an interesting result showing that increasing the efficiency and increasing the maximum damping force goes hand in hand, see also (17). This has to do with the fact that half the maximum damping force comes from copper losses and half becomes output power. The iron losses, which also contribute with damping force, are omitted here.

The calculations of the maximum damping force are approximations and does not take saturation into account, which could be expected at this operational mode. However, the values can be used for comparison between different design choices. The maximum damping force implies quite high current density and potentially high temperature. If the force is only required for short periods of time, this might be acceptable. Otherwise, the limiting factor for the maximum damping force for a CTA controlled PM generator is more likely thermal than the size of the inherent maximum damping force.

5. Simulation Method

5.1. The Original Generator Design

For the parameter study in this paper a given generator design is used as a starting point. The chosen generator has been designed, built, and tested offshore. Its main parameters are summarized in Table 1 and the generator and wave energy converter is further presented in [9,10]. The generator has surface-mounted magnets on the translator. The stator has circular conductors and is wound with multi-strand cables with a copper area of 25 mm². The generator has a surprisingly large inner resistance which affects the performance as discussed above. This is due to long end windings but probably also caused by connections of windings. Experimental results on output power for different wave climates for the original generator connected to a resistive load, are found in [23].

Table 1. Rated design parameters of the original generator, data adapted from [9,10].

Parameter	Value
Speed (m/s)	0.70
Stator height, H (m)	1.2
Translator height (m)	2.1
Free stroke length (m)	2.0
Number of sides, s	4
Stator side length (m)	0.4
Total stator length, l_s (m)	1.6
Pole pitch, τ_p (mm)	40
Conductors per slot, n_s	8
Electrical frequency, f_{el} (Hz)	8.8
Current density, I_{dens} (A/mm^2)	1.52
Inner resistance, R_i (Ω)	0.64
Inner inductance (mH)	20
Output power, P_{out} (kW)	17.1
Voltage L-L (V)	257
Current, I (A)	38.4
Number of phases	3
Load	Resistive

5.2. MATLAB Simulation Method

Simulations based on the formulas derived in Sections 3 and 4, have been performed by using MATLAB. CTA control has been used for all simulations. The MATLAB simulation method used here can be used to compare different design choices to each other. However, it cannot substitute full-physics simulations with FEM. The original generator (see Section 5.1) has been used as a starting point for the simulations and to verify the simulation method.

In the design study, some design parameters have been kept fixed such as the magnetic flux density in the airgap, which is kept fixed at 0.75 T for all loading conditions. The generator design speed is 0.7 m/s. The number of slots per pole and phase, q, is a fixed design choice, usually between 1 and 3 for slow machines and preferably a fraction for a PM synchronous generator to decrease cogging. Here it is kept fixed at $q = 5/4$. The winding factor, k_w, expected to be between 0.9 and 1 is kept at 1 for simplicity. The number of parallel current paths, c, is 1. Rectangular conductors, with twice the side length in the radial direction is considered instead of circular conductors, as in the original generator. A stacking factor of 1 is used for simplicity.

In the simulations it is assumed that the stator is always completely filled with translator, i.e., partial stator-translator overlap is not considered. The free stroke length is kept fixed at 1.998 m and the translator height for the original design is increased to 3.198 m, to ensure full stator-translator overlaps at all times. The translator height is set at $(H + 1.998)$ m, and varies with changing H. The number of poles in the original design is 30 which is the number of electrical poles on a 1.2 m long stator with a pole pitch of 40 mm. However, in reality the translator is longer than the stator and has more poles than 30. In the parameter study, the number of poles, pole pitch, number of conductors per slot, stator length, stator height, and current density are varied. The inherent speed difference in a linear generator for wave power, is not considered in the comparative study, so simulations are performed at rated speed. However, additional simulations for an annual speed cycle is also performed to find the average efficiency.

The original generator was connected to a resistive load during experiments, which implies a power factor of 1. When the generator is run at the same speed and with the same current, the output power for the CTA control is 17.7 kW and with a resistive load 17.1 kW, i.e., the power output is 3.5% higher with CTA control. To find the power for the resistive load, the inner inductance of 20 mH from Table 1 was used. If the CTA control is used on the same generator to get the power 17.1 kW, the copper losses decrease to 92% of the losses for the same output power and resistive load.

The original generator has a surprisingly large inner resistance which affects the performance as discussed above. This is mainly caused by long end windings but probably also caused by connections of windings. By assuming that somewhat shorter end windings and more optimized connections can be used, the inner resistance can be decreased. Therefore, a slightly smaller inner resistance is used as a starting point for the parameter study here. The end windings are set to a fixed length of 0.8 m which might seem long but includes end windings of all four sides, which gives an inner resistance for the original generator with shortened end windings of 0.48 Ω. The rated output power of the generator is then increased from 17.7 kW to 18.4 kW for CTA control. In the derivations it was convenient to let the end windings be a function of l_s. However, in the parameter study the end windings are kept at the fixed value of 0.8 m for all cases.

5.3. Verification with FEM Simulations

The original generator design has been used as a basis for the MATLAB simulations. With a magnetic flux density in the airgap of 0.75 T in the simulations, all the parameters reached in the simulations match the original generator design parameters of Table 1 apart from the inductance, which is not included in the simulations.

A simulation tool based on field equations coupled with circuit equations and solved with FEM has also been used for verification. The simulation method is further explained and has been experimentally verified for a rotating generator in [17]. A FEM simulation was performed for the original generator design. All geometrical parameters from Table 1 was inserted in the FEM program. The end winding length was adjusted to reach the inner resistance of 0.64 Ω. The generator was run at rated speed and rated current density when connected to a resistive load. The rated voltage and power was achieved for the given generator dimensions. However, a slightly higher magnetic flux density amplitude in the airgap of 0.92 T was needed to achieve the rated voltage. In the MATLAB calculations the magnetic flux density was 0.75 T. Some simplifications in the MATLAB model are that the winding factor is set to 1, the stacking factor is set to 1, and that no saturation is taken account for. If the winding factor of 0.91 (for a q of 5/4) and a stacking factor of 0.96, is taken into account, the magnetic flux density in the calculations corresponds to 0.86 T, which is still slightly lower than the FEM simulations. This is a reasonable result as no leakage flux is included in the MATLAB simulations. To compensate for this, a higher magnetic flux density can be achieved by adding more PM material. The FEM simulations verified that no parts in the generator were saturated as the maximum magnetic flux density was 1.74 T and occurred in the stator teeth. Therefore, the MATLAB simulations have been verified against FEM simulations and gives acceptable results at least for a starting point for design work. However, FEM simulations are needed to account for all effects when designing a generator. The geometry and the magnetic flux density for the original generator from FEM simulations can be seen in Figure 3.

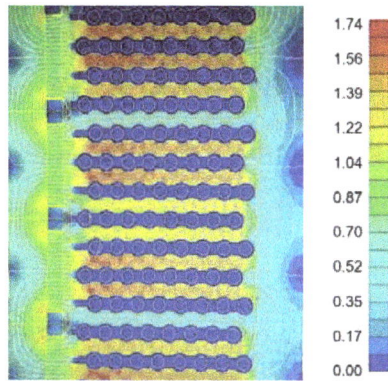

Figure 3. Field plot of the generator based on the original design from a FEM simulation, showing magnetic flux density in Tesla.

The main limitation with the presented simulation method is not including saturation in the calculations of the maximum damping force. Therefore, the maximum torque, from FEM simulations of a similar but larger generator as the one used here [24], of 200 kNm (3.1 p.u.) has been compared to circuit equation-based calculations of the maximum torque giving 219 kNm. The difference is 9.3% so it can be assumed that the maximum damping force calculations presented here also are over-estimated with roughly 10%, which gives an acceptable estimation.

5.4. Experimental Verification

To verify the simulation method, the induced voltage at no load has been compared to measurements. The experiment was performed on a multi-pole, low speed rotating generator designed for a wind turbine and presented in [25]. The tested generator is similar to the tested linear generators studied here as it has a large diameter (so for one pole pitch it is close to linear), a low speed in the airgap and a relatively large number of conductors per slot, n_s, of six. No changes in the simulations were needed to simulate the no-load voltage for a rotating generator instead of a linear generator. The magnetic flux density in the airgap was measured on the prototype and fitted very well with FEM simulations with an amplitude of the fundamental of 0.93 T [25]. The experimental generator is rated at 225 kW at 33 rpm corresponding to an electrical frequency of 9.9 Hz and an airgap speed of 3.5 m/s. It has a stator length of 0.884 m, a diameter of 2.054 m, 3.75 turns per pole and phase and 36 poles consisting of surface-mounted NdFeB-magnets. The stacking factor was measured in the prototype to 0.954 and the winding factor is calculated to 0.915. In the experiment, the rotational speed of the generator was lower than the rated speed. It was run with an electrical frequency of 5.5 Hz corresponding to a speed in the airgap of 2.0 m/s, which is closer to the operating frequency and speed of the linear generator. An oscilloscope (Tektronix TDS2014) and a high voltage probe (Tektronix P5120) were used to measure the voltage. The simulation was run with the properties of the experimental generator and gives a result as the rms value of the voltage at the given speed. The voltage has then been plotted as a sinusoidal signal with the simulated frequency.

The induced phase voltage at no load from the measurement and from the simulation are compared in Figure 4. The shapes of the curves are slightly different, which shows a limitation with the simulations, as only the fundamental of the voltage is considered. However, for a multi-pole machine with a fractional number of slots per pole and pitch, the harmonics are normally low and it can be seen that the results from experiment and simulation are quite similar. The experimental verification further validates the use of the method as a design comparison tool.

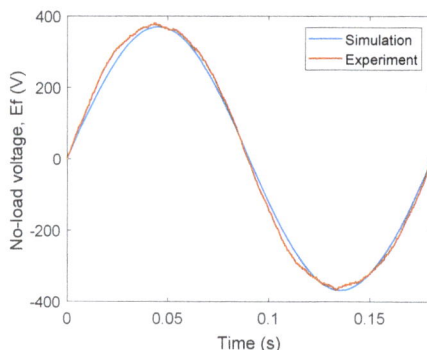

Figure 4. Comparison of results from the simulation method and from experiments. Data from experiments are reused from [25] licenced under CC BY 3.0 https://creativecommons.org/licenses/by/3.0/.

5.5. Variable Speed Operation

Waves are an inherently variable source of energy. To evaluate a generator design and to compare generators to each other, the generators should be run at variable speed and load, i.e., the annual energy absorption from the waves should be evaluated. Therefore, the average efficiency has been evaluated for different generator designs using wave data for a year.

Wave data was taken from Site 1 in Figure 3 in [26] presenting annual data of hours per year of a certain wave climate, i.e., as a function of energy period and significant wave height. The generator speed is assumed to vary sinusoidally with the wave frequency and the translator is assumed long enough for all waves. The energy flux in the waves is a function of the energy period and the significant wave height squared. The power into the generator is assumed directly proportional to the power from the waves, where the efficiency of the buoy absorption and the buoy size is adjusted to fit the rated power of the generator for an energy period of 5.5 s and a significant wave height of 1.925 m, which is one of the wave climates that contributes with the most energy during one year for this particular site.

6. Parameter Study

A parameter study has been performed and the effect of a few different parameters on the generator performance has been evaluated. All designs are based on the original generator design, see Section 5.1, with decreased end winding length at a fixed value of 0.8 m and a rated power of 18.4 kW. A comparison has been made based on efficiency, maximum damping force, and relative cost. In the wave power generator presented in [9], the number of conductors per slot, n_s is 8, whereas for large synchronous generators n_s is usually 2. Therefore, the influence of different values for n_s on the generator design, has been studied. In addition, the other parameters affecting the power from Equation (16) are varied, i.e., the generator height, number of poles, pole pitch, and current density. For all changed parameters, the stator length, l_s is adjusted to reach a rated power of 18.4 kW.

In Table 2, the effect of varying different parameters for six different generators with the same rated power is shown. Two generators with different l_s and n_s but the same power level is compared in Table 2, Case 1 and 3, showing that increasing the stator length gives a much higher cost. Case 2 represents a case where n_s has been halved but the increase in stator length and cost are not as extreme as in Case 3 due to fewer poles.

The rated efficiency and the annual average efficiency, η_{av}, for a specific site are also compared for the six generators in Table 2. The average efficiency was found from the method described in Section 5.5 . It can be seen that the average efficiency for all cases are much lower than the rated efficiency. However, when comparing the generators, it is seen that the generator with the highest rated efficiency has the highest average efficiency, so the rated efficiency seems to be an acceptable parameter for comparing the generators. For the original generator design, Case 1, the mean output power is 3.89 kW for the particular site considered.

Table 2. Comparison of six different generators.

Case	1	2	3	4	5	6
Varied parameters						
l_s (m)	1.60	2.32	3.14	1.70	1.92	1.07
n_s (m)	8	4	4	4	2	6
H (m)	1.2	1.2	1.2	1.2	1.2	1.47
p	30	22	30	16	16	22
I_{dens} (A/mm²)	1.52	1.52	1.52	1.52	3	1.52
Resulting parameters						
P_{out} (kW)	18.4	18.4	18.4	18.4	18.4	18.4
E_f (V)	178	95	175	50	29	65
I (A)	38.4	71.5	38.4	135	267	107
η	88.4	89.7	89.8	89.2	80.3	87.3
η_{av}	78.6	80.9	81.4	79.9	66.5	76.7
Rel. cost	1	1.20	1.34	1.09	0.90	1.00
F_{max}^{pu} (p.u.)	4.8	5.4	5.8	4.9	2.6	4.2

Figure 5 shows the effect of varying n_s and l_s simultaneously to achieve the same output power of 18.4 kW. It can be seen that the cost and required total stator length, l_s is decreased with increasing number of cables per slot. However, the maximum damping force is also decreasing with increasing number of cables per slot. The efficiency appears constant but is actually slowly decreasing with increasing number of cables. It is shown that only using one cable per slot is probably unrealistic both from a space and cost perspective.

Figure 6 shows how changing the stator height affects the generator. Changing the stator height will also change the translator height so it has a quite large effect on the generator volume and therefore on the relative cost. In the left figure the pole pitch is fixed so the number of poles is adjusted to the stator height. In the right figure the number of poles is kept constant at 30 so the pole pitch is adjusted to the new stator height. It can be seen that varying the pole pitch has a larger effect on the generator. A minimum value can be seen for the relative cost at a stator height of about 1.5 m giving a pole pitch of 50 mm.

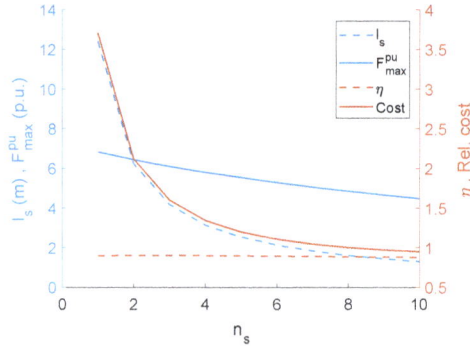

Figure 5. Stator length, maximum damping force in per unit, efficiency and relative cost for the original generator design with varied number of conductors per slot, n_s and an adjusted stator length, l_s to achieve an output power of 18.4 kW.

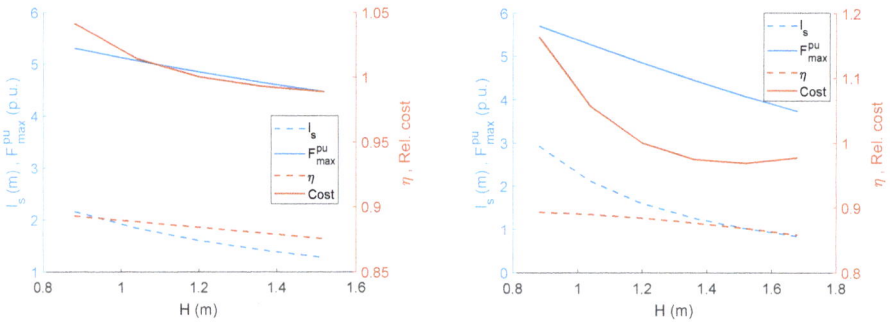

Figure 6. Stator length, maximum damping force in per unit, efficiency and relative cost for the original generator design with varied stator height, H and an adjusted stator length, l_s to achieve an output power of 18.4 kW. **Left** figure: The pole pitch is fixed. **Right** figure: The number of poles is fixed, so the pole pitch varies.

Figure 7 shows the effect of changing the number of poles and therefore also the pole pitch since the stator height is fixed. The relative cost has a clear minimum for about 18 poles. However, the efficiency has decreased quite a lot for this case, so it is not recommended to minimize the relative cost in this case. A minimum is also present if studying the ratio of maximum damping force to relative cost, which is a common design parameter to choose for optimization. None of the changes studied so far has any significant impact on the efficiency as was also shown in the analytical derivations. Figure 8 shows the effect of changing the rated current density and adjusting the stator length to reach the same rated power. As previously discussed, the efficiency decreases almost linearly with increased rated current density. The maximum damping force and cost decreases even faster with increasing current density. From this figure, it can be seen that the design is a trade-off between decreasing the cost and increasing both the efficiency and the maximum damping force. Case 5 in Table 2 shows an example with increased rated current density comparable to the original design, giving lower cost but much lower efficiency and lower maximum damping force. Case 6 in Table 2 shows a generator with increased stator height and fewer poles. It has a relative cost comparable to the original design, Case 1, but the efficiency and maximum damping force are lower. From Table 2 it can be concluded that Case 1, i.e., the original design with slightly shorter end windings, is a quite good design with a high efficiency, a large maximum damping force and a low cost compared to most of the other cases.

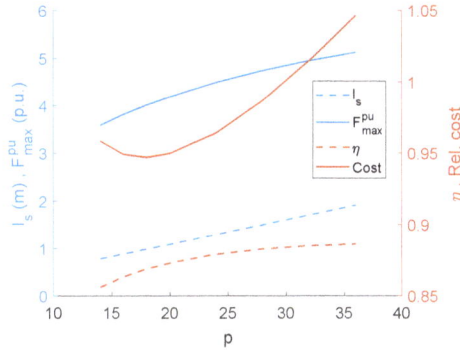

Figure 7. Stator length, maximum damping force in per unit, efficiency and relative cost for the original generator design with varied number of poles, p and an adjusted stator length, l_s to achieve an output power of 18.4 kW.

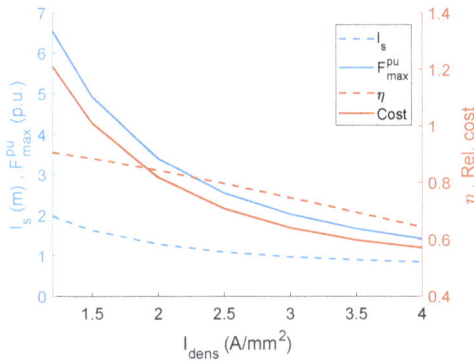

Figure 8. Stator length, maximum damping force in per unit, efficiency and relative cost for the original generator design with different values of the rated current density, I_{dens} and an adjusted stator length, l_s to achieve an output power of 18.4 kW.

In this study, design aspects have been compared for operation at rated speed. This is not the case for a linear generator where the speed is continuously changing. Apart from comparing different parameters as in this study, a study for a proposed design must be tested at different speed and loading situations. Figure 9 shows the maximum damping force and efficiency for varying speed as an example of variable speed operation for the original generator design. How the efficiency changes with varying speed depends on how the generator is controlled and in this example the generator is run with the damping force increasing linearly with increasing speed. The copper losses constitute 87% of the total losses in the original design for rated speed of 0.7 m/s and clearly dominate the losses in all studied designs. However, at very low speed the iron losses will dominate, and the efficiency therefore drops fast, as seen in Figure 9.

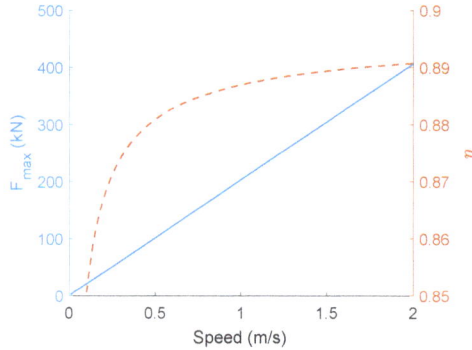

Figure 9. Maximum damping force and efficiency for different speeds for the original generator design operated with CTA control. The efficiency is here shown for a generator run with the damping force increasing linearly with increasing speed.

7. Discussion

The results in this study have shown the effect different variables have on performance of the generator, which is valuable for generator design. For a complete design work the generator behavior should be analyzed for variable speed and for the specific wave climate expected as well as for specific requirements on damping force. Partial stator-translator overlap was not considered here and geometrical constraints on for instance the stator length, l_s have not been considered.

The comparison shows that the original design performs quite well. However, this design with eight conductors per slot has quite deep slots which might increase leakage flux and give somewhat lower magnetic induction for the conductors in the bottom of the slot. These issues should be studied in a FEM simulation as the MATLAB calculations omit all such effects.

As the generator is operated with CTA control, the voltage needs to be boosted by the electrical system. This places demand on the electrical system and its rating. The rated current varies a lot between the different generators presented in Table 2. In this study, thermal issues have not been considered. To reach the theoretical maximum damping force, a high current is needed, and the associated high copper losses will temporarily heat the generator. Therefore, this type of operation should be time limited for a passively cooled generator or carefully controlled by a cooling system. The maximum force will also be affected by saturation which has been omitted here for simplification.

The MATLAB simulations gave the same results as the original generator design when a magnetic flux density in the airgap of 0.86 T was used and the stacking factor and winding factor was taken into account. The magnetic flux density required in the FEM simulations was 0.92 T. This is an acceptable result since no saturation or leakage flux in the stator are taken into account in the simplified MATLAB simulations. The experimental verification showed only a small voltage difference between the simulation and measurement. However, as only the fundamental of the voltage signal is considered in the model, the method is only suitable for generators expected to have a quite sinusoidal induced voltage.

The average efficiencies are lower than the rated efficiencies, but they follow the rated efficiencies when comparing the generators, i.e., the generator that has the highest rated efficiency also has the highest average efficiency. This means that the rated efficiency is an acceptable parameter to use when comparing designs, even though variable speed operation also must be considered. The efficiency for the design with the higher rated current density decreases the most when operated at variable speed. This gives one more indication of that the rated current density is one of the most important design parameters.

8. Conclusions

A novel simulation method for design comparison of direct-drive linear generators run with CTA control has been presented. The simulation method enables generator designers to quickly evaluate many different design parameters before deciding on a design to further investigate through more time-demanding FEM simulations. A tool such as this is especially useful in the field of linear generator design for wave power as no consensus exists on how to optimize these machines. It can be concluded that the CTA control simplifies the analytical expressions so the efficiency, maximum damping force, and relative cost for the generator can be analyzed without knowledge of the reactance or load angle. The simulation method has been verified against FEM simulations and experiments. The maximum damping force is over-estimated since saturation is neglected, for one example roughly 10% which is still acceptable for design comparison. The fast simulation method can be used for design evaluation and comparison but cannot replace full-physics simulations of generators during design.

An expression of the copper loss to output power quotient has been derived. It shows that for CTA control the quotient is only dependent on airgap magnetic flux density, translator speed, current density, and the quotient between end winding length and stator length. A slow-moving direct-drive synchronous generator has inherently high copper losses and it was shown difficult to make design choices to decrease these substantially. Only decreasing the current density or shortening the end windings can decrease the copper losses. The amount of copper losses also affects the achievable maximum damping force. Therefore, special consideration should be taken to design and production technology to decrease the end winding length as much as possible. A low rated current density gives a large generator. The generator design is therefore a trade-off between cost versus high efficiency and high maximum damping force. It can be concluded that the desired maximum damping force should be considered early in the design stage of the generator.

The parameter study showed that the original generator design, used as a basis in the simulation, represents a quite good design for a PM linear generator for wave power. It can be concluded that a design with a minimized relative cost can be found. However, the low cost might be chosen at the expense of efficiency and damping force. The pole pitch and the rated current density were shown to affect the studied parameters quite a lot, so care should be taken to optimize those parameters when designing a generator. The results indicate that rated speed is acceptable when comparing the efficiency for different generator designs. It will not represent the average efficiency but it can be used for comparison.

As future work it is suggested to include the magnetic circuit in the simulations, preferably as a reluctance circuit. Generator designs with different magnet materials such as ferrite magnets and rare-earth metal-based magnets could be compared. In addition, a more advanced economical model coupled to the annual energy output could be included in the simulations.

Funding: This research was supported by the Swedish Energy Agency grant number 42243-1, Carl Trygger Foundation grant number 15:152 and STandUP for Energy.

Acknowledgments: The author gratefully acknowledges the contribution of Arne Wolfbrandt and Urban Lundin for developing the FEM simulation tool.

Conflicts of Interest: The author declares no conflict of interest. The founding sponsors had no role in the design of the study; in the collection, analyses, or interpretation of data; in the writing of the manuscript, and in the decision to publish the results.

Energies **2019**, *12*, 1312

Abbreviations

The following abbreviations are used in this manuscript:

CTA	Constant-torque-angle
EMF	Electromotive force
FEM	Finite element method
PI	Proportional Integral
PM	Permanent-magnet

References

1. Ozkop, E.; Altas, I.H. Control, power and electrical components in wave energy conversion systems: A review of the technologies. *Renew. Sustain. Energy Rev.* **2017**, *67*, 106–115.
2. Polinder, H.; Damen, M.E.C.; Gardner, F. Linear PM Generator system for wave energy conversion in the AWS. *IEEE Trans. Energy Convers.* **2004**, *5*, 583–589. [CrossRef]
3. Hodgins, N.; Keysan, O.; McDonald, A.S.; Mueller, M.A. Design and Testing of a Linear Generator for Wave-Energy Applications. *IEEE Trans. Ind. Electron.* **2012**, *59*, 2094–2103. [CrossRef]
4. Danielsson, O.; Eriksson, M.; Leijon, M. Study of a longitudinal flux permanent magnet linear generator for wave energy converters. *Int. J. Energy Res.* **2006**, *30*, 1130–1145. [CrossRef]
5. Piscopo, V.; Benassai, G.; Morte, R.D.; Scamardella, A. Cost-Based Design and Selection of Point Absorber Devices for the Mediterranean Sea. *Energies* **2018**, *11*, 946. [CrossRef]
6. Faiz, J.; Nematsaberi, A. Linear permanent magnet generator concepts for direct-drive wave Energy converters: A comprehensive review. In Proceedings of the 12th IEEE Conference on Industrial Electronics and Applications (ICIEA), Siem Reap, Cambodia, 18–20 June 2017; pp. 618–623.
7. Leijon, J.; Sjölund, J.; Ekergård, B.; Boström, C.; Eriksson, S.; Temiz, I.; Leijon, M. Study of an Altered Magnetic Circuit of a Permanent Magnet Linear Generator for Wave Power. *Energies* **2018**, *11*, 84. [CrossRef]
8. Frost, A.E.; Ulvgård, L.; Sjökvist, L.; Eriksson, S.; Leijon, M. Partial Stator Overlap in a Linear Generator for Wave Power: An Experimental Study. *J. Mar. Sci. Eng.* **2017**, *5*, 53. [CrossRef]
9. Strömstedt, E.; Savin, A.; Heino, H.; Antbrams, K.; Haikonen, K.; Götschl, T. Project WESA (Wave Energy for a Sustainable Archipelago)—A Single Heaving Buoy Wave Energy Converter Operating and Surviving Ice Interaction in the Baltic Sea. In Proceedings of the 10th European Wave and Tidal Conference (EWTEC), Aalborg, Denmark, 2–5 September 2013.
10. Lejerskog, E.; Strömstedt, E.; Savin, A.; Boström, C.; Leijon, M. Study of the operation characteristics of a point absorbing direct driven permanent magnet linear generator deployed in the Baltic Sea. *IET Renew. Power Gener.* **2016**, *10*, 1204–1210. [CrossRef]
11. Parel, T.S.; Rotaru, M.D.; Sykulski, J.K.; Hearn, G.E. Optimisation of a tubular linear machine with permanent magnets for wave energy extraction. *COMPEL Int. J. Comput. Math. Electr. Electron. Eng.* **2011**, *30*, 1056–1068. [CrossRef]
12. Rao, K.R.; Sunderan, T.; Adiris, M.R. Performance and design optimization of two model based wave energy permanent magnet linear generators. *Renew. Energy* **2017**, *101*, 196–203. [CrossRef]
13. Shibaike, A.; Sanada, M.; Morimoto, S. Suitable Configuration of Permanent Magnet Linear Synchronous Generator for Wave Power Generation. In Proceedings of the 2007 Power Conversion Conference, Nagoya, Japan, 2–5 April 2007; pp. 210–215.
14. Ngu, S.S.; Dorrell, D.G.; Cossar, C. Design and Operation of Very Slow Speed Generators for a Bristol Cylinder Sea Wave Generating Device. *IEEE Trans. Ind. Appl.* **2014**, *50*, 2749–2759. [CrossRef]
15. Cancelliere, P.; Marignetti, F.; Colli, V.D.; Stefano, R.D.; Scarano, M. A tubular generator for marine energy direct drive applications. In Proceedings of the IEEE International Conference on Electric Machines and Drives, San Antonio, TX, USA, 15–18 May 2005; pp. 1473–1478.
16. Wu, F.; Zhang, X.P.; Ju, P.; Sterling, M.J.H. Optimal Control for AWS-Based Wave Energy Conversion System. *IEEE Trans. Power Syst.* **2009**, *24*, 1747–1755.

17. Eriksson, S.; Solum, A.; Bernhoff, H.; Leijon, M. Simulations and experiments on a 12 kW direct driven PM synchronous generator for wind power. *Renew. Energy* **2008**, *33*, 674–681. [CrossRef]

18. Lejerskog, E.; Leijon, M. Detailed Study of Closed Stator Slots for a Direct-Driven Synchronous Permanent Magnet Linear Wave Energy Converter. *Machines* **2014**, *2*, 73–86. [CrossRef]

19. Ekström, R.; Ekergård, B.; Leijon, M. Electrical damping of linear generators for wave energy converters—A review. *Renew. Sustain. Energy Rev.* **2014**, *42*, 116–128. [CrossRef]

20. Purwadi, A.; Hutahaean, R.; Rizqiawan, A.; Heryana, N.; Heryanto, N.A.; Hindersah, H. Comparison of maximum torque per Ampere and Constant Torque Angle control for 30 kW Interior Interior Permanent Magnet Synchronous Motor. In Proceedings of the Joint International Conference on Electric Vehicular Technology and Industrial, Mechanical, Electrical and Chemical Engineering, Surakarta, Indonesia, 4–5 November 2015; pp. 253–257.

21. Polinder, H.; Slootweg, J.G.; Hoeijmakers, M.J.; Compter, J.C. Modeling of a linear PM Machine including magnetic saturation and end effects: Maximum force-to-current ratio. *IEEE Trans. Ind. Appl.* **2003**, *39*, 1681–1688. [CrossRef]

22. Bülow, F.; Eriksson, S.; Bernhoff, H. No-load core loss prediction of PM generator at low electrical frequency. *Renew. Energy* **2012**, *43*, 389–392. [CrossRef]

23. Savin, A.; Temiz, I.; Strömstedt, E.; Leijon, M. Statistical analysis of power output from a single heaving buoy WEC for different sea states. *Mar. Syst. Ocean Technol.* **2018**, *13*, 103–110. [CrossRef]

24. Eriksson, S.; Bernhoff, H. Rotor design for PM generators reflecting the unstable neodymium price. In Proceedings of the 2012 XXth International Conference on Electrical Machines, Marseille, France, 2–5 September 2012; Volume 33, pp. 1419–1423.

25. Eriksson, S.; Bernhoff, H.; Leijon, M. A 225 kW Direct Driven PM Generator Adapted to a Vertical Axis Wind Turbine. *Adv. Power Electron.* **2011**, *2011*, 239061. [CrossRef]

26. Waters, R.; Engström, J.; Isberg, J.; Leijon, M. Wave climate off the Swedish west coast. *Renew. Energy* **2009**, *34*, 1600–1606. [CrossRef]

Article

Fast and Accurate Model of Interior Permanent-Magnet Machine for Dynamic Characterization

Klemen Drobnič [1,*] , **Lovrenc Gašparin** [2] and **Rastko Fišer** [1]

[1] Faculty of Electrical Engineering, University of Ljubljana, Tržaška 25, SI-1000 Ljubljana, Slovenia; rastko.fiser@fe.uni-lj.si

[2] Mahle Electric Drives Slovenia, Polje 15, SI-5290 Šempeter pri Gorici, Slovenia; lovrenc.gasparin@mahle.si

* Correspondence: klemen.drobnic@fe.uni-lj.si

Received: 31 January 2019; Accepted: 22 February 2019; Published: 26 February 2019

Abstract: A high-fidelity two-axis model of an interior permanent-magnet synchronous machine (IPM) presents a convenient way for the characterization and validation of motor dynamic performance during the design stage. In order to consider a nonlinear IPM nature, the model is parameterized with a standard dataset calculated beforehand by finite-element analysis. From two possible model implementations, the current model (CM) seems to be preferable to the flux-linkage model (FLM). A particular reason for this state of affairs is the rather complex and time-demanding parameterization of FLM in comparison with CM. For this reason, a procedure for the fast and reliable parameterization of FLM is presented. The proposed procedure is significantly faster than comparable methods, hence providing considerable improvement in terms of computational time. Additionally, the execution time of FLM was demonstrated to be up to 20% shorter in comparison to CM. Therefore, the FLM should be used in computationally intensive simulation scenarios that have a significant number of iterations, or excessive real-time time span.

Keywords: digital simulation; motor drives; interior permanent-magnet machines; finite-element analysis; modeling; automotive applications; electric vehicle (EV); hybrid electric vehicle (HEV); mathematical model; saturation

1. Introduction

Permanent-magnet synchronous machines excel in high torque density and high efficiency, so they have been widely investigated in recent years [1–3]. From the two permanent-magnet synchronous-machine topologies, i.e., surface permanent-magnet machines (SPM) and interior permanent-magnet machines (IPM), the latter are especially suitable for transport applications, where a wide speed range is a key requirement [4,5]. The specific feature that enables such an operation is their pronounced flux-weakening capability. This is an immediate consequence of IPM design, where the magnetic field due to the armature current is considerable with respect to the permanent magnet field [6]. Hence, IPMs are substantially more saturation-prone than their SPM counterparts, rendering any assumption of magnetic linearity at high flux levels untenable.

In modern electric-machine design, finite-element analysis (FEA) is widely used and has become an industry standard [7–9]. By taking into account the specific geometry and material, FEA produces an accurate magnetic-field model, which is an excellent foundation for the reliable prediction of machine operational characteristics. In this way, the expensive and time-consuming prototype stage can be completely left out [10,11]. This fact is especially important for any designer in the predevelopment design stage, who is daily faced with specific consumer demands that must be promptly served with fast and reliable design offers. An additional benefit of using FEA is that existent FEA results can be readily converted in state-space models, which are computationally much more efficient than FEA [12]. These models can be used as a platform for the verification of various types of machine dynamics. An illustrative example is a symmetrical short circuit, which can be simulated within seconds. Otherwise, dynamic FEA would have to be employed, resulting in about 10 h of computation time on a typical workstation [7]. Furthermore, the state-space model can be turned into a black box, which can directly be deployed to consumers for their system simulation. As an example, the state-space model of an IPM drive can serve as a high-fidelity real-time emulator in Hardware-in-the-Loop (HIL) tests [13], or as a platform for system-efficiency investigation [14].

Among the family of state-space models, a two-axis (*dq*) model, defined in synchronous (rotor) co-ordinates, is by far the most frequently used [15]. Two basic modeling paradigms depending on chosen state variables (i.e., currents or flux linkages) are possible, resulting in the current model (CM) and flux-linkage model (FLM). For linear models with constant parameters, both choices are equally suitable in terms of simplicity, even though the FLM is computationally slightly less demanding (one integrator instead of two), but this difference is negligible with today's computational power [16]. In spite of that, practitioners in many subfields where a simulation model is essential (e.g., control design, system design) overwhelmingly prefer the CM over FLM. The reason for this is that CM-based computations directly yield quantities, some of which can be directly monitored, e.g., stator currents.

On the other hand, the *dq*-model can also reproduce various nonlinear phenomena, such as magnetic saturation, cross-coupling, spatial harmonics, and iron losses [12,17–24]. In this case, the parameters of the *dq*-model are estimated with FEA results. The merits of a state-space model and a magnetic-field model are thus advantageously combined [22,25]. However, a fundamental distinction arises between CM and FLM in terms of model complexity. In CM, a nonlinear current-flux relation causes inductance to be split and reintroduced as apparent and incremental inductance [6]. At least three consequences come to light: (a) the physical meaning of inductance is blurred, (b) the CM structure becomes involved, and (c) in order to preserve accuracy, an additional set of FEA with a special technique (e.g., the frozen permeability method [26]) should be made to separate the spatial effect of permanent magnets and the stator current. Nothing of the above applies to FLM, where the model structure stays the same and existent FEA results suffice [19].

Even though there are clear advantages for choosing FLM over CM, the practical application of FEA-parameterized FLM is still comparatively rare [18]. The main reason probably lies in the standard static FEA setup. The stator current being an input, FEA produces flux-linkage maps where the stator current is treated as an independent variable (direct form). As FEA-parameterized FLM actually requires an inverse relation between current and flux linkage (inverse form), the designer is typically induced to rather follow CM paradigm [6]. Nevertheless, few authors have recently tackled the subject of inverting flux-linkage maps [18–21]. Various strategies were employed in order to find the inverse form, such as the interior-point method [18] and radial-basis function [20], but scarce implementation details were provided. In Reference [21], the lookup table-inversion process is outlined but too vaguely for easy third-party implementation. All procedures listed above are significantly more computationally demanding in comparison with the straightforward parameterization of CM. However, one should note that the principal objective of the aforementioned papers was to develop a high-fidelity state-space model per se, regardless of the possible increase in complexity of some sort (mathematical and/or computational).

This paper intends to fill this gap, and specifically proposes a straightforward parameterization procedure that adapts FEA results into a form suitable for FLM parameterization. Parameterization algorithm features are clear implementation, computational efficiency, and high reliability. Only a standard set of FEA results are considered as an input, ruling out any need for additional precomputation.

In this paper, spatial effects and and iron losses are not considered, because they are negligible for the studies we are performing, i.e., evaluation of motor performance in a certain drive cycle and a prediction of short-circuit currents, although they can be readily included by using effective and proven techniques [18,27–29]. However, readers should be aware that consideration of spatial effects is an integral part of a true high-fidelity IPM model [30]. Namely, concentrated stator windings, different rotor geometries, or any other variation of magnetic permeance along the air gap are a common sight in modern practices. In this way, the effects of air-gap harmonics on voltages, currents, flux linkages, and torque can be properly described. This is essential for a user, if he intents to use the IPM model for, e.g., developing control (including sensorless and field-weakening operations) [22], analyzing torque ripple [31], and studying acoustic characteristics [19].

2. Two-Axis IPM Models

2.1. General Equations

Restricting the study to the case of perfect field orientation, the voltage equation of IPM is defined as

$$v_s = R_s i_s + \frac{d\psi_s}{dt} + J\omega_r \psi_s, \tag{1}$$

where $v_s = [v_d \; v_q]^T$, $i_s = [i_d \; i_q]^T$ and $\psi_s = [\psi_d \; \psi_q]^T$ are real vectors defined in the rotor reference frame, R_s is stator resistance (scalar), ω_r is rotor electrical speed, and J the matrix equivalent to complex unit j

$$J = \begin{bmatrix} 0 & -1 \\ 1 & 0 \end{bmatrix}.$$

Electromagnetic torque T_e is defined as a vector product of flux linkage and current

$$T_e = \frac{3}{2} p_p (\psi_s \times i_s), \tag{2}$$

where p_p is the number of pole pairs. Dynamic equation links the electromagnetic and mechanical domains

$$T_e - T_l = \Gamma \frac{d\omega_M}{dt}, \tag{3}$$

where T_l is the load torque, Γ the moment of inertia, and ω_M rotor mechanical speed.

Equations (1)–(3) are globally valid for any working condition. In order to obtain a full set of IPM equations, the relation between flux linkage and current (flux–current relation, flux map) $\psi_s = f(i_s)$ needs to be defined. Specifically addressing two-axis IPM representation in the rotor reference frame, the model is linear only if saturation is not present. Therefore, function $f : \mathbf{R}^2 \to \mathbf{R}^2$ is affine if the system is linear, and nonlinear if otherwise.

2.2. Linear Model

2.2.1. CM

If saturation is neglected, the system is linear, and the stator flux linkage may be expressed as an affine function of i_s

$$\psi_s = f(i_s) = L_s i_s + \psi_R, \tag{4}$$

where $\psi_R = [\psi_R \; 0]^T$ is a *constant* real vector of the rotor flux linkage defined in the rotor reference frame, and L_s inductance matrix with *constant* terms $L_d = (\psi_d - \psi_R)/i_d$ and $L_q = \psi_q/i_q$

$$L_s = \begin{bmatrix} L_d & 0 \\ 0 & L_q \end{bmatrix}. \tag{5}$$

L_s is a diagonal matrix that satisfies the fundamental assumption of the decoupled model. Plugging Equation (4) into Equation (1) and considering $d\psi_R/dt = 0$ gives voltage equation

$$v_s = (R_s I + J\omega_r L_s) i_s + L_s \frac{di_s}{dt} + J\omega_r \psi_R, \tag{6}$$

where I is the 2×2 identity matrix. In this way, IPM dynamics is fully described by Equations (6), (2), (3), and (4). As the stator current is state-variable in Equation (6), this particular model is called CM.

2.2.2. FLM

Note that L_s in Equation (4) is invertible; hence, inverse relationship f^{-1} can be readily obtained

$$i_s = f^{-1}(\boldsymbol{\psi}_s) = L_s^{-1}(\boldsymbol{\psi}_s - \boldsymbol{\psi}_R),\tag{7}$$

where L_s^{-1} is an inverse inductance matrix

$$L_s^{-1} = \begin{bmatrix} 1/L_d & 0 \\ 0 & 1/L_q \end{bmatrix}.$$

Plugging Equation (7) into Equation (1), we obtain

$$v_s = (R_s L_s^{-1} + J\omega_r)\boldsymbol{\psi}_s + \frac{d\boldsymbol{\psi}_s}{dt} - R_s L_s^{-1}\boldsymbol{\psi}_R.\tag{8}$$

Now, Equation (8) can be combined with Equations (2), (3), and (7) to form an FLM, with stator flux linkage as a state-variable. The particular form of Equation (8) clearly indicates that the parameterization of FLM is exactly the same as for a CM, as only four parameters, i.e., R_s, L_d, L_q, and $\boldsymbol{\psi}_R$, are needed.

Remark 1. *Torque Equation (2) for either of linear models can be rewritten in familiar form*

$$T_e = \frac{3}{2} p_p \left(\psi_R i_q + (L_d - L_q) i_d i_q \right).\tag{9}$$

2.3. Nonlinear Model

2.3.1. CM

In the presence of saturation, the flux–current relation becomes nonlinear. Nevertheless, we can still write it in the form of Equation (4), provided that $L_s(i_s)$ and $\boldsymbol{\psi}_R(i_s)$ alter to variables dependent on the stator current

$$\boldsymbol{\psi}_s = f(i_s) = L_s(i_s)i_s + \boldsymbol{\psi}_R(i_s).\tag{10}$$

In this way, the stator current remains a state-space variable. The rotor flux-linkage vector in Equation (10)

$$\boldsymbol{\psi}_R = [\psi_R(i_s) \ 0]^T$$

preserves perfect field orientation $\psi_R = ||\boldsymbol{\psi}_R||$ by definition. On the other hand, its amplitude is influenced by the stator current, by both current components i_d and i_q in general. The 2×2 matrix $L_s(i_s)$ has four nonzero elements:

$$L_s(i_s) = \begin{bmatrix} L_{dd}(i_s) & L_{dq}(i_s) \\ L_{qd}(i_s) & L_{qq}(i_s). \end{bmatrix}$$

Each element is defined as a ratio of appropriate flux linkage and the current component; therefore, the following form is obtained [6]

$$L_s(i_s) = \begin{bmatrix} L_{dd}(i_s) & L_{dq}(i_s) \\ L_{qd}(i_s) & L_{qq}(i_s) \end{bmatrix} = \begin{bmatrix} \frac{\psi_d - \psi_R}{i_d} & \frac{\psi_d - \psi_R}{i_q} \\ \frac{\psi_q}{i_d} & \frac{\psi_q}{i_q} \end{bmatrix},\tag{11}$$

where explicit dependence of flux linkages on stator currents was dropped to avoid clutter. In a nonlinear context, L_s is called an apparent inductance matrix. Elements of L_s are not constant as they are all dependent on the stator current. In comparison with Equation (5), the matrix also has two nondiagonal terms, a clear indication of the cross-saturation phenomena. It is interesting to notice that coupling between direct and quadrature axes arises due to the effects of main-flux saturation.

Substituting Equation (10) in Equation (1), but leaving the derivative $d\boldsymbol{\psi}_s/d\boldsymbol{i}_s$ in its original form yields

$$v_s = (R_s + J\omega_r L_s)i_s + L_i \frac{di_s}{dt} + J\omega_r \boldsymbol{\psi}_R, \tag{12}$$

where L_i is the incremental inductance matrix, defined as

$$L_i = \frac{d\boldsymbol{\psi}_s}{d\boldsymbol{i}_s} = \begin{bmatrix} \frac{\partial \psi_d}{\partial i_d} & \frac{\partial \psi_d}{\partial i_q} \\ \frac{\partial \psi_q}{\partial i_d} & \frac{\partial \psi_q}{\partial i_q} \end{bmatrix} = \begin{bmatrix} l_{dd}(\boldsymbol{i}_s) & l_{dq}(\boldsymbol{i}_s) \\ l_{qd}(\boldsymbol{i}_s) & l_{dd}(\boldsymbol{i}_s) \end{bmatrix}. \tag{13}$$

In general, apparent and incremental inductance matrices are not equal ($L_s \neq L_i$) unless the machine is linear, which can easily be verified by plugging Equation (4) in Equation (13). The full CM set of equations is defined by Equations (2), (3), (10), and (12).

2.3.2. FLM

Because the model is nonlinear, there is no benefit in rewriting the inverse relation as in Equation (8). Instead, we simply define current-flux relation

$$i_s = g(\boldsymbol{\psi}_s) = F_s(\boldsymbol{\psi}_s)\boldsymbol{\psi}_s, \tag{14}$$

where F_s is a generalized "reluctance" matrix obtained in an analogous way to Equation (11)

$$F_s(\boldsymbol{\psi}_s) = \begin{bmatrix} \frac{i_d}{\psi_d} & \frac{i_d}{\psi_q} \\ \frac{i_q}{\psi_d} & \frac{i_q}{\psi_q} \end{bmatrix} = \begin{bmatrix} F_{dd}(\boldsymbol{\psi}_s) & F_{dq}(\boldsymbol{\psi}_s) \\ F_{qd}(\boldsymbol{\psi}_s) & F_{qq}(\boldsymbol{\psi}_s) \end{bmatrix}. \tag{15}$$

Matrix F_s is composed of elements that are strictly ratios of the stator-current and stator flux-linkage components. Note, that rotor flux linkage is explicitly left out.

Inserting Equation (14) in Equation (1), the FLM voltage equation is obtained

$$v_s = R_s F_s \boldsymbol{\psi}_s + \frac{d\boldsymbol{\psi}_s}{dt} + J\omega_r \boldsymbol{\psi}_s, \tag{16}$$

where stator flux linkage is preserved as state-variable. The full FLM set of equations is defined by Equations (2), (3), (14), and (16).

2.4. Simulation Form

A digital simulation requires the state-space form of state equations $\dot{x} = f(x)$. Even though rotor speed ω_r is state-variable in both models as well, only the state variables in voltage equations are of interest. Voltage Equations (12) and (16) can be rearranged into

$$\frac{di_s}{dt} = L_i^{-1}[v_s - (R_s + J\omega_r L_s)i_s - J\omega_r \boldsymbol{\psi}_R] \tag{17}$$

$$\frac{d\boldsymbol{\psi}_s}{dt} = v_s - R_s F_s \boldsymbol{\psi}_s - J\omega_r \boldsymbol{\psi}_s. \tag{18}$$

Figures 1 and 2 depict the block diagrams of the CM and FLM, respectively. It is worth noting that CM diagram is clearly more complex. Comparison between the two diagrams shows that CM requires two matrices, L_s and L_i^{-1}, whereas the FLM only matrix F_s. In addition, the simultaneous presence of two inductance matrices, L_s and L_i^{-1}, obscures the meaning of inductance.

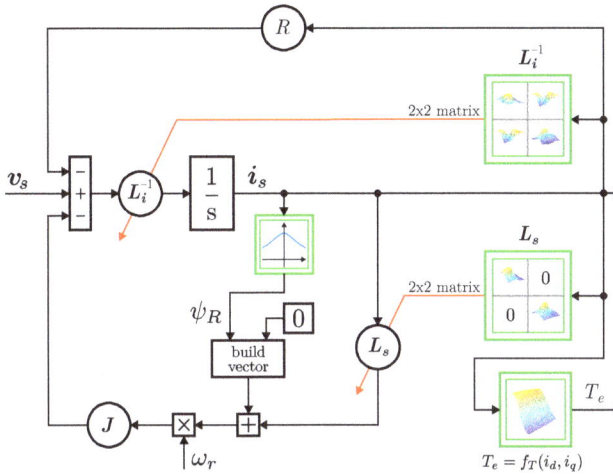

Figure 1. Block diagram of nonlinear current model (CM).

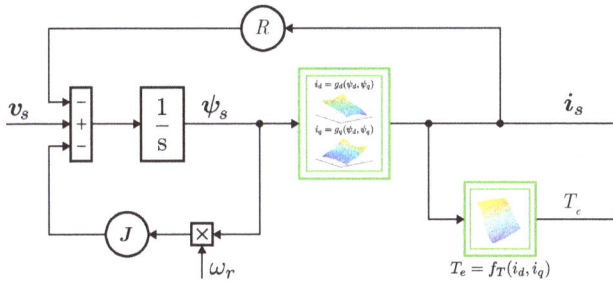

Figure 2. Block diagram of nonlinear flux-linkage model (FLM).

3. Model Parameterization

3.1. Static FEA Batch Simulation

Parameterization of a nonlinear model can be established in two distinct ways, by means of FEA or measurements [32,33]. FEA is preferred when reasonably accurate information about geometry and magnetic characteristics of materials is close at hand, which is especially true for the predesign stage. However, if reliable input data are not at one's disposal, a systematic measurement procedure can provide fairly accurate parameterization as well [33,34].

Once the designer specifies the magnetostatic finite-element method (FEM) model of IPM, a series of simulations are performed in order to obtain the fundamental electromagnetic relations of the machine. Firstly, the d- and q-axis stator-current vectors with N values over a reasonable range are specified

$$I_d^{\text{orig}} = \begin{bmatrix} I_{d1} & \cdots & I_{dN} \end{bmatrix}^T \qquad I_q^{\text{orig}} = \begin{bmatrix} I_{q1} & \cdots & I_{qN} \end{bmatrix}^T. \tag{19}$$

Then, the batch process iterates over all possible combinations of the d- and q-axis stator currents while performing FEA for each combination. Finally, the batch job returns three matrices $\mathbf{\Psi}_d^{\text{orig}}$, $\mathbf{\Psi}_q^{\text{orig}}$ and T_e

$$\mathbf{\Psi}_d^{\text{orig}} = \begin{bmatrix} \Psi_{d11} & \cdots & \Psi_{d1N} \\ \vdots & \ddots & \vdots \\ \Psi_{dN1} & \cdots & \Psi_{dNN} \end{bmatrix} \qquad \mathbf{\Psi}_q^{\text{orig}} = \begin{bmatrix} \Psi_{q11} & \cdots & \Psi_{q1N} \\ \vdots & \ddots & \vdots \\ \Psi_{qN1} & \cdots & \Psi_{qNN} \end{bmatrix} \tag{20}$$

$$T_e = \begin{bmatrix} T_{e11} & \cdots & T_{e1N} \\ \vdots & \ddots & \vdots \\ T_{eN1} & \cdots & T_{eNN} \end{bmatrix} \tag{21}$$

Matrices $\mathbf{\Psi}_d^{\text{orig}}$ and $\mathbf{\Psi}_q^{\text{orig}}$ are commonly known as flux-linkage maps, while T_e is a torque map. The matrices share the same form with straightforward interpretation: it is a table of numbers that depend on two indices, i and j, which correspond to entries in vectors I_d^{orig} and I_q^{orig}, respectively. For example, element Ψ_{d24} gives the value of the d-axis stator flux linkage at specific stator current values $I_d^{\text{orig}}(2)$ and $I_q^{\text{orig}}(4)$.

Figure 3 shows the original flux-linkage maps for an IPM with pronounced nonlinearity than can be visually checked from figures. For example, in case of the d-axis flux linkage (Figure 3a) the surface curvature along d-axis indicates the main saturation, whereas the curvature along the q-axis is indicative of cross-saturation (Figure 3b). Figure 4 shows the electromagnetic torque of the same IPM featuring a curved surface. As expected, the nonlinear effect is more pronounced with higher stator currents in the respective axes.

In this paper, an IPM with data in Table 1 was used as a case study. There are two reasons that all maps are defined for a full (symmetrical) range of currents. The short-circuit test (SCT), which is an important application of the model, is a specific operating state where stator current are not under control. Furthermore, our intention was to illustrate a complete treatment of flux-map inversion. Readers should be aware that, in general, there is no need for a positive i_d current if only a normal IPM operation is considered.

Flux and torque maps are built for current values $I_d^{\text{orig}} = I_q^{\text{orig}} = [-1860 \ \ldots \ 1860]^T$ A in 33 equidistant steps, which results in $33 \times 33 = 1089$ data points. The appropriate range of stator-current values used for flux- and torque-map calculation depends on the specific IPM. In the authors' experience, the limits should be at least 2.2 larger than IPM's maximal current $I_{\max} = 778$ A. However, if it turns out that the range is insufficient, reiteration is needed. We also found that further increase in grid density is not justified in light of increasing FE computational costs. An equidistant grid was chosen for convenience.

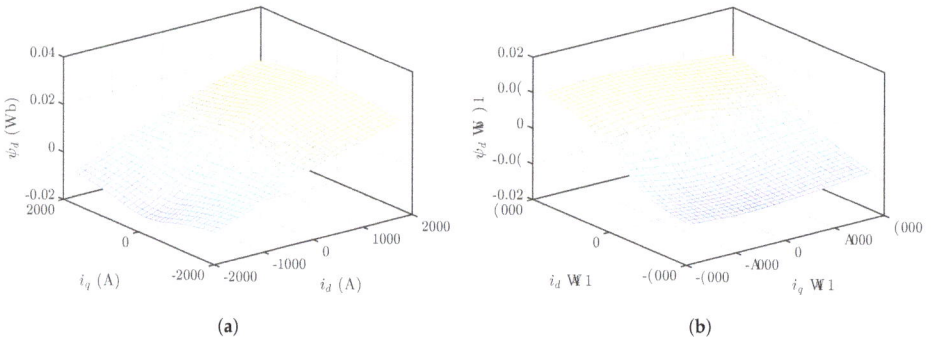

Figure 3. Original flux-linkage map for (**a**) d-axis and (**b**) q-axis flux linkage.

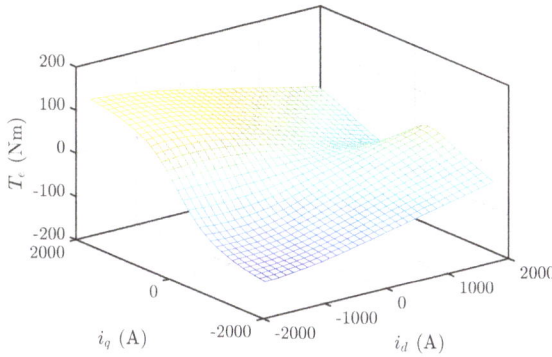

Figure 4. Original torque map for electromagnetic torque.

3.2. CM Parameterization

Once Ψ_d^{orig}, Ψ_q^{orig}, and T_e are known for a specific design, it is possible to define their respective interpolation functions f_d, f_q, and f_T:

$$\psi_d = f_d(i_d, i_q), \quad \psi_q = f_q(i_d, i_q), \quad T_e = f_T(i_d, i_q). \tag{22}$$

CM parameterization requires the calculation of matrices L_s and L_i^{-1}. Because L_i is completely oblivious (see Equation (17)), L_i^{-1} can be directly calculated using

$$L_i^{-1} = \left(\frac{\partial \psi_d}{\partial i_d} \frac{\partial \psi_q}{\partial i_q} - \frac{\partial \psi_d}{\partial i_q} \frac{\partial \psi_q}{\partial i_d} \right)^{-1} \begin{bmatrix} \frac{\partial \psi_q}{\partial i_q} & -\frac{\partial \psi_d}{\partial i_q} \\ -\frac{\partial \psi_q}{\partial i_d} & \frac{\partial \psi_d}{\partial i_d} \end{bmatrix}, \tag{23}$$

where an inversion formula for a 2×2 matrix was applied to Equation (13).

On the other hand, apparent inductance matrix L_s defined in Equation (11) suffers from major drawback. For example, the first term in L_s

$$L_{dd}(i_d, i_q) = \frac{\psi_d(i_d, i_q) - \psi_R(i_d, i_q)}{i_d}$$

besides $\psi_d(i_d, i_q)$ also requires $\psi_R = f_R(i_d, i_q)$, which is not available from the original set of simulation results. A straightforward solution would be to define ψ_R as a constant $\psi_R = f_d(0, 0)$, but then singularity at $i_d = 0$ is unavoidable. Namely, the cross-saturation effect of i_q onto ψ_d results in a slight reduction of ψ_d. In consequence, numerator $\psi_d(0, i_q) - \psi_R$ does not evaluate to 0, hence resulting in a singularity. A similar reasoning can be adopted for second term $L_{dq}(i_d, i_q)$, where numerator $\psi_d(i_d, 0) - \psi_R \neq 0$ for $i_q = 0$.

A practical solution is twofold: neglecting cross-saturation terms in (11)

$$L_s(i_s) = \begin{bmatrix} \frac{\psi_d - \psi_R}{i_d} & 0 \\ 0 & \frac{\psi_q}{i_q} \end{bmatrix} = \begin{bmatrix} L_{dd}(i_s) & 0 \\ 0 & L_{qq}(i_s) \end{bmatrix}, \tag{24}$$

and defining $\psi_R = f_d(0, i_q)$ as an explicit function of q-axis current. Then, L_{dd} can be calculated for all current combinations without incurring a singularity

$$L_{dd}(i_d, i_q) = \frac{\psi_d(i_d, i_q) - \psi_d(0, i_q)}{i_d}.$$

It is clear from the above argument that parameterization with the original set of FEA data requires CM simplification. Whether its impact is considerable or marginal depends on the grade of nonlinearity of the particular machine. Machines with pronounced nonlinearity, as is the case for IPMs, may therefore require a full inductance matrix. At this point, it should be mentioned that approaches that can overcome the limits in calculating original Equation (11) were reported, e.g., frozen-permeances method. However, this technique requires an additional FEA set, effectively doubling the computational burden.

3.3. FLM Parameterization

In comparison to CM, FLM simulation form Equation (18) is simpler, as only matrix F_s has to be parameterized. Even though the four terms in F_s are elementary ratios of the stator current and flux linkage, e.g., a first term in Equation (15)

$$F_{dd}(\psi_d, \psi_q) = \frac{i_d(\psi_d, \psi_q)}{\psi_d},$$

they cannot be readily calculated using previously defined functions f_d and f_q. The underlying reason is that the simulation flow of FLM treats flux linkage as the independent variable as far as current-flux relation is concerned. Therefore, an inverse relation between current and flux linkage must be determined beforehand. Drawing an analogy between CM and FLM, we define corresponding interpolating functions

$$i_d = g_d(\psi_d, \psi_q), \quad i_q = g_q(\psi_d, \psi_q). \tag{25}$$

The relation between two sets of interpolating Functions (22) and (25) can be thought of as a mathematical inverse of two variate functions

$$i_d = g_d(f_d(i_d, i_q), f_q(i_d, i_q)) \tag{26}$$
$$i_q = g_q(f_d(i_d, i_q), f_q(i_d, i_q)). \tag{27}$$

An important question to be answered is whether inverse functions g_d and g_q actually exist. Consider mapping $f : \mathbf{R}^2 \rightarrow \mathbf{R}^2$, defined by

$$f(i_d, i_q) = \begin{bmatrix} f_d(i_d, i_q) \\ f_q(i_d, i_q) \end{bmatrix}. \tag{28}$$

If the derivative of a mapping f is invertible at some point (i_{d0}, i_{q0}), then mapping f is locally invertible in some neighborhood of point $f(i_{d0}, i_{q0})$ [35]. The derivative of f is just a Jacobian matrix

$$J_f = \begin{bmatrix} \frac{\partial f_d}{\partial i_d} & \frac{\partial f_d}{\partial i_q} \\ \frac{\partial f_q}{\partial i_d} & \frac{\partial f_q}{\partial i_q} \end{bmatrix}, \tag{29}$$

which is invertible exactly when its determinant is nonzero:

$$\det(J_f) = \frac{\partial f_d}{\partial i_d} \frac{\partial f_q}{\partial i_q} - \frac{\partial f_d}{\partial i_q} \frac{\partial f_q}{\partial i_d} \neq 0. \tag{30}$$

Mapping f is invertible over the domain of current vectors I_d^{orig} and I_q^{orig} only if the corresponding $\det(J_f)$ is strictly nonzero at every point in the domain. Figure 5 shows the Jacobian determinant for the IPM in question (see Table 1). For this particular machine, it is clear that the determinant is strictly positive; therefore, the IPM's flux maps are invertible over the whole domain.

It is interesting to inspect Equation (30) more closely in order to gain some insight in its behavior for a general machine. The first term is always strictly positive, as f_d and f_q are both monotone-increasing, which can easily be seen by inspection. This property is also physically sound because increasing the current always results in increased flux linkage in the respective axis. The inspection of the second term is more involved, but careful analysis shows that it is also strictly positive, as can be seen in Figure 6. For this reason, the first term must always be bigger than the second so as to maintain invertible mapping f. If we associate the two terms with main and cross magnetization, respectively, we can see that this is an entirely reasonable expectation for a practical machine. Typically, the level of main magnetization is about one order of magnitude higher than that of cross-magnetization.

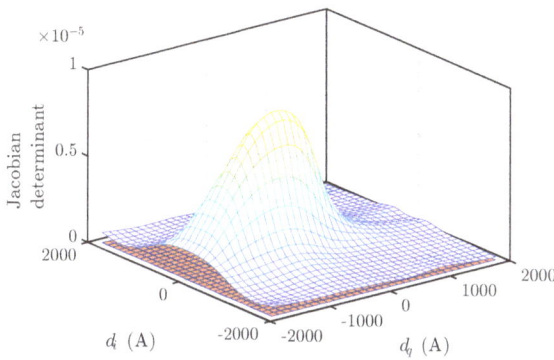

Figure 5. Jacobian determinant J_f over the domain of current vectors I_d^{orig} and I_q^{orig}.

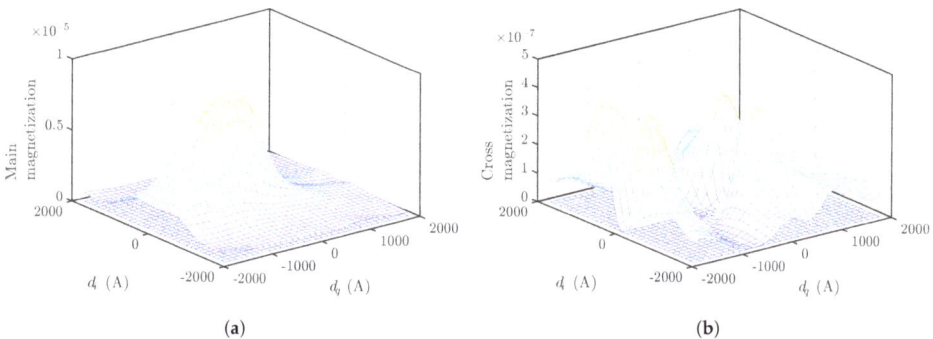

Figure 6. (**a**) Main and (**b**) cross-magnetization term in equation for Jacobian determinant. Note the different order of magnitude.

If interpolation functions g_d and g_q exist, they can be found in various ways. In accordance to Equation (19), we generate d- and q-axis stator flux-linkage vectors with N-equidistant, monotone-increasing entries

$$\boldsymbol{\Psi}_d^{\mathrm{new}} = [\Psi_{d1} \ \cdots \ \Psi_{dN}]^T \qquad \boldsymbol{\Psi}_q^{\mathrm{new}} = [\Psi_{q1} \ \cdots \ \Psi_{qN}]^T,$$

where their range is specified by the minimal and maximal value of original maps $\boldsymbol{\Psi}_d^{\mathrm{orig}}$ and $\boldsymbol{\Psi}_d^{\mathrm{orig}}$, respectively.

3.3.1. Inversion via Minimization

One way to find inverse maps I_d^{new} and I_q^{new} is to minimize the residual between interpolation points and the value of interpolating function over whole set $i, j = 1 \ldots N$. This translates to a multiobjective nonlinear least-squares problem with two objectives, one for each component

$$J_{d,ij}(i_d, i_q) = ||\Psi_d^{new}(i) - f_d(i_d, i_q)||^2$$
$$J_{q,ij}(i_d, i_q) = ||\Psi_q^{new}(j) - f_q(i_d, i_q)||^2.$$

For each iteration, a unique solution (i_d^{sol}, i_q^{sol}) exists, which is interpreted as an entry in inverse maps $I_d^{new}(i, j) = i_d^{sol}$ and $I_q^{new}(i, j) = i_q^{sol}$. Formally, we can express the idea as

$$\text{minimize} \quad ||r_{ij}(i_d, i_q)||^2, \tag{31}$$

where the residual is defined as $r_{ij}(i_d, i_q) = J_{d,ij}(i_d, i_q) + J_{q,ij}(i_d, i_q)$. Once inverse maps I_d^{new} and I_q^{new} are obtained, one can easily define interpolation functions g_d and g_d.

3.3.2. Inversion via Intersections

An equivalent solution for determining inverse maps I_d^{new} and I_q^{new} is via intersections. Original maps Ψ_d^{orig} and Ψ_q^{orig} are first interpolated (functions f_d and f_q) and then sliced into predefined number of contour isolines at flux-linkage levels $\Psi_d^{new}(i)$ or $\Psi_q^{new}(j)$. These lines can be interpreted as curves in the current co-ordinates, where each of these curves encodes all possible combinations (i_d, i_q) for a particular flux-linkage level. Algorithm 1 then proceeds to find an intersection between curves at $\Psi_d^{new}(i)$ and $\Psi_q^{new}(j)$ for all combinations of the stator flux-linkage. If Condition (30) holds, then a unique intersection exists for each pair of $\Psi_d^{new}(i)$ and $\Psi_q^{new}(j)$. These solutions are not only more intuitive due to a clear geometrical interpretation, but also significantly faster.

Algorithm 1 Inverse flux map via intersection.

1: Define number of flux isolines n (levels)
2: Slice flux map $\psi_d = f_d(i_d, i_q)$ (3D) into n isolines (2D)
3: Slice flux map $\psi_q = f_q(i_d, i_q)$ (3D) into n isolines (2D)
4: Obtain $2n$ isolines $\psi_{d,i}$ and $\psi_{q,j}$
5: Initialize empty $n \times n$ matrices Ψ_d^{new}, Ψ_q^{new}, I_d^{new}, and I_q^{new}
6: **for** $i \leftarrow 1 \ldots n$ **do**
7:     **for** $j \leftarrow 1 \ldots n$ **do**
8:         Find intersection (x_d, x_q) between $\psi_{d,i}$ and $\psi_{q,j}$
9:         Update $\Psi_d^{new}(i, j) = \psi_{d,i}$ and $\Psi_q^{new}(i, j) = \psi_{q,j}$
10:         Update $I_d^{new}(i, j) = x_d$ and $I_q^{new}(i, j) = x_q$
11:     **end for**
12: **end for**
13: Extrapolate I_d^{new} and I_q^{new} at corners
14: **return** Ψ_d^{new}, Ψ_q^{new}, I_d^{new}, and I_q^{new}

Figure 7 shows the resulting inverse flux maps using the intersection algorithm. In order to validate inversion accuracy, flux-linkage errors

$$\Delta_d = f_d(g_d(\psi_d, \psi_q), g_q(\psi_d, \psi_q)) - \psi_d \tag{32}$$
$$\Delta_q = f_q(g_d(\psi_d, \psi_q), g_q(\psi_d, \psi_q)) - \psi_q \tag{33}$$

are checked for the whole range of flux-linkage values. For each flux-linkage pair (ψ_d, ψ_q), their respective currents (i_d, i_q) are calculated using interpolating functions g_d and g_q. Then, these currents are plugged into f_d and f_q, resulting in a new flux-linkage pair that should be as close as possible to the original. Figure 8 confirms that the maximum error did not exceed 0.1% in either axis.

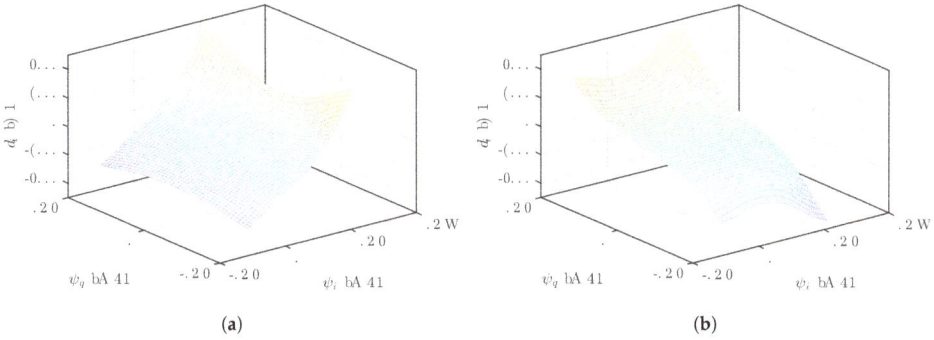

Figure 7. Inverse flux map for (**a**) d-axis and (**b**) q-axis current.

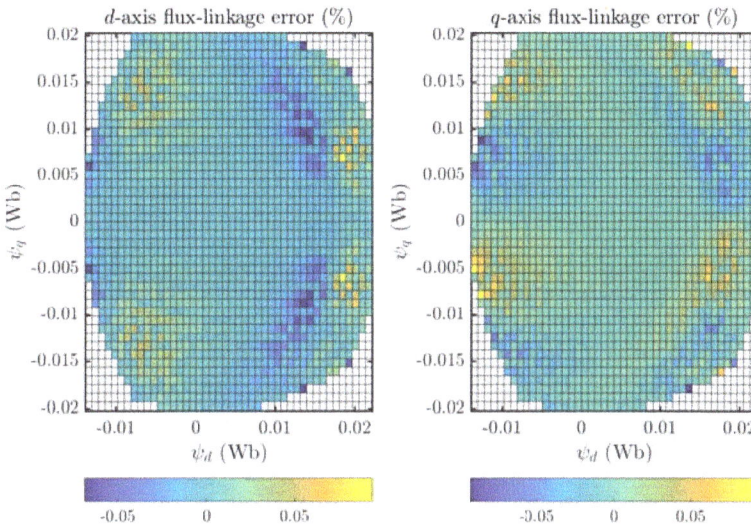

Figure 8. Flux-linkage error for (**Left**) d-axis and (**Right**) q-axis.

4. CM and FLM Performance Comparison

4.1. Model Verification

First, we compare the performance of CM and FLM in a dynamic simulation. A three-phase short-circuit test (SCT), a well-established IPM test procedure in the industry [36], was chosen because high transient currents were expected to push the IPM deep into saturation. In this way, both models could conveniently be assessed in terms of nonlinearity. The chosen IPM (Table 1) is intended for small urban electric vehicles. As the IPM nominal voltage level was low ($V_{dc} = 48$ V), the maximal current was rather large $I_{max} = 778$ A. Both inductances were valid for a nominal operation point.

Table 1. Motor data.

Nominal data	
Power P_n	25 kW
Dc-link voltage V_{dc}	48 V
Maximal phase current I_{max}	778 A
Characteristic current I_{ch}	550 A
Number of pole pairs p_p	4
Moment of inertia Γ	0.003 kgm^2
Parameters	
Stator resistance R_s	3.3 mΩ
d-axis inductance L_d	0.013 mH
q-axis inductance L_q	0.029 mH
Rotor flux-linkage ψ_R	12.1 mWb

During the SCT, the IPM rotor is driven by an external machine at a constant speed, when stator windings (at first open) are suddenly short-circuited. Figure 9 depicts the stator currents in dq co-ordinates for CM and FLM. There was practically a perfect match between currents, the sole difference being a slightly higher oscillation frequency for CM. One can conclude that CM and FLM are equivalent in terms of performance.

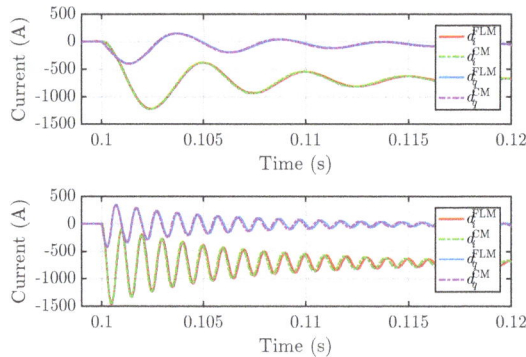

Figure 9. Comparison of FLM and CM for a three-phase symmetric short circuit at (**top**) 3000 min^{-1} and (**bottom**) 15,000 min^{-1}.

Then, FLM simulation results of a three-phase SCT were compared to measurements on a prototype machine (Figure 10). The time-domain response of the phase currents to the sudden short circuit depends on the starting position of the rotor. In order to compare measurements with simulations, the time traces of phase currents must be synchronized. Unfortunately, the measurement setup did not allow to choose the exact instant of the short-circuit maneuver. Therefore, the measurement was performed first and the appropriate time instant of the short circuit in simulation was determined afterwards. It is noteworthy to observe that FLM is capable or forecasting true maximal phase current (in this case red signal at $t = 0.013$ s).

Nevertheless, a direct comparison of individual phase currents (Figure 11) reveals discrepancies between measurement and simulation. All three measured phase currents had smaller peak-to-peak values, which can be attributed to nonzero contact resistance. The DC component in measurements died out slower than predicted by the simulations.

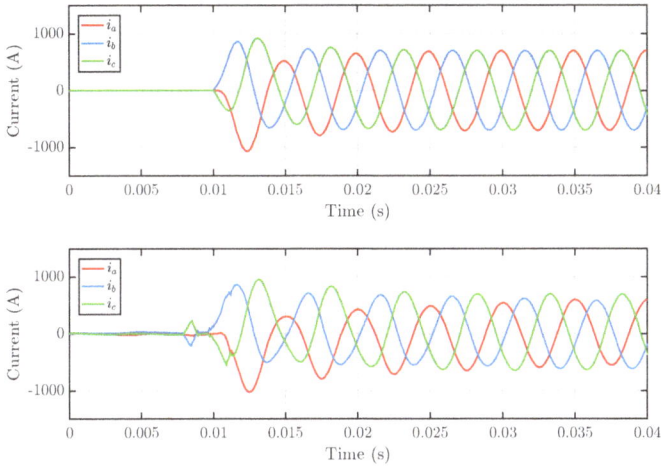

Figure 10. Three-phase short circuit at $3000\,\text{min}^{-1}$: (**top**) simulations and (**bottom**) measurements.

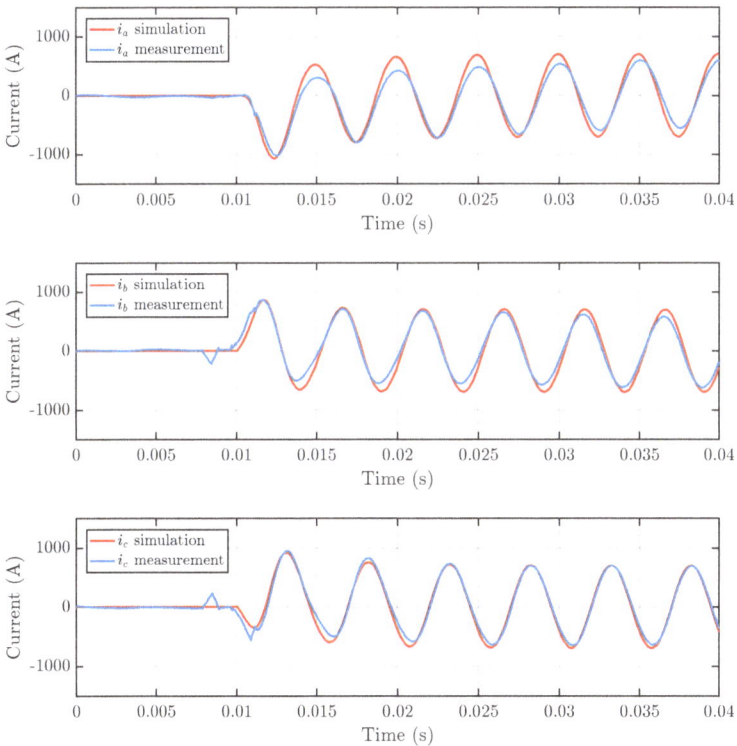

Figure 11. Three-phase short circuit at $3000\,\text{min}^{-1}$: a direct comparison of simulated and measured phase currents (**top**) i_a, (**middle**) i_b, and (**bottom**) i_c.

4.2. Parameterization Time

CM and FLM were implemented in Matlab/Simulink and run on a typical workstation with 3.40 GHz and 16 GB RAM. Even though both models were parameterized with the same input data, the process itself differed. CM parameterization includes the calculation of interpolation functions for the inverse of incremental inductance, apparent inductance, and rotor flux linkage. On the other hand, the FLM requires an inversion of flux maps and subsequent determination of interpolation functions. Parameterization time denotes the computational time required for full-model parameterization before the simulation can be actually run. Table 2 shows that, in the case of FLM parameterization, the proposed intersection method is superior to the minimization method by an order of magnitude. The FLM parameterization time is thus comparable to CM, even though the latter still performs twice faster. However, the said differences fade completely, as execution time of simulation is nearly always significantly longer.

Table 2. Comparison of parameterization times.

Model Type	Parameterization Time
CM	0.24 s
FLM via intersections	0.49 s
FLM via minimization	58.0 s

4.3. Execution Time

Next, we compared the performance of the CM and FLM in terms of their execution time. Two rather basic simulation tasks were chosen as examples: three-phase short-circuit and full acceleration from standstill to deep field weakening, where the motor model is part of the speed control loop. The real-time time frame of the latter task was set at 10 s. The results in Table 3 clearly show that the FLM was faster than CM in both cases. It is interesting to note that, in terms of the control loop simulation, the FLM has a considerable margin of almost 20% over CM.

Table 3. Comparison of execution times.

Simulation Task	FLM	CM	Improvement
Three-phase short circuit	12.5 s	13.8 s	9.4%
Control with field weakening	34.6 s	42.4 s	18.4%
Demagnetization risk assessment	375 min	414 min	9.4%
WLTC driving cycle	104 min	127 min	18.1%

As there are numerous possible situations for which dynamical simulations are typically performed, we particularly focused on scenarios where significant computational effort is inevitable. We were interested in the question of whether time-consuming tasks could be significantly reduced by using the FLM instead of CM. Two computationally intensive simulation tasks with high practical value were chosen as representative scenarios: (a) assessment of demagnetization risk during a three-phase short circuit and (b) WLTC driving cycle. The computational effort of the former scenario stems from an excessive number of iterations. In contrast, the latter scenario demands comparatively a long real-time time frame of 30 min. It should be added that both scenarios require only one parameterization at the beginning.

The assessment of demagnetization risk during a three-phase short circuit requires many iterations of the basic three-phase short-circuit for different operating points (T, n) in the torque-speed diagram. Every steady-state operating point (T, n) is associated with the specific value of I_d (Figure 12) and I_q (not shown). Here, the torque-speed diagram contains $30 \times 60 = 1800$ operating points. Each current pair (I_d, I_q) is fed as an initial value $(I_d^{init}, I_q^{init}) = (I_d, I_q)$ into the space-state model, which calculates respective transient short-circuit currents $i_d(t)$ and $i_q(t)$. For example, Figure 13 depicts transient for 9 operation point (56 Nm, 5000 min^{-1}). The critical time instant t_{crit}, where d-axis current reaches its global negative peak, is determined by

$$t_{\text{crit}} = \arg\min_{t} i_d(t).$$

The negative peak value of d-axis current $I_d^{\text{crit}} = i_d(t_{\text{crit}})$ and the coincidental q-axis current $I_q^{\text{trans}} = i_q(t_{\text{crit}})$ are then identified and stored (Figure 13). In this way, a complete prediction of $(I_d^{\text{crit}}, I_q^{\text{crit}})$ for the whole torque-speed diagram is obtained.

The d-axis current I_d^{crit} is the worst-case instantaneous value for specific operating point (T, n) and is directly associated with demagnetization risk. Figure 14 depicts the predicted I_d^{crit} for the torque-speed diagram. We can observe that the highest negative values of I_d^{crit} are expected when the IPM is delivering peak power (big i_q). It is interesting to note that there is less of a demagnetization risk when the machine operates in deep flux weakening. Moreover, the I_d^{crit} in generating mode is larger than its counterpart in motoring mode.

Operating points, where values of I_d^{crit} are critically low, are identified as candidates for demagnetization. Therefore, precise re-evaluation of this operating point with FEA was performed. The corresponding current pair $(I_d^{\text{crit}}, I_q^{\text{crit}})$ is fed into the magnetostatic simulation. Precise calculation of the magnetic field inside magnets enables reliable demagnetization risk assessment.

As this particular scenario is a simple iteration of the basic simulation task described above, we could estimate the execution time for both models. The FLM enabled 9.4% faster execution in comparison to CM, which can be considered a significant improvement (Table 3).

The torque-speed diagram in Figures 12 and 14 requires comment. The envelope of the diagram is not symmetrical for motoring (top) and generating (bottom), as the transition speed between a constant torque operation and flux-weakening operation differs for motoring and generating mode, which is mainly due to the 48 V car battery being used as a DC source. A significant relative difference in bus voltage for motoring (48 V) and generating (52 V) mode is further aggravated by the effect of losses and stator resistance, which explains the difference in the torque-speed envelope.

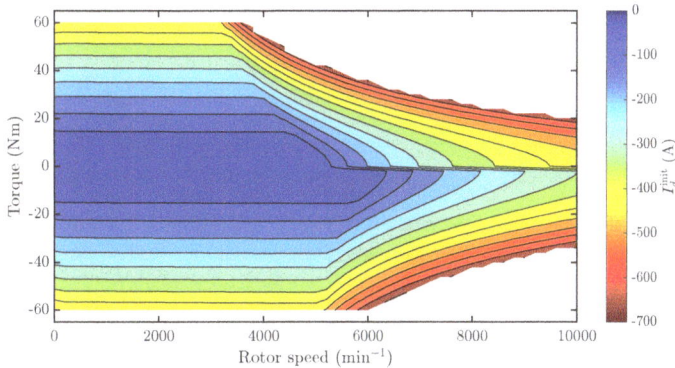

Figure 12. Initial current I_d^{init} for a three-phase short circuit.

Analogously, the WLTC driving cycle scenario is very similar to another basic simulation task described above, namely, control with field weakening. WLTC prescribes speed values for each second in the 30 min time frame. Consequently, the time series of speed values is treated as a reference for the control loop drive. As we can observe from Table 3, the FLM-based simulation model was 18.1% faster than its counterpart.

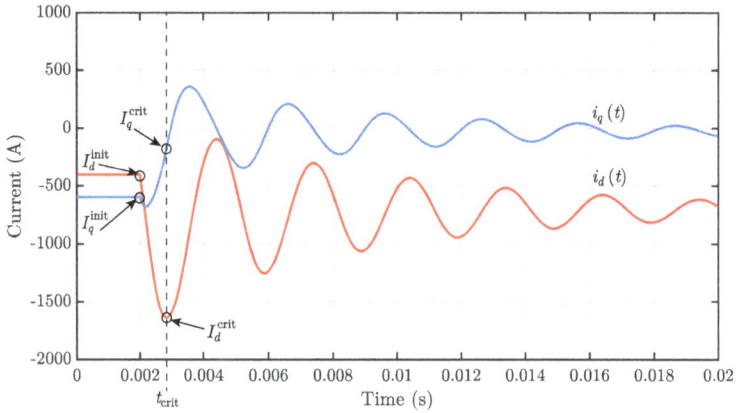

Figure 13. Three-phase short circuit for worst-case operation point (56 Nm, 5000 min^{-1}), where $I_d^{\text{init}} = -405$ A and $I_q^{\text{init}} = -599$ A.

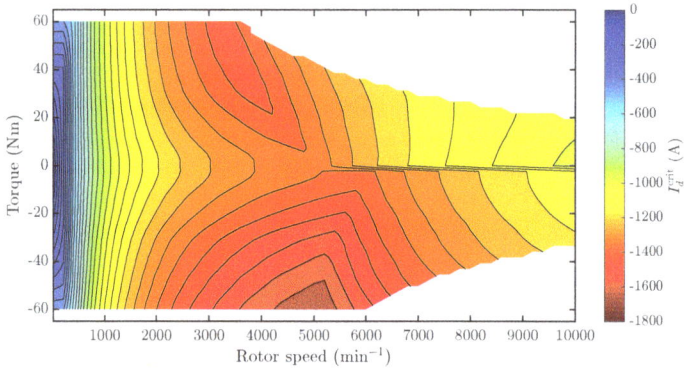

Figure 14. Peak transient current I_d^{crit} for three-phase short circuit.

5. Conclusions

In this paper, an FEM-parameterized two-axis model of IPM was thoroughly analyzed, and its two implementations (i.e., CM and FLM) were compared from the point of view of accuracy and computational demand. Since parameterization of FLM involves the inversion of flux maps, it is mathematically more complex than analog CM tasks, thus resulting in significantly higher computational and, by extension, time demand. For this reason, a procedure for fast and reliable parameterization FLM was presented. The proposed procedure is one order of magnitude faster than the comparable methods, hence providing considerable improvement in terms of computational time. In this way, the major drawback of using FLM were overcome. It was established that the error of the inversion process was under 0.02%.

Additionally, we showed that the execution time of FLM was up to 20% shorter in comparison to CM. As the parameterization time of FLM is now comparable to CM, we strongly advocate the use of FLM in computationally intensive simulation scenarios, which include a significant number of iterations or have an excessive time span. This is why FLM should be used in the Hardware-in-the-Loop experiment platform as well, since the real-time computational burden should be as low as possible.

Author Contributions: Conceptualization, K.D.; methodology, K.D.; software, K.D.; validation, K.D. and L.G.; formal analysis, K.D.; investigation, K.D.; resources, K.D. and L.G.; data curation, L.G.; writing–original draft preparation, K.D.; writing–review and editing, K.D. and R.F.; visualization, K.D.; supervision, R.F.; project administration, R.F.

Funding: This research received no external funding.

Acknowledgments: This work was supported by the Slovenian Research Agency (research core funding No. P2-0258).

Conflicts of Interest: The authors declare no conflict of interest.

Abbreviations

The following abbreviations are used in this manuscript:

IPM	Interior permanent magnet synchronous machine
FLM	Flux-linkage model
CM	Current model
FEA	Finite-element analysis
SCT	Short-circuit test
WLTC	Worldwide harmonized light vehicles test cycle

References

1. EL-Refaie, A.M. Fractional-Slot Concentrated-Windings Synchronous Permanent Magnet Machines: Opportunities and Challenges. *IEEE Trans. Ind. Electron.* **2010**, *57*, 107–121. [CrossRef]
2. Hu, Y.; Zhu, S.; Liu, C.; Wang, K. Electromagnetic Performance Analysis of Interior PM Machines for Electric Vehicle Applications. *IEEE Trans. Energy Convers.* **2018**, *33*, 199–208. [CrossRef]
3. Carraro, E.; Bianchi, N. Design and comparison of interior permanent magnet synchronous motors with non-uniform airgap and conventional rotor for electric vehicle applications. *IET Electr. Power Appl.* **2014**, *8*, 240–249. [CrossRef]
4. Pellegrino, G.; Vagati, A.; Guglielmi, P.; Boazzo, B. Performance Comparison Between Surface-Mounted and Interior PM Motor Drives for Electric Vehicle Application. *IEEE Trans. Ind. Electron.* **2012**, *59*, 803–811. [CrossRef]
5. Torrent, M.; Perat, J.I.; Jiménez, J.A. Permanent Magnet Synchronous Motor with Different Rotor Structures for Traction Motor in High Speed Trains. *Energies* **2018**, *11*. [CrossRef]
6. Pellegrino, G.; Jahns, T.M.; Bianchi, N.; Soong, W.; Cupertino, F. *The Rediscovery of Synchronous Reluctance and Ferrite Permanent Magnet Motors: Tutorial Course Notes*; Springer: Basel, Switzerland, 2016. [CrossRef]
7. Germishuizen, J.J.; Kamper, M.J. IPM Traction Machine With Single Layer Non-Overlapping Concentrated Windings. *IEEE Trans. Ind. Appl.* **2009**, *45*, 1387–1394. [CrossRef]
8. Germishuizen, J.J.; Stanton, S.; Delafosse, V. Integrating FEM in an everyday design environment to accurately calculate the performance of IPM motors. In Proceedings of the 2009—XIV International Symposium on Electromagnetic Fields in Mechatronics, Electrical and Electronic Engineering (ISEF), Arras, France, 10–12 September 2009.
9. Li, L.; Li, W.; Li, D.; Li, J.; Fan, Y. Research on Strategies and Methods Suppressing Permanent Magnet Demagnetization in Permanent Magnet Synchronous Motors Based on a Multi-Physical Field and Rotor Multi-Topology Structure. *Energies* **2018**, *11*. [CrossRef]
10. Liu, X.; Lin, Q.; Fu, W. Optimal Design of Permanent Magnet Arrangement in Synchronous Motors. *Energies* **2017**, *10*, 1700. [CrossRef]
11. Liu, Y.X.; Li, L.Y.; Cao, J.W.; Gao, Q.H.; Sun, Z.Y.; Zhang, J.P. The Optimization Design of Short-Term High-Overload Permanent Magnet Motors Considering the Nonlinear Saturation. *Energies* **2018**, *11*, 3272. [CrossRef]
12. Pries, J.; Burress, T. High fidelity D-Q modeling of synchronous machines using spectral interpolation. In Proceedings of the 2017 IEEE Transportation Electrification Conference and Expo (ITEC), Chicago, IL, USA, 26–28 June 2017; pp. 779–785. [CrossRef]

13. Alvarez-Gonzalez, F.; Griffo, A.; Sen, B.; Wang, J. Real-Time Hardware-in-the-Loop Simulation of Permanent-Magnet Synchronous Motor Drives Under Stator Faults. *IEEE Trans. Ind. Electron.* **2017**, *64*, 6960–6969. [CrossRef]

14. Zhang, C.; Guo, Q.; Li, L.; Wang, M.; Wang, T. System Efficiency Improvement for Electric Vehicles Adopting a Permanent Magnet Synchronous Motor Direct Drive System. *Energies* **2017**, *10*, 2030. [CrossRef]

15. Krause, P.; Wasynczuk, O.; Sudhoff, S.; Pekarek, S. *Analysis of Electric Machinery and Drive Systems*; Wiley-IEEE Press: Hoboken, NJ, USA, 2013.

16. Vas, P. *Electrical Machines and Drives: A Space-Vector Theory Approach*; Clarendon Press: Oxford, UK, 1993.

17. Štumberger, B.; Štumberger, G.; Dolinar, D.; Hamler, A.; Trlep, M. Evaluation of saturation and cross-magnetization effects in interior permanent-magnet synchronous motor. *IEEE Trans. Ind. Appl.* **2003**, *39*, 1264–1271. [CrossRef]

18. Chen, X.; Wang, J.; Sen, B.; Lazari, P.; Sun, T. A High-Fidelity and Computationally Efficient Model for Interior Permanent-Magnet Machines Considering the Magnetic Saturation, Spatial Harmonics, and Iron Loss Effect. *IEEE Trans. Ind. Electron.* **2015**, *62*, 4044–4055. [CrossRef]

19. Boesing, M.; Niessen, M.; Lange, T.; Doncker, R.D. Modeling spatial harmonics and switching frequencies in PM synchronous machines and their electromagnetic forces. In Proceedings of the 2012 XXth International Conference on Electrical Machines, Marseille, France, 2–5 September 2012; pp. 3001–3007. [CrossRef]

20. Weidenholzer, G.; Silber, S.; Jungmayr, G.; Bramerdorfer, G.; Grabner, H.; Amrhein, W. A flux-based PMSM motor model using RBF interpolation for time-stepping simulations. In Proceedings of the 2013 International Electric Machines Drives Conference, Chicago, IL, USA, 12–15 May 2013; pp. 1418–1423. [CrossRef]

21. Pinto, D.E.; Pop, A.C.; Kempkes, J.; Gyselinck, J. dq0-modeling of interior permanent-magnet synchronous machines for high-fidelity model order reduction. In Proceedings of the 2017 International Conference on Optimization of Electrical and Electronic Equipment (OPTIM) 2017 Intl Aegean Conference on Electrical Machines and Power Electronics (ACEMP), Fundata, Romania, 25–27 May 2017; pp. 357–363. [CrossRef]

22. Luo, G.; Zhang, R.; Chen, Z.; Tu, W.; Zhang, S.; Kennel, R. A Novel Nonlinear Modeling Method for Permanent-Magnet Synchronous Motors. *IEEE Trans. Ind. Electron.* **2016**, *63*, 6490–6498. [CrossRef]

23. Fasil, M.; Antaloae, C.; Mijatovic, N.; Jensen, B.B.; Holboll, J. Improved *dq*-Axes Model of PMSM Considering Airgap Flux Harmonics and Saturation. *IEEE Trans. Appl. Supercond.* **2016**, *26*, 1–5. [CrossRef]

24. Li, S.; Han, D.; Sarlioglu, B. Modeling of Interior Permanent Magnet Machine Considering Saturation, Cross Coupling, Spatial Harmonics, and Temperature Effects. *IEEE Trans. Transp. Electrif.* **2017**, *3*, 682–693. [CrossRef]

25. Stipetić, S.; Goss, J.; Žarko, D.; Popescu, M. Calculation of Efficiency Maps Using a Scalable Saturated Model of Synchronous Permanent Magnet Machines. *IEEE Trans. Ind. Appl.* **2018**, *54*, 4257–4267. [CrossRef]

26. Chu, W.Q.; Zhu, Z.Q. Average Torque Separation in Permanent Magnet Synchronous Machines Using Frozen Permeability. *IEEE Trans. Magn.* **2013**, *49*, 1202–1210. [CrossRef]

27. Dück, P.; Ponick, B. A novel iron-loss-model for permanent magnet synchronous machines in traction applications. In Proceedings of the 2016 International Conference on Electrical Systems for Aircraft, Railway, Ship Propulsion and Road Vehicles International Transportation Electrification Conference (ESARS-ITEC), Toulouse, France, 2–4 November 2016; pp. 1–6. [CrossRef]

28. Mellor, P.H.; Wrobel, R.; Holliday, D. A computationally efficient iron loss model for brushless AC machines that caters for rated flux and field weakened operation. In Proceedings of the 2009 IEEE International Electric Machines and Drives Conference, Miami, FL, USA, 3–6 May 2009; pp. 490–494. [CrossRef]

29. Bramerdorfer, G.; Andessner, D. Accurate and Easy-to-Obtain Iron Loss Model for Electric Machine Design. *IEEE Trans. Ind. Electron.* **2017**, *64*, 2530–2537. [CrossRef]

30. Miller, T.J.E.; Popescu, M.; Cossar, C.; McGilp, M. Performance estimation of interior permanent-magnet brushless motors using the voltage-driven flux-MMF diagram. *IEEE Trans. Magn.* **2006**, *42*, 1867–1872. [CrossRef]

31. Bianchi, N.; Alberti, L. MMF Harmonics Effect on the Embedded FE Analytical Computation of PM Motors. *IEEE Trans. Ind. Appl.* **2010**, *46*, 812–820. [CrossRef]

32. Liu, K.; Feng, J.; Guo, S.; Xiao, L.; Zhu, Z.Q. Identification of Flux Linkage Map of Permanent Magnet Synchronous Machines Under Uncertain Circuit Resistance and Inverter Nonlinearity. *IEEE Trans. Ind. Inf.* **2018**, *14*, 556–568. [CrossRef]

33. Armando, E.; Bojoi, R.; Guglielmi, P.; Pellegrino, G.; Pastorelli, M. Experimental methods for synchronous machines evaluation by an accurate magnetic model identification. In Proceedings of the 2011 IEEE Energy Conversion Congress and Exposition, Phoenix, AZ, USA, 17–22 September 2011; pp. 1744–1749. [CrossRef]

34. Marčič, T.; Štumberger, G.; Štumberger, B. Analyzing the Magnetic Flux Linkage Characteristics of Alternating Current Rotating Machines by Experimental Method. *IEEE Trans. Magn.* **2011**, *47*, 2283–2291. [CrossRef]

35. Hubbard, J.H.; Burke Hubbard, B. *Vector Calculus, Linear Algebra, and Differential Forms: A Unified Approach*; Prentice Hall: Upper Saddle River, NJ, USA, 1998.

36. Choi, G.; Jahns, T.M. Investigation of Key Factors Influencing the Response of Permanent Magnet Synchronous Machines to Three-Phase Symmetrical Short-Circuit Faults. *IEEE Trans. Energy Convers.* **2016**, *31*, 1488–1497. [CrossRef]

energies

MDPI

Article

Robust Nonlinear Predictive Current Control Techniques for PMSM

Mingcheng Lyu, Gongping Wu, Derong Luo *, Fei Rong and Shoudao Huang

College of Electrical and Information Engineering, Hunan University, Changsha 410082, China;
lmc2016@hnu.edu.cn (M.L.); gongping_wu@163.com (G.W.); rf_hunu@126.com (F.R.);
Hsd_1962@hnu.edu.cn (S.H.)
* Correspondence: hdldr@sina.com; Tel.: +86-139-0748-8608

Received: 30 November 2018; Accepted: 16 January 2019; Published: 30 January 2019

Abstract: This paper proposes a robust nonlinear predictive current control (RNPCC) method for permanent magnet synchronous motor (PMSM) drives, which can optimize the current control loop performance of the PMSM system with model parameter perturbation. First, the disturbance caused by parameter perturbation was considered in the modeling of PMSM. Based on this model, the influence of parameter perturbation on the conventional predictive current control (PCC) was analyzed. The composite integral terminal sliding mode observer (SMO) was then designed to estimate the disturbance caused by the parameter perturbation in real time. Finally, a RNPCC method is developed without relying on the mathematical model of PMSM, which can effectively eliminate the influence of parameter perturbation by injecting the estimated disturbance value. Simulations and experiments verified that the proposed RNPCC method was able to remove the current error caused by the parameter perturbation during steady state operation.

Keywords: permanent magnet synchronous motor (PMSM); sliding mode observer (SMO); parameter perturbation; predictive current control (PCC)

---

1. Introduction

Field-oriented control (FOC) have been widely used in permanent magnet synchronous motor (PMSM) drives due to the fast and fully decoupled control of torque and flux [1–4]. In a FOC-based PMSM drive, the inner double-loop control is usually adopted. The internal current loop is designed for current tracking, while the external speed loop regulates the rotor speed [5]. The dynamic response and stability of the internal current loop are the key factors that determine the performance of the whole drive system.

To achieve a high performance PMSM drive system, the predictive current control (PCC) method has been adopted in a wide range of applications [6]. The PCC can calculate the required command voltage based on the discrete mathematical model of PMSM, and achieves an accurate current tracking [7]. Furthermore, the PCC improves the bandwidth of current control loops and the dynamic performance of motors theoretically [8–10]. However, the performance of PCC is largely dependent on the accuracy of the PMSM model, which means that the parameter perturbation will deteriorate its performance.

Neglecting the parameter perturbation in the PCC design results in a significant steady-state current error and oscillation. To overcome this shortage, a robust PCC technique is proposed in [11], which can reduce the current error caused by motor parameter variation. In [12], an extension of the PCC method is presented to improve the prediction accuracy, which can not only reduce the current ripple, but also improve the robustness of the PCC against parameter perturbation. In [13], a real-time predictive control scheme with a parallel integral loop is presented for Pulse Width Modulation (PWM) nverter-fed PMSM drives, which can compensate for the variations in motor parameters.

However, to achieve an acceptable transient response without overshooting, the integral gain needs to be properly designed.

Some other robust PCC methods based on a disturbance observer have been investigated to enhance the predictive control performance under parameter perturbations. The authors in [14] propose a flux immunity robust PCC for PMSM drives which can operate without knowing the rotor flux. In order to avoid the divergence caused by stator inductance mismatch, the proposed robust PCC adopts an extended state observer to enhance inductance robustness. In [15], a composite predictive control method based on stator current and a disturbance observer is developed, which can achieve perfect current control performance of the PMSM with model parameter mismatch. In [16], a simple disturbance observer is designed to increase the robustness of the proposed deadbeat predictive control algorithm against parameter uncertainties of the Permanent Magnetic Synchronous Generator (PMSG). In [17], a robust fault-tolerant predictive current control algorithm is proposed based on a composite observer, which can enhance robustness against parameter perturbation and permanent magnet demagnetization by adding compensation voltage. In [18], a continuous-time offset-free model predictive control approach based on a disturbance observer is developed for a general disturbed system, which can correct the errors caused by disturbances and uncertainties. By designing a novel sliding surface based on the disturbance estimation, a disturbance observer method was developed to counteract the mismatched disturbance in [19]. In [20], a generalized nonlinear model predictive control augmented with a disturbance observer is proposed, which can solve the disturbance problem of nonlinear systems. The disturbance observers in the aforementioned methods are able to achieve precise disturbance compensation. Moreover, it is possible to enhance robustness against parameter perturbation. In industrial applications, the flux linkage parameter is often accompanied by a change of inductance parameter. However, it is difficult to use these disturbance observers to simultaneously compensate for the resistance, inductance and flux linkage parameter perturbation.

In order to eliminate motor parameter perturbation effects, with improved robustness against inductance and flux linkage parameter mismatch, this paper proposes a robust nonlinear predictive current control (RNPCC) algorithm, which has three main contributions. Firstly, the influence of parameter perturbation on the PCC was analyzed. PCC is mostly sensitive to flux linkage and inductance parameters, whereas the influence of the resistance parameter can be ignored. Secondly, a composite integral terminal observer based on the Luenberger observer and sliding mode observer (SMO) was designed, which can estimate the external disturbance caused by parameter perturbation. Thirdly, the proposed RNPCC with online disturbance estimation can overcome the weakness of the steady-state current error of the conventional PCC. Thus it can be readily applicable in practical engineering.

This paper is organized as follows. A nonlinear PMSM model is developed in Section 2. The influence of parameter perturbation on conventional PCC is analyzed in Section 3. The RNPCC method is proposed in Section 4. The composite integral terminal SMO is designed in Section 5. The simulations and experiments are set up in Sections 6 and 7, respectively. Section 8 concludes this paper.

2. Nonlinear PMSM Model

Machine Model Description

The voltage equations of the PMSM in a synchronous rotating reference frame are written as in [10,21]:

$$\begin{cases} \frac{di_d}{dt} = -\frac{R}{L_d}i_d + \frac{L_q}{L_d}\omega i_q + \frac{1}{L_d}u_d \\ \frac{di_q}{dt} = -\frac{R}{L_q}i_q - \frac{L_d}{L_q}\omega i_d - \frac{\psi_r \omega}{L_q} + \frac{1}{L_q}u_q \end{cases} \tag{1}$$

where u_d and u_q denotes the d- and q-axis stator voltages, respectively; i_d and i_q present the d- and q-axis currents, respectively; R, L_d, and L_q denotes the nominal values of the stator resistance, the d-axis inductance and the q-axis inductance, respectively; ω is the electrical rotor speed; and ψ_r is the flux linkage established by the permanent magnet.

For surface-mounted PMSM (SPMSM), $L_{do} = L_{qo} = L_o$. If we define $\Delta R = R - R_o$, $\Delta L = L - L_o$ and $\Delta \psi_r = \psi_r - \psi_{ro}$, the voltage equations of the SPMSM are expressed as follows, according to (1), when parameter perturbations are considered.

$$\begin{bmatrix} \dot{i}_d \\ \dot{i}_q \end{bmatrix} = \begin{bmatrix} -\frac{R_o}{L_o} & \omega \\ -\omega & -\frac{R_o}{L_o} \end{bmatrix}\begin{bmatrix} i_d \\ i_q \end{bmatrix} + \begin{bmatrix} \frac{1}{L_o} & 0 \\ 0 & \frac{1}{L_o} \end{bmatrix}\begin{bmatrix} u_d \\ u_q \end{bmatrix} + \begin{bmatrix} 0 \\ -\frac{\psi_{ro}}{L_o}\omega \end{bmatrix} - \begin{bmatrix} \frac{1}{L_o} & 0 \\ 0 & \frac{1}{L_o} \end{bmatrix}\begin{bmatrix} \delta_d \\ \delta_q \end{bmatrix} \tag{2}$$

where R_o, L_o, and ψ_{ro} are the nominal values and ΔR, ΔL, and $\Delta \psi_r$ are the perturbation values of the corresponding model parameters; and δ_d and δ_q represent the disturbances caused by parameter perturbations. The δ_d and δ_q can be expressed as:

$$\begin{aligned} \delta_d &= \frac{\Delta L}{(L_o + \Delta L)}(u_d - R_o i_d - \Delta R i_d + L_o \omega i_q + \Delta L \omega i_q) + (\Delta R i_d - \Delta L \omega i_q) \\ \delta_q &= \frac{\Delta L}{(L_o + \Delta L)}(u_q - R_o i_q - \Delta R i_q - L_o \omega i_d - \Delta L \omega i_d - \omega \psi_{ro} - \omega \Delta \psi_r) + (\Delta R i_q + \Delta L \omega i_d + \omega \Delta \psi_r) \end{aligned} \tag{3}$$

According to (2), the nonlinear mathematical model of SPMSM is established as:

$$\begin{cases} \dot{x} = Ax + Bu + Df_\psi - B\delta \\ y = Ex \end{cases} \tag{4}$$

where $x = \begin{bmatrix} i_d \\ i_q \end{bmatrix}$, $u = \begin{bmatrix} u_d \\ u_q \end{bmatrix}$, $y = \begin{bmatrix} i_d \\ i_q \end{bmatrix}$, and $\delta = \begin{bmatrix} \delta_d \\ \delta_q \end{bmatrix}$ are state variables, system inputs, system outputs, and disturbances, respectively.

The coefficient matrixes of the state equations are:

$$A = \begin{bmatrix} -\frac{R_o}{L_o} & \omega \\ -\omega & -\frac{R_o}{L_o} \end{bmatrix}, B = \begin{bmatrix} \frac{1}{L_o} & 0 \\ 0 & \frac{1}{L_o} \end{bmatrix}, f_\psi = \begin{bmatrix} 0 \\ \psi_{ro} \end{bmatrix}, D = \begin{bmatrix} 0 & 0 \\ 0 & -\frac{\omega}{L_o} \end{bmatrix}, E = \begin{bmatrix} 1 & 0 \\ 0 & 1 \end{bmatrix}$$

3. Parameter Sensitivity Analysis of Conventional PCC

According to [15], the output voltage vectors of conventional PCC are expressed by:

$$u(k) = F^{-1}[i^{ref}(k+1) - H(k)i(k) - P(k)] \tag{5}$$

where, T_s is the sampling period: $i^{ref}(k+1) = \begin{bmatrix} i_d^{ref}(k+1) \\ i_q^{ref}(k+1) \end{bmatrix}$, $i(k) = \begin{bmatrix} i_d(k) \\ i_q(k) \end{bmatrix}$, $u(k) = \begin{bmatrix} u_d(k) \\ u_q(k) \end{bmatrix}$,

$F = \begin{bmatrix} \frac{T_s}{L_o} & 0 \\ 0 & \frac{T_s}{L_o} \end{bmatrix}$, $P(k) = \begin{bmatrix} 0 \\ -\frac{\psi_{ro}}{L_o}T_s\omega(k) \end{bmatrix}$, $H(k) = \begin{bmatrix} 1 - \frac{R_o}{L_o}T_s & T_s\omega(k) \\ -T_s\omega(k) & 1 - \frac{R_o}{L_o}T_s \end{bmatrix}$.

When the sampling period (T_s) is small enough, we can obtain $\frac{i_{dq}}{dt} = \frac{i_{dq}(k+1) - i_{dq}(k)}{T_s}$. According to (2), the discrete model of the SPMSM under parameter perturbations can be expressed by,

$$i(k+1) = H(k) \cdot i(k) + F \cdot u(k) + P(k) - F \cdot f(k) \tag{6}$$

where: $f(k) = \begin{bmatrix} \delta_d(k) \\ \delta_q(k) \end{bmatrix}$.

The conventional PCC method belongs to one beat delay control, and the voltage vector $u(k)$ is applied to the SPMSM at the $(k+1)T_s$ moment. Therefore, substituting (5) into (6) yields:

$$\Delta i(k+1) = i^{ref}(k+1) - i(k+1) = F \cdot f(k) \tag{7}$$

with,

$$\Delta i(k+1) = F \cdot f(k) = \begin{bmatrix} \Delta i_d \\ \Delta i_q \end{bmatrix} = \frac{T_s}{L_o} \begin{bmatrix} \delta_d(k) \\ \delta_q(k) \end{bmatrix} \tag{8}$$

where Δi_d and Δi_q are the d- and q-axis current errors, respectively.

The stationary equations are given as follows [22]:

$$\begin{cases} u_d(k) = R_o i_d(k) - L_o w i_q(k) \\ u_q(k) = R_o i_q(k) + L_o w i_d(k) + w\psi_{ro} \end{cases} \tag{9}$$

In this paper, i_d is set to 0. Thus, the current on d-axis is neglected. Combining (3), (8) and (9), the current errors on the d- and q-axis can be simplified as:

$$\begin{cases} \Delta i_d = -\frac{\Delta L}{(L_o + \Delta L)} T_s w i_q(k) \\ \Delta i_q = \frac{\Delta R}{(L_o + \Delta L)} T_s i_q(k) + \frac{\Delta \psi_r}{(L_o + \Delta L)} T_s w \end{cases} \tag{10}$$

From (10), the d- and q-axis current errors can be deduced when there is a resistance parameter error ΔR between actual value R and nominal value R_o:

$$\begin{cases} \Delta i_d = 0 \\ \Delta i_q = \frac{\Delta R}{L_o} T_s i_q(k) \end{cases} \tag{11}$$

Similarly, the d- and q-axis current errors can be obtained by (12) under flux linkage parameter error $\Delta \psi_{rd}$:

$$\begin{cases} \Delta i_d = 0 \\ \Delta i_q = \frac{\Delta \psi_r}{L_o} T_s w \end{cases} \tag{12}$$

The d- and q-axis current errors can be obtained by (13) under inductance parameter error ΔL:

$$\begin{cases} \Delta i_d = -\frac{\Delta L}{(L_o + \Delta L)} T_s w i_q(k) \\ \Delta i_q = 0 \end{cases} \tag{13}$$

The motor parameters are provided in Table 1. According to (11), (12), and (13) the d- and q-axis current errors are plotted, as in Figure 1. Figure 1a shows the q-axis current error under resistance parameter mismatch. It can be observed from Figure 1a and Equation (11) that the resistance parameter perturbation does not affect the d-axis current, and resistance parameter perturbation has little effect on the q-axis current. Figure 1b shows the q-axis current error under flux linkage parameter mismatch. From Figure 1b and Equation (12), it is known that the flux linkage parameter mismatch does not affect the d-axis current, but the influence on the q-axis current is very strong. Figure 1c shows the d-axis current error under inductance parameter mismatch. From Figure 1c and Equation (13), it is known that the q-axis current error is zero, and the d-axis current error increases with the increase of the speed and q-axis current. Through the above analysis, the inductance and flux linkage parameter mismatches deteriorate the performance of the PCC, while the influence of the resistance parameter mismatch can be ignored.

Table 1. Main parameters of surface permanent magnet synchronous motor (SPMSM).

Parameters	Value
Rated power	125 kW
Rated speed	2000 r/min
Rate torque	600 N·m
Stator phase resistance (R_o)	0.02 Ω
Number of pole pairs (n_p)	4
Inductance (L_o)	1 mH
Flux linkage of PM (Ψ_{ro})	0.892 Wb
Rotational inertia (J)	1.57 kg·m²

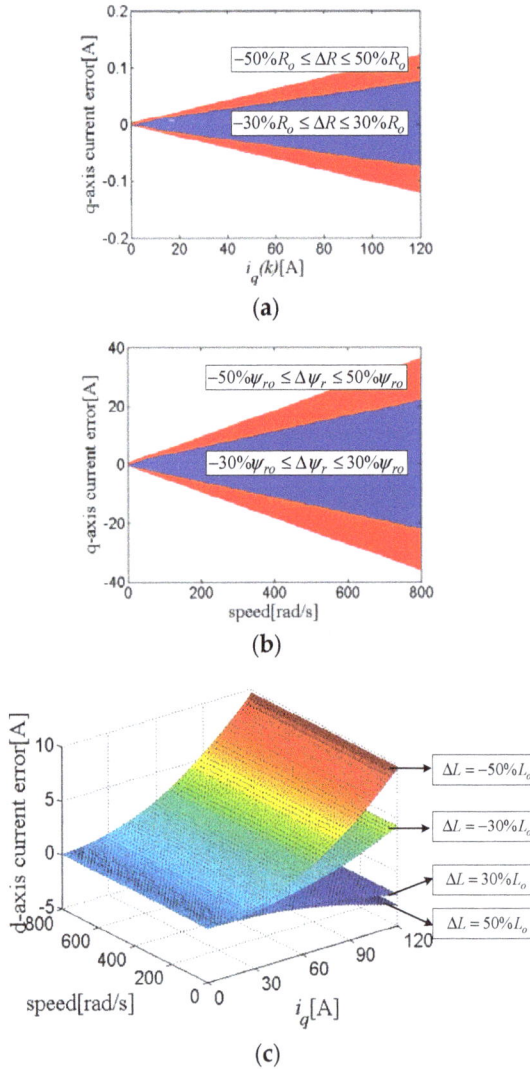

(a)

(b)

(c)

Figure 1. *d*- and *q*-axis current errors under parameter mismatches. (**a**) *q*-axis current error under resistance parameter mismatch; (**b**) *q*-axis current error under flux linkage parameter mismatch; (**c**) *d*-axis current error under inductance parameter mismatch.

4. Design of the RNPCC

4.1. Design of the Optimal Control Law

The reference value and predictive value of the output currents are respectively defined as:

$$y_r(t+\tau) = \begin{bmatrix} i_d^{ref}(t+\tau) \\ i_q^{ref}(t+\tau) \end{bmatrix}, y(t+\tau) = \begin{bmatrix} i_d(t+\tau) \\ i_q(t+\tau) \end{bmatrix} \tag{14}$$

To simplify the calculations, these two values are expanded to 1th-order Taylor series:

$$y_r(t+\tau) = \begin{bmatrix} E & M \end{bmatrix} \begin{bmatrix} y_r(t) \\ \dot{y}_r(t) \end{bmatrix}, y(t+\tau) = \begin{bmatrix} E & M \end{bmatrix} \begin{bmatrix} y(t) \\ \dot{y}(t) \end{bmatrix} \tag{15}$$

where τ is the time constant: $y_r(t) = \begin{bmatrix} i_d^{ref}(t) \\ i_q^{ref}(t) \end{bmatrix}, y(t) = \begin{bmatrix} i_d(t) \\ i_q(t) \end{bmatrix}, E = \begin{bmatrix} 1 & 0 \\ 0 & 1 \end{bmatrix}, M = \begin{bmatrix} \tau & 0 \\ 0 & \tau \end{bmatrix}.$

In Figure 2, the control sequence logic diagram of the proposed RNPCC method is illustrated in a discrete-time framework. The design of the cost function depends on the requirements of the control system performance. The servo control system belongs to the tracking control system, and it is hoped that the output can track the input reference accurately. Therefore, the cost function is defined as:

$$P = \frac{1}{2}\int_0^{T_s} e^T(t+\tau)e(t+\tau)d\tau \tag{16}$$

with

$$e(t+\tau) = y_r(t+\tau) - y(t+\tau) = \begin{bmatrix} E & M \end{bmatrix} \begin{bmatrix} e(t) \\ \dot{e}(t) \end{bmatrix} \tag{17}$$

where: $e(t) = y_r(t) - y(t) = \begin{bmatrix} e_d \\ e_q \end{bmatrix} = \begin{bmatrix} i_d^{ref}(t) - i_d(t) \\ i_q^{ref}(t) - i_q(t) \end{bmatrix}$

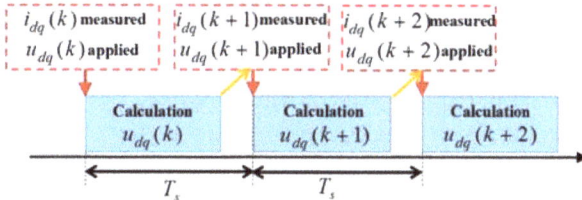

Figure 2. Logic diagram of a beat delay compensation using robust nonlinear predictive current control (RNPCC) method.

Substituting (17) into (16), the cost function is written as:

$$\begin{aligned} P &= \frac{1}{2}\int_0^{T_s} \begin{bmatrix} e^T(t) & \dot{e}^T(t) \end{bmatrix} \begin{bmatrix} E^T \\ M^T \end{bmatrix} \begin{bmatrix} E & M \end{bmatrix} \begin{bmatrix} e(t) \\ \dot{e}(t) \end{bmatrix} d\tau \\ &= e^T(t)P_1 e(t) + \dot{e}^T(t)P_2 e(t) + e^T(t)P_3\dot{e}(t) + \dot{e}^T(t)P_4\dot{e}(t) \end{aligned} \tag{18}$$

where: $P_1 = \frac{1}{2}\int_0^{T_s} E^T E d\tau, P_2 = \frac{1}{2}\int_0^{T_s} M^T E d\tau, P_3 = \frac{1}{2}\int_0^{T_s} E^T M d\tau, P_4 = \frac{1}{2}\int_0^{T_s} M^T M d\tau.$

The constraint condition of the optimal control law is given by:

$$\frac{\partial P}{\partial u(t)} = 0 \tag{19}$$

Substituting (18) into (19) yields:

$$-\frac{\partial \dot{\boldsymbol{y}}^T(t)}{\partial \boldsymbol{u}(t)}\boldsymbol{P}_2\boldsymbol{e}(t) - \boldsymbol{e}^T(t)\boldsymbol{P}_3\frac{\partial \dot{\boldsymbol{y}}(t)}{\partial \boldsymbol{u}(t)} + \frac{\partial \dot{\boldsymbol{y}}^T(t)}{\partial \boldsymbol{u}(t)}\boldsymbol{P}_4\dot{\boldsymbol{y}}(t) + \dot{\boldsymbol{y}}^T(t)\boldsymbol{P}_4\frac{\partial \dot{\boldsymbol{y}}(t)}{\partial \boldsymbol{u}(t)} = 0 \tag{20}$$

Thus, the optimal control law can be obtained according to (19), as given by:

$$\boldsymbol{u}(t) = (\boldsymbol{B})^{-1}\left[\frac{\boldsymbol{P}_2}{\boldsymbol{P}_4}\boldsymbol{e}(t) - \boldsymbol{A}\boldsymbol{x}(t) - \boldsymbol{D}\boldsymbol{f}_\psi + \boldsymbol{B}\boldsymbol{\delta}(t)\right] \tag{21}$$

4.2. Design of the RNPCC

In order to improve the robustness against the parameter perturbation, a composite integral terminal SMO is designed. Taking u_d^{ref} and u_q^{ref} as the references of u_d and u_q, respectively, the predictive current controller is designed as:

$$\begin{cases} u_d^{ref} = \left[\frac{3L_o}{2T_s}\hat{e}_d + R_o\hat{i}_d - L_o\hat{i}_q\omega\right] + \hat{\delta}_d \\ u_q^{ref} = \left[\frac{3L_o}{2T_s}\hat{e}_q + L_o\hat{i}_d\omega + R_o\hat{i}_q + \psi_{ro}\omega\right] + \hat{\delta}_q \\ \hat{e}_d = i_d^{ref} - \hat{i}_d \\ \hat{e}_q = i_q^{ref} - \hat{i}_q \end{cases} \tag{22}$$

where $\hat{\delta}_d$, $\hat{\delta}_q$ are the estimation values of δ_d, δ_q, respectively; $\hat{i}_d$, $\hat{i}_q$ are the estimation values of i_d, i_q, respectively; and u_d^{ref} and u_q^{ref} are the output references of the RNPCC. The block diagram of RNPCC method with a composite integral terminal SMO is shown in Figure 3.

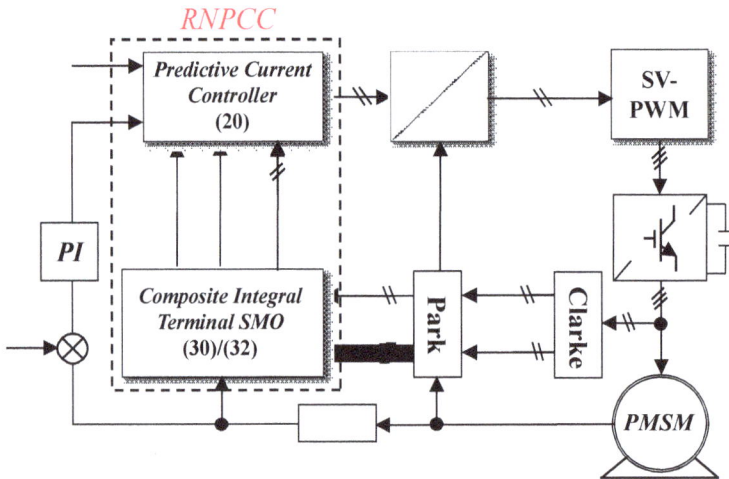

Figure 3. Block diagram of the RNPCC method with composite integral terminal sliding mode observer (SMO) for permanent magnet synchronous motor (PMSM).

5. Design of the Composite Integral Terminal SMO

The key to realizing the RNPCC method is to estimate the disturbance caused by parameter perturbation. Based on (4), the composite integral terminal SMO is designed as:

$$\dot{\hat{x}} = A_1\hat{x} + \omega A_2\hat{x} + Bu + Df_\psi + Q\omega(\hat{x} - x) - U_o \tag{23}$$

where $\hat{x}$ is the observed value of x, Q is the designed gain matrix of Luenberger observer, $U_o = \begin{bmatrix} U_{od} & U_{oq} \end{bmatrix}^T$ is the sliding mode control function, $A = A_1 + \omega A_2$, $A_1 = \begin{bmatrix} -\frac{R_o}{L_o} & 0 \\ 0 & -\frac{R_o}{L_o} \end{bmatrix}$, and $A_2 = \begin{bmatrix} 0 & 1 \\ -1 & 0 \end{bmatrix}$.

Consider the following integral terminal sliding surface vector:

$$s_o = e_o + \lambda \int_0^t sgn(e_o)d\tau \tag{24}$$

In industrial applications, chattering suppression is necessary when designing observers because the chattering of the sign function easily causes the chattering of the system. Therefore, the hyperbolic tangent function with smooth continuity is used instead of the sin function. The hyperbolic tangent function is expressed as:

$$sgn(v) = tanh(v) = \frac{e^{2v} - 1}{e^{2v} + 1} \tag{25}$$

where $s_o = \begin{bmatrix} s_{od} & s_{oq} \end{bmatrix}^T$, $\lambda > 0$; $sgn()$ is the sign function; and e_o is defined as:

$$e_o = \hat{x} - x = \begin{bmatrix} e_{od} \\ e_{oq} \end{bmatrix} = \begin{bmatrix} \hat{i}_d - i_d \\ \hat{i}_q - i_q \end{bmatrix} \tag{26}$$

Taking the time derivative for (24), and combining it with (4) and (23), the error equation of the composite integral terminal SMO can be obtained as:

$$\dot{s}_o = A_1 e_o + B\delta + (A_2 + Q)\omega e_o + \lambda sgn(e_o) - U_o \tag{27}$$

To prevent the influence of the motor speed on the error equation, the gain matrix of the Luenberger observer must be designed as $Q = -A_2$. Thus, Equation (27) can be simplified to:

$$\dot{s}_o = A_1 e_o + B\delta + \lambda sgn(e_o) - U_o \tag{28}$$

Considering the integral terminal sliding surface Equation (24) and the error Equation (27), to guarantee the convergence of error e_o, the sliding mode control function is designed:

$$U_o = A_1 e_o + \lambda sgn(e_o) + k s_o + k_s sgn(s_o) \tag{29}$$

where $k = \begin{bmatrix} k_1 & 0 \\ 0 & k_2 \end{bmatrix}$ is a positive diagonal matrix, and $k_s = \begin{bmatrix} k_{s1} & 0 \\ 0 & k_{s2} \end{bmatrix}$ is the observer sliding gain.

Consider the Lyapunov function candidate as follows:

$$V = \frac{1}{2} s_o^T s_o \tag{30}$$

Differentiating Equation (30) and combining with Equation (28) yields:

$$\dot{V} = s_o^T (A_1 e_o + B\delta + \lambda sgn(e_o) - U_o) \tag{31}$$

Applying Equations (29)–(31) yields:

$$
\begin{aligned}
\dot{V} &= s_o^T (B\delta - ks_o - k_s sgn(s_o)) \\
&= -k_1 s_{od}^2 - k_2 s_{oq}^2 + s_o^T B\delta - s_o^T k_s sgn(s_o) \\
&\leq s_o^T B\delta - k_{s1}\|s_{od}\| - k_{s2}\|s_{oq}\| \\
&= (\tfrac{\delta_d}{L_o}s_{od} - k_{s1}\|s_{od}\|) + (\tfrac{\delta_q}{L_o}s_{oq} - k_{s2}\|s_{oq}\|)
\end{aligned}
\tag{32}
$$

In industrial applications, the disturbances δ should be bounded, that is, normal values exist, satisfying $|\delta_d| \leq N_1$, $|\delta_q| \leq N_2$. If the gain matrix of composite integral terminal SMO satisfies $k_{s1} \geq N_1 \geq \left|\tfrac{\delta_d}{L_o}\right|$, $k_{s2} \geq N_2 \geq \left|\tfrac{\delta_q}{L_o}\right|$, then $\dot{V} \leq 0$. Therefore, the stability and convergence of the composite observer are guaranteed.

From (23), the composite integral terminal SMO can be represented by:

$$
\begin{cases}
\dot{\hat{i}}_d = -\tfrac{R_o}{L_o}\hat{i}_d + \omega\hat{i}_q + \tfrac{1}{L_o}u_d - \omega(\hat{i}_q - i_q) - U_{od} \\
\dot{\hat{i}}_q = -\tfrac{R_o}{L_o}\hat{i}_q - \omega\hat{i}_d + \tfrac{1}{L_o}u_q - \tfrac{\omega}{L_o}\psi_{ro} + \omega(\hat{i}_d - i_d) - U_{oq} \\
U_{od} = -\tfrac{R_o}{L_o}(\hat{i}_d - i_d) + \lambda sgn(\hat{i}_d - i_d) + k_1(\hat{i}_d - i_d) + k_1\lambda\int_0^t sgn(\hat{i}_d - i_d)d\tau \\
\quad + k_{s1}sgn(\hat{i}_d - i_d) + k_{s1}sgn\lambda\int_0^t sgn(\hat{i}_d - i_d)d\tau \\
U_{oq} = -\tfrac{R_o}{L_o}(\hat{i}_q - i_q) + \lambda sgn(\hat{i}_q - i_q) + k_2(\hat{i}_q - i_q) + k_2\lambda\int_0^t sgn(\hat{i}_q - i_q)d\tau \\
\quad + k_{s2}sgn(\hat{i}_q - i_q) + k_{s2}sgn\lambda\int_0^t sgn(\hat{i}_q - i_q)d\tau
\end{cases}
\tag{33}
$$

According to the sliding mode equivalent principle, when the system reaches the sliding mode surface, that is, $\dot{e}_o = e_o = 0$, Equation (28) can be simplified to:

$$
B\delta = U_o
\tag{34}
$$

From (34), the estimated disturbances $\hat{\delta}$ can be represented by:

$$
\begin{cases}
\hat{\delta} = B^{-1}U_o = \begin{bmatrix} L_o & 0 \\ 0 & L_o \end{bmatrix}\begin{bmatrix} U_{od} \\ U_{oq} \end{bmatrix} = \begin{bmatrix} L_o U_{od} \\ L_o U_{oq} \end{bmatrix} \\
\hat{\delta} = \begin{bmatrix} \hat{\delta}_d \\ \hat{\delta}_q \end{bmatrix}
\end{cases}
\tag{35}
$$

With:

$$
\begin{cases}
\hat{\delta}_d = L_o U_{od} \\
\hat{\delta}_q = L_o U_{oq}
\end{cases}
\tag{36}
$$

6. Simulations

The parameters of SPMSM used in the simulation are given in Table 1. The sampling frequency for current control loop was 10 kHz; the proposed composite integral terminal SMO parameters were $\lambda = 800$, $K_1 = K_2 = K_3 = 5000$, and $K_{s1} = K_{s2} = K_{s3} = 100$, respectively. At 0 s, the speed reference increased from 0 to 800 rad/s, and the load torque increased suddenly from no-load to rated load.

6.1. Performance Comparison of Conventional PCC and Proposed RNPCC under Inductance Parameter Perturbation

In this simulation, the inductance parameter perturbation value ΔL was set to zero initially. At 0.5 s, ΔL changed to $\Delta L = 50\% L_o$. The simulation results are shown in Figures 4 and 5.

When taking the conventional PCC method, the d-axis current response value cannot track the current reference after 0.5 s, which is shown clearly in Figure 4a. The d-axis current response value was 11 A, which was bigger than its current reference, that is, 0 A, and the steady error was -11 A.

The reason for this was that the conventional PCC depends on the inductance parameter of the motor. Figure 5c presents the simulation results of the disturbance estimation when using the RNPCC method. The results reflect that the proposed composite integral terminal SMO can accurately estimate the disturbance under inductance parameter perturbation. From Figure 5a, it can be seen that the current response can track its reference accurately by using the proposed composite integral terminal SMO. Figure 4b shows the three-dimension rotor flux trajectories of the conventional PCC method under inductance parameter perturbation. Figure 5b shows the three-dimensional rotor flux trajectories of the RNPCC method under inductance parameter perturbation. The influence of inductance perturbation on the α- and β-axis flux linkage can be neglected because of the small d-axis current error, which can be clearly seen in Figures 4b and 5b.

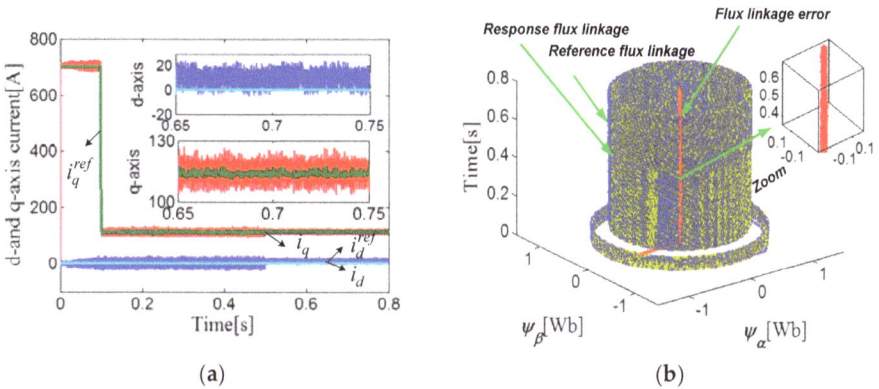

(a)   (b)

Figure 4. Simulation results of the conventional predictive current control (PCC) method under inductance parameter perturbation. (**a**) d- and q-axis current responses and references; (**b**) three-dimension rotor flux trajectories.

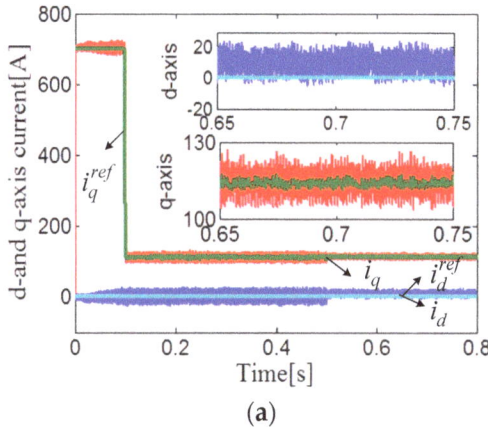

(a)

Figure 5. *Cont.*

(b)

(c)

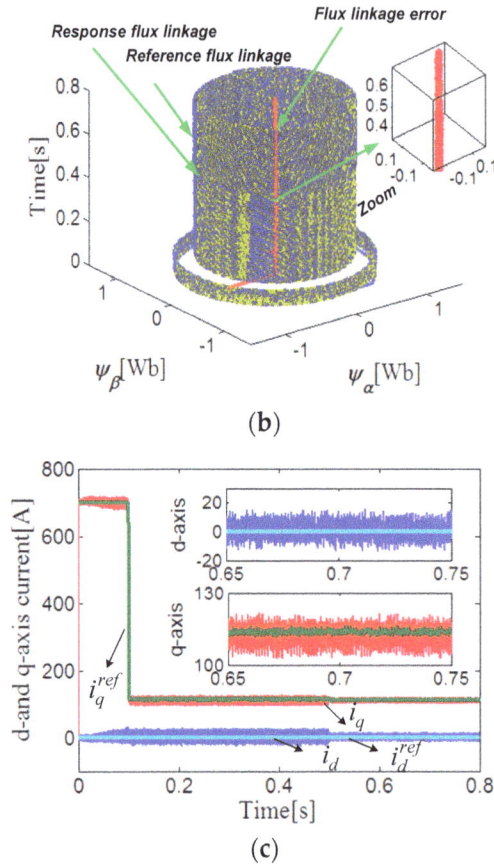

Figure 5. Simulation results of the RNPCC method under inductance parameter perturbation. (**a**) *d*- and *q*-axis current responses and references; (**b**) three-dimensional rotor flux trajectories; (**c**) disturbance estimation results.

6.2. Performance Comparison of Conventional PCC and Proposed RNPCC under Flux Linkage Parameter Perturbation

In this simulation, the flux linkage parameter perturbation value $\Delta\psi_r$ was set initially to zero. At 0.5 s, $\Delta\psi_r$ changed to $\Delta\psi_r = -50\%\psi_{ro}$. The simulation results are shown in Figures 6 and 7.

From Figure 6a, it is known that the *q*-axis current response values cannot track current references under flux linkage parameter perturbation when using the conventional PCC method. After 0.5 s, the *q*-axis current response value and reference value were respectively 225 A and 185 A, and the steady error was −40 A. When using the RNPCC method, the disturbance was accurately estimated using the proposed composite integral terminal SMO, as can be seen in Figure 7. From Figure 7a, it can be seen that the steady state current error can effectively be removed by injecting the estimated external disturbance value. Figure 6b shows the three-dimensional rotor flux trajectories of the conventional PCC method under flux linkage parameter perturbation. Figure 7b shows the three-dimensional rotor flux trajectories of the RNPCC method under flux linkage parameter perturbation. Note that there are large *α*- and *β*-axis flux linkage errors when the flux linkage parameter is changed suddenly at 0.5 s. The reason for this is that the conventional PCC produces a large *q*-axis current steady error under flux linkage parameter perturbation.

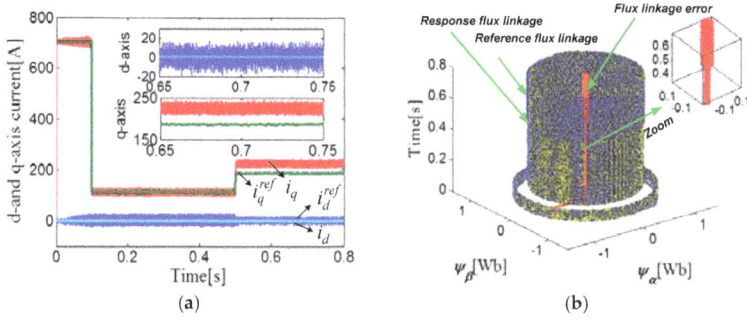

Figure 6. Simulation results of the conventional PCC method under flux linkage parameter perturbation. (**a**) *d*- and *q*-axis current responses and references; (**b**) three-dimensional rotor flux trajectories.

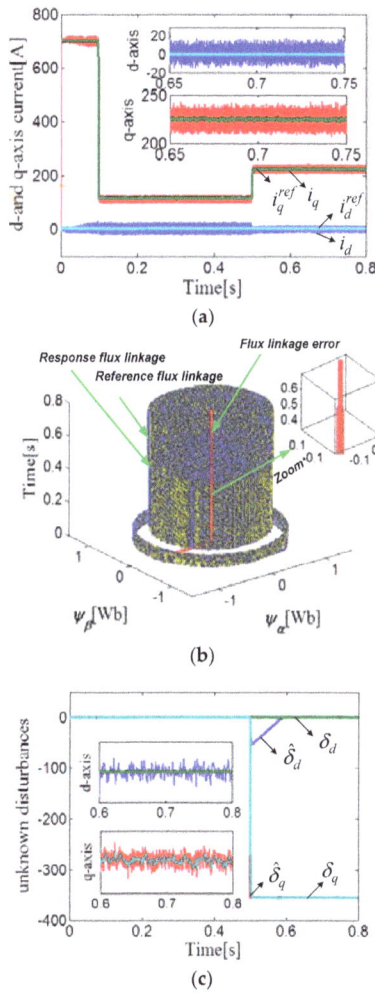

Figure 7. Simulation results of the RNPCC method under flux linkage parameter perturbation. (**a**) *d*- and *q*-axis current responses and references; (**b**) three-dimension rotor flux trajectories; (**c**) disturbance estimation results.

6.3. Performance Comparison of Conventional PCC and Proposed RNPCC under Inductance and Flux Linkage Parameter Perturbation

In this simulation, the inductance and flux-linkage parameter perturbation values were both set to zero initially, and changed to $\Delta L = 50\% L_o$ and $\Delta \psi_r = -50\% \psi_{ro}$ at 0.5 s, respectively. The simulation results are shown in Figures 8 and 9.

From Figure 8a, it can be seen that the *d*- and *q*-axis current steady error were respectively 12 A and -43 A after 0.5 s when using the conventional PCC, which were larger than those of simulations (a) and (b). The disturbance caused by the inductance and flux linkage parameter perturbations was also able to be accurately estimated when using the RNPCC, as can be seen in Figure 9. Furthermore, the *d*-axis and *q*-axis current steady error was effectively removed using the proposed composite integral terminal SMO. Figure 8b shows the three-dimensional rotor flux trajectories of the conventional PCC method under inductance and flux linkage parameter perturbation. Figure 9b shows the three-dimensional rotor flux trajectories of the RNPCC method under inductance and flux linkage parameter perturbation. The errors of *α*- and *β*-axis flux linkage were more serious than those of simulation (b) under inductance and flux linkage parameter perturbations, while still being nearly zero when using the RNPCC, as is clearly shown in Figures 8b and 9b. The Comparison of the current tracking between the conventional PPC and RNPCC are provided in Table 2.

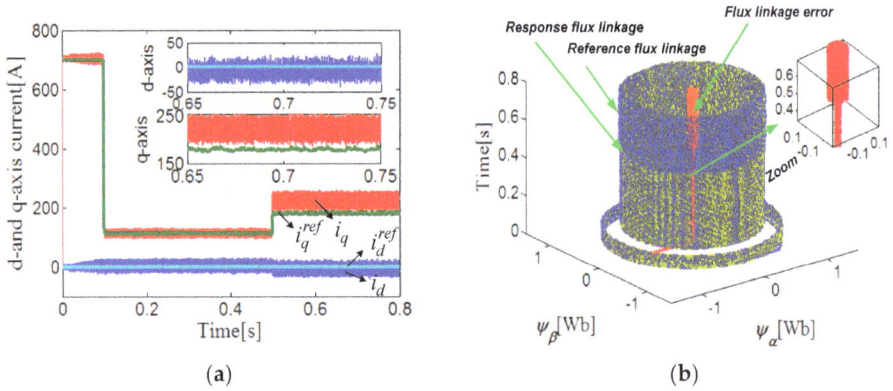

(a)

(b)

Figure 8. Simulation results of the conventional PCC method under inductance and flux linkage parameter perturbation. (**a**) *d*- and *q*-axis current responses and references; (**b**) three-dimensional rotor flux trajectories.

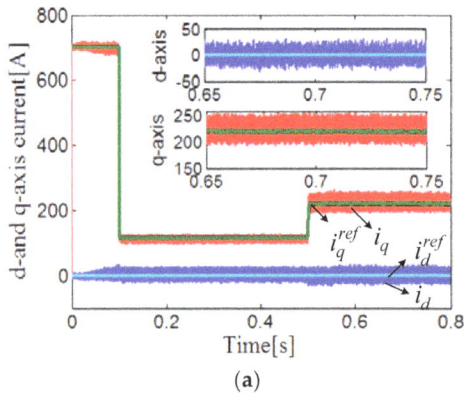

(a)

Figure 9. *Cont.*

(b)

(c)

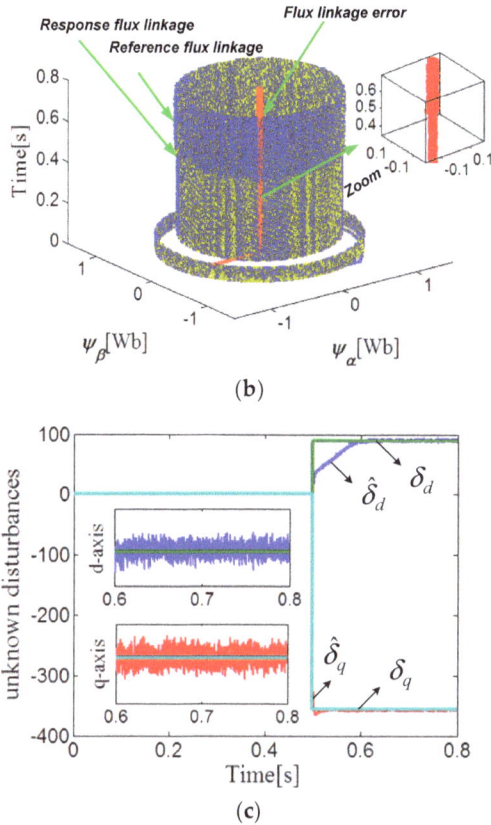

Figure 9. Simulation results of the RNPCC method under inductance and flux linkage parameter perturbations. The comparative performance of the current tracking between the conventional PCC and RNPCC methods was provided as follows. (**a**) *d*- and *q*-axis current responses and references.; (**b**) three-dimensional rotor flux trajectories; (**c**) disturbance estimation results.

Table 2. Comparison of the current tracking between the conventional PPC and RNPCC.

Steady State	Current Errors	Controller Type	
		Conventional PCC	Proposed RNPCC
Parameter Perturbation	$\pm(i_d{}^{ref} - i_d)$	±17 A	±0.8 A
Inductance	$\pm(i_q{}^{ref} - i_q)$	±1.4 A	±1.2 A
Parameter Perturbation	$\pm(i_d{}^{ref} - i_d)$	±0.7 A	±0.4 A
Flux Linkage	$\pm(i_q{}^{ref} - i_q)$	±55 A	±2 A
Parameter Perturbation	$\pm(i_d{}^{ref} - i_d)$	±18 A	±1.3 A
Inductance and Flux Linkage			
Parameter Perturbation	$\pm(i_q{}^{ref} - i_q)$	±47 A	±0.7 A

$i_d{}^{ref}$ is the reference value of i_d.

7. Experimental Results

To further verify the proposed RNPCC method, experiments were conducted using a PMSM laboratory platform, as is shown in Figure 10. It was composed of two motors (drive and load machines). The per-unit (p.u.) values of 10kW PMSM parameters are consistent with the simulation models. The current clamp was used to detect the current during the experiment. A torque sensor

(Beijing Sanjing Creation Science & technology Group Co., LTD., Beijing, China) was used to detect motor torque. Finally, the experimental results were obtained by Tek oscilloscope.

Figure 10. Experimental platform of the permanent magnet synchronous motor (PMSM)

The experimental results of the conventional PCC and proposed RNPCC under inductance parameter perturbation are shown in Figures 11 and 12, respectively. From Figure 11, it can be seen that the *d*-axis current value was greater than the reference value, but the inductance parameter perturbation had little effect on the *q*-axis current when using the conventional PCC. Figure 12 shows that the *d*-axis current steady state error could be removed by using the proposed RNPCC. Figures 13 and 14 show the control performance comparisons under flux linkage parameter mismatch. Figure 13 shows that the flux linkage parameter mismatch exerted an effect on the *q*-axis current response in the conventional PCC method. The *q*-axis current response value was greater than the current reference under flux linkage parameter mismatch. When using the RNPCC method, it can be observed from Figure 14 that the *q*-axis current response could track the current reference accurately and quickly. Figures 15 and 16 show the control performance comparisons under inductance and flux linkage parameter mismatch. From Figure 15, it can be seen that neither the *d*-axis nor the *q*-axis current response could track the current reference. The *q*-axis value of the current response was greater than the current reference, whereas the *d*-axis current response value was less than the current reference when using the conventional PCC. Figure 16 shows that the proposed RNPCC could remove the steady state current error caused by the inductance and flux linkage parameter perturbations during steady state operation.

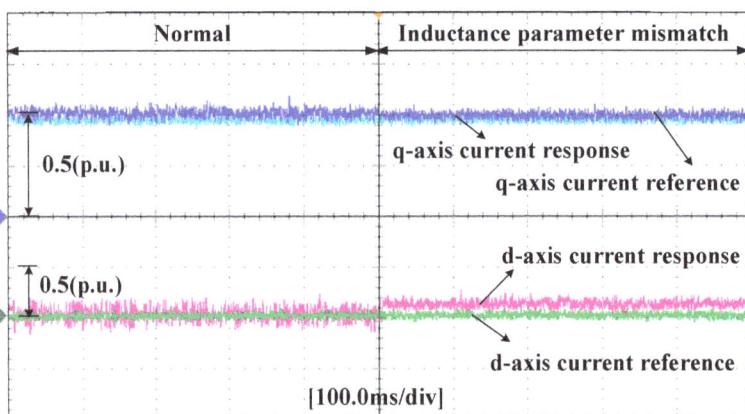

Figure 11. Experimental results of the conventional PCC method under inductance parameter perturbation.

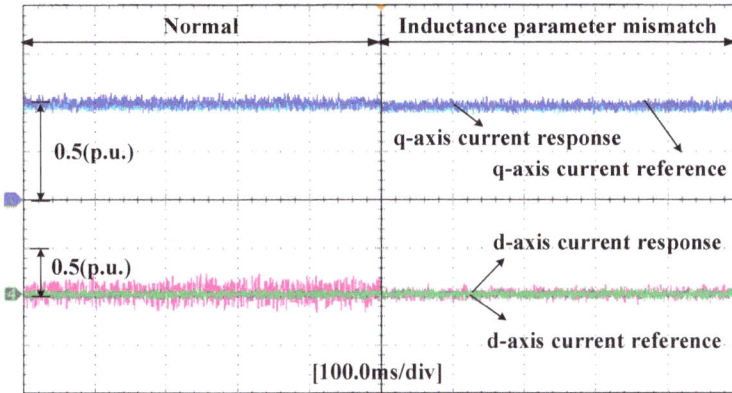

Figure 12. Experimental results of the RNPCC method under inductance parameter perturbation.

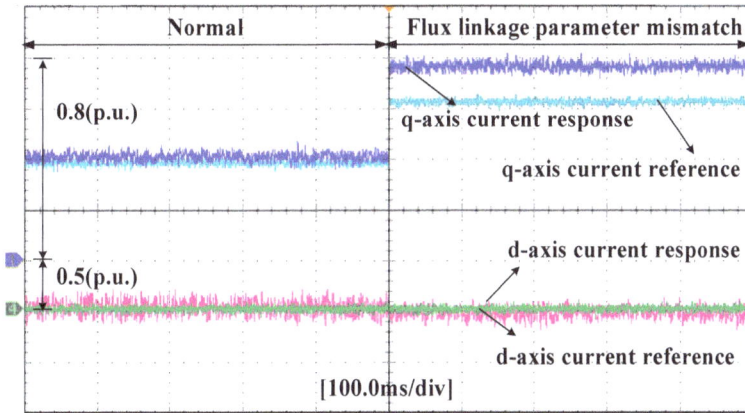

Figure 13. Experimental results of the conventional PCC method under flux linkage parameter perturbation.

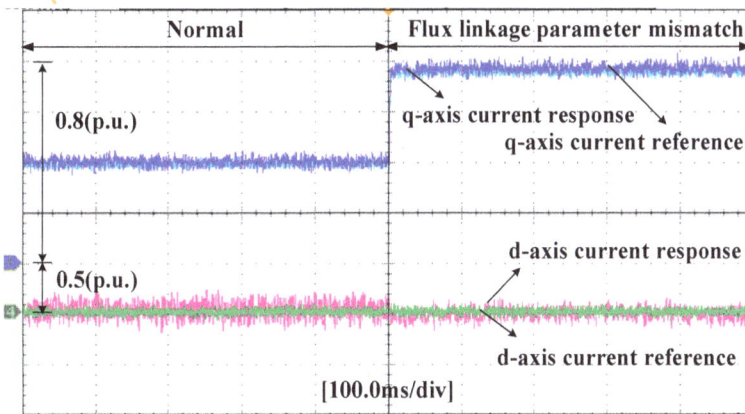

Figure 14. Experimental results of the RNPCC method under flux linkage parameter perturbation.

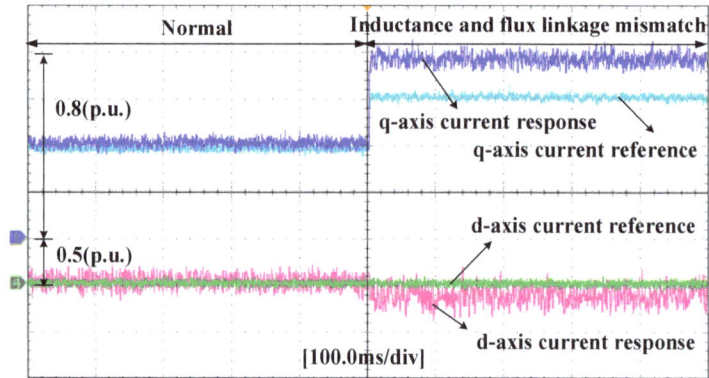

Figure 15. Experimental results of the conventional PCC method under inductance and flux linkage parameter perturbations.

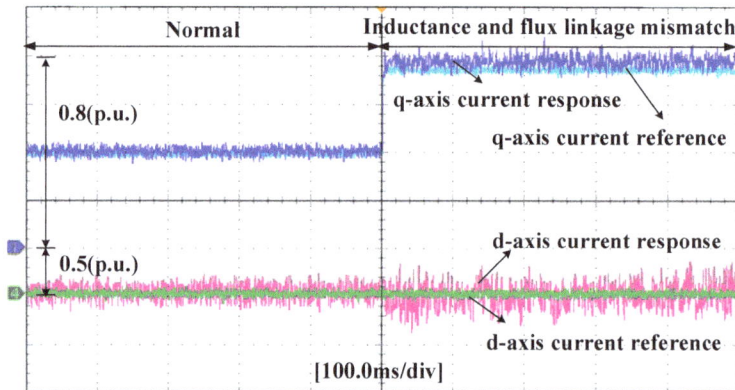

Figure 16. Experimental results of the RNPCC method under inductance and flux linkage parameter perturbations.

8. Conclusions

This paper proposed an RNPCC method based on a composite integral terminal SMO for current loop of the PMSM control system, without relying on the mathematical mode. The steady state current tracking errors caused by parameter perturbations were eliminated by the introduction of the disturbance estimation into the proposed control law. Moreover, the proposed RNPCC was able to effectively guarantee the control precision of the current loop regardless of the inductance and flux linkage parameter perturbations. The simulation and experimental results validated the performance of the proposed RNPCC. Compared with the traditional PCC, the proposed RNPCC shows its superiority in control precision and disturbance rejection. Additionally, the proposed RNPCC presents strong robustness and an excellent dynamic response in the case of parameter perturbations, where the weakness of the conventional PCC was successfully overcome. However, the proposed control method ignores the effect caused by rotor position detection error. The rotor position of the PMSM may deviate from the true position due to the position detection device fault, which can provide a mistaken response current to the proposed control method. As a result, the control performance of the proposed method deteriorated. A further research direction is to propose a novel robust predictive control algorithm to overcome the influence of inaccurate rotor position detection.

Energies **2019**, *12*, 443

Author Contributions: M.L., G.W. and D.L. designed the proposed control strategy, M.L., G.W. and D.L. conducted experimental works, modeling and simulation, S.H. and F.R. gave help of paper writing.

Funding: This work was supported in part by the National Key R&D Program of China (2016YFF0203400).

Conflicts of Interest: The authors declare no conflicts of interest.

References

1. Consoli, A.; Scarcella, G.; Testa, A. Slip-frequency detection for indirect field-oriented control drives. *IEEE Trans. Ind. Electron.* **2004**, *40*, 194–201. [CrossRef]
2. Qian, W.; Panda, S.K.; Xu, J.X. Torque ripple minimization in PM synchronous motors using iterative learning control. *IEEE Trans. Power Electron.* **2004**, *19*, 272–279. [CrossRef]
3. Wang, B.; Chen, X.; Yu, Y.; Wang, G.; Xu, D. Robust predictive current control with online disturbance estimation for induction machine drives. *IEEE Trans. Power Electron.* **2017**, *32*, 4663–4674. [CrossRef]
4. Zoghlami, M.; Kadri, A.; Bacha, F. Analysis and Application of the Sliding Mode Control Approach in the Variable-Wind Speed Conversion System for the Utility of Grid Connection. *Energies* **2018**, *11*, 720. [CrossRef]
5. Stojic, D.M.; Milinkovic, M.; Veinovic, S.; Klasnic, I. Stationary frame induction motor feed forward current controller with back EMF compensation. *IEEE Trans. Power Electron.* **2015**, *30*, 1356–1366. [CrossRef]
6. Yan, L.; Dou, M.; Hua, Z. Disturbance Compensation-Based Model Predictive Flux Control of SPMSM with Optimal Duty Cycle. *IEEE J. Emerg. Sel. Top. Power Electron.* **2018**. [CrossRef]
7. Carpiuc, S.C.; Lazar, C. Fast real-time constrained predictive current control in permanent magnet synchronous machine-based automotive traction drives. *IEEE Trans. Tran. Electrif.* **2015**, *1*, 326–335. [CrossRef]
8. Cortes, P.; Rodriguez, J.; Silva, C.; Flores, A. Delay Compensation in Model Predictive Current Control of a Three-Phase Inverter. *IEEE Trans. Ind. Electron.* **2012**, *59*, 1323–1325. [CrossRef]
9. Lim, C.S.; Levi, E.; Jones, M.; Rahim, N.A.; Hew, W.P. FCSMPC- Based Current Control of a Five-Phase Induction Motor and its Comparison with PI-PWM Control. *IEEE Trans. Ind. Electron.* **2016**, *61*, 149–163. [CrossRef]
10. Chen, Z.; Qiu, J.; Jin, M. Prediction-Error-Driven Position Estimation Method for Finite Control Set Model Predictive Control of Interior Permanent Magnet Synchronous Motors. *IEEE J. Emerg. Sel. Top. Power Electron.* **2018**. [CrossRef]
11. Wipasuramonton, P.; Zhu, Z.Q.; Howe, D. Predictive current control with current-error correction for PM brushless AC drives. *IEEE Trans. Ind. Appl.* **2006**, *42*, 1071–1079. [CrossRef]
12. Siami, M.; Khaburi, D.A.; Abbaszadeh, A.; Rodríguez, J. Robustness improvement of predictive current control using prediction error correction for permanent-magnet synchronous machines. *IEEE Trans. Ind. Electron.* **2016**, *63*, 3458–3466. [CrossRef]
13. Le-Huy, H.; Slimani, K.; Viarouge, P. Analysis and implementation of a real-time predictive current controller for permanent-magnet synchronous servo drives. *IEEE Trans. Ind. Electron.* **1994**, *41*, 110–117. [CrossRef]
14. Yang, M.; Lang, X.; Long, J.; Xu, D. Flux immunity robust predictive current control with incremental model and extended state observer for PMSM drive. *IEEE Trans. Power Electron.* **2017**, *32*, 9267–9279. [CrossRef]
15. Zhang, X.; Hou, B.; Mei, Y. Deadbeat Predictive Current Control of Permanent-Magnet Synchronous Motors with Stator Current and Disturbance Observer. *IEEE Trans. Power Electron.* **2017**, *32*, 3818–3834. [CrossRef]
16. Abdelrahem, M.; Hackl, C.M.; Zhang, Z.; Kennel, R. Robust Predictive Control for Direct-Driven Surface-Mounted Permanent-Magnet Synchronous Generators Without Mechanical Sensors. *IEEE Trans. Energy Conv.* **2018**, *33*, 179–189. [CrossRef]
17. Zhang, C.; Wu, G.; Fei, R.; Feng, J.H.; Jia, L. Robust Fault-Tolerant Predictive Current Control for Permanent Magnet Synchronous Motors Considering Demagnetization Fault. *IEEE Trans. Ind. Electron.* **2018**, *65*, 5324–5334. [CrossRef]
18. Yang, J.; Zheng, W.X.; Li, S.; Wu, B.; Cheng, M. Design of a prediction-accuracy-enhanced continuous-time MPC for disturbed systems via a disturbance observer. *IEEE Trans. Ind. Electron.* **2015**, *62*, 5807–58164. [CrossRef]
19. Yang, J.; Li, S.; Yu, X. Sliding-mode control for systems with mismatched uncertainties via a disturbance observer. *IEEE Trans. Ind. Electron.* **2013**, *60*, 160–169. [CrossRef]

Energies **2019**, *12*, 443

20. Yang, J.; Zheng, W.X. Offset-free nonlinear MPC for mismatched disturbance attenuation with application to a static var compensator. *IEEE Trans. Circuits Syst. II Express Br.* **2014**, *61*, 49–53. [CrossRef]

21. Hu, F.; Luo, D.; Luo, C.; Long, Z.; Wu, G. Cascaded Robust Fault-Tolerant Predictive Control for PMSM Drives. *Energies* **2018**, *11*, 3087. [CrossRef]

22. Richter, J.; Doppelbauer, M. Predictive trajectory control of permanent-magnet synchronous machines with nonlinear magnetic. *IEEE Trans. Ind. Electron.* **2016**, *63*, 3915–3924. [CrossRef]

![energies logo] *energies*

MDPI

Article

Commutation Error Compensation Strategy for Sensorless Brushless DC Motors

Xuliang Yao, Jicheng Zhao, Guangxu Lu, Hao Lin and Jingfang Wang *

College of Automation, Harbin Engineering University, Harbin 150001, China; yaoxuliang@hrbeu.edu.cn (X.Y.); worryfree@hrbeu.edu.cn (J.Z.); michel@hrbeu.edu.cn (G.L.); linhao@hrbeu.edu.cn (H.L.)
* Correspondence: jingfangwang@hrbeu.edu.cn; Tel.: +86-189-4603-2562

Received: 27 November 2018; Accepted: 4 January 2019; Published: 9 January 2019

Abstract: Sensorless brushless DC (BLDC) motor drive systems often suffer from inaccurate commutation signals, which result in current fluctuation and high conduction loss. To improve precision of commutation signals, this paper presents a novel commutation error compensation strategy for BLDC motors. First, the relationship between the line voltage difference integral in 60 electrical degree conduction interval and the commutation error is analyzed. Then, in terms of the relationship derived, a feedback compensation strategy based on the line voltage difference integral is proposed to regulate commutation signals by making three-phase back electromotive force (EMF) integral to zero, and the effect of the freewheeling process on the line voltage difference integral is considered. Moreover, an incremental PI controller is designed to achieve closed-loop compensation for the commutation error automatically. Finally, experiment results verify feasibility and effectiveness of the proposed strategy.

Keywords: brushless DC (BLDC) motor; sensorless motor; commutation error compensation; free-wheeling period

---

1. Introduction

Brushless DC (BLDC) motors are widely used in aerospace, electric vehicle, and household applications due to their inherent advantages, which include high-power density, high efficiency, and simple structure [1–3]. Position sensors, such as Hall-effect sensors, are often employed to acquire rotor position signals for proper commutation, however, misalignment in the mechanical installation of the sensors, extreme temperature conditions, or electromagnetic interference may induce errors in the rotor position information, which severely limit the application scope of BLDC motors [4]. To overcome these disadvantages, many sensorless BLDC motor techniques have been proposed in recent years. The zero-crossing points (ZCPs) of the back electromotive force (EMF) are easily detected and widely used in sensorless control. Among the various sensorless strategies currently available, these presented ZCP detection techniques show satisfactory results in specific application scopes; unfortunately, these strategies are still disturbed by various commutation errors [5–7]. The ZCPs of back-EMF are generally abstracted from the terminal voltage, but the terminal voltage is vulnerable to the high-frequency components due to the pulse-width modulation (PWM) switching. A low pass filter is usually applied to filter high-frequency switching noise of the terminal voltage, whereas the filters will introduce phase delay for ZCPs detection of back-EMF and the phase delay varies with the operating speed. The lag angle will increase as the motor speed rises. The distorted terminal voltage caused by the freewheeling process also will generate the commutation error. The disturbed terminal voltage may make position detection signal phase ahead, deviating from the ideal commutation instants. The inaccurate commutation signals will induce current pulsation and increase conduction loss.

In reference [8], a hysteresis comparator composed of discrete components is presented to compensate phase delay from the low-pass filter. In reference [9], the authors analyze the errors caused by signal extraction circuits and machine parameters in the sensorless control technique based on average line to line voltage. In reference [10], commutation errors due to the armature reaction and low-pass filters are analyzed, and an error compensation circuit is proposed to achieve proper commutation. In reference [11], the effects of the low-pass filter, noise, and nonideal properties of the estimated back-EMF on the commutation instants are considered. To adjust commutation instants for BLDC motors, a modified digital phase shifter is designed to implement compensation of commutation errors. In the research mentioned previously [8–11], some specific error factors are discussed and investigated, and the accuracy of the commutation signal is improved accordingly. To achieve more precise commutation signals of sensorless BLDC motors, the commutation error compensation strategies which consider more comprehensive error factors should be studied.

Many estimation and observer methods of commutation error compensation have been proposed to achieve high-precision commutation [12–18]. In reference [12], the torque constant is employed as the reference signal of the rotor position to obtain accurate commutation signals. In reference [13], the rotor position is acquired from the back-EMF difference, which is estimated from the disturbance observer structure. Discrete-time sliding mode observer [14] and iterative sliding mode observer [15] have been introduced to obtain precise commutation signals for sensorless permanent magnet synchronous motor control. Estimation strategies based on Kalman filters [16] and extended Kalman filters [17,18] have also been presented to estimate information of rotor position accurately. However, these algorithms are complicated and require a large amount of calculation.

Commutation error factors are reflected in the commutation signals, and thus voltages or currents under inaccurate commutation will contain the commutation error information. Many compensation strategies of commutation errors based on the phenomenon of voltages or currents produced under inaccurate commutation have been proposed. In reference [19], an intelligent self-tuning strategy is presented to finely adjust commutation instants, and regulation of the commutation instants is realized by minimizing the stator current. In reference [20], a self-compensation strategy of the commutation angle based on the DC-link current is proposed; in this technique, the commutation instants are corrected by regulating the estimated commutation time error to zero. However, this method is based on the linearized approximation between the DC-link current difference and commutation time error in low-inductance motors. In reference [21], commutation errors are compensated by keeping the freewheeling current of the unenergized phase symmetrical. However, when dealing with some PWM strategies featuring suppression of the freewheeling current [22], such as PWM_ON_PWM mode, the method is unsuitable. The research in [23] proposed a current index optimal approach to adjust commutation errors. However, the back-EMF is assumed to be a trapezoidal waveform with less than 120° flat-top width, which limits the applications of the proposed method. In reference [24], the symmetrical characteristics of the unenergized phase terminal voltage are presented to regulate commutation instants, but the effect of the freewheeling period is not considered.

This paper presents a novel compensation strategy for commutation errors of sensorless BLDC motors. In this paper, the relationship between the commutation error and the line voltage difference integral in 60 electrical degrees interval is analyzed, and a feedback compensation strategy based on the line voltage difference integral is presented to regulate commutation instants and the effect of the freewheeling period is considered. In addition, a PI controller is designed to achieve closed-loop compensation for commutation errors. The remainder of the paper is organized as follows. The proposed commutation error compensation strategy is proposed in Section 2. Section 3 provides experimental results to support the validity and effectiveness of the proposed method. The conclusions are presented in Section 4.

2. Proposed Commutation Error Compensation Strategy

This paper presents a commutation error compensation strategy based on the line voltage difference integral. In this section, the relationship between the line voltage difference integral and the commutation error is theoretically analyzed. Then, a feedback compensation strategy with the line voltage difference integral is presented. Finally, an incremental PI controller is designed to regulate the commutation error.

2.1. Mathematical Model of BLDC Motor

The equivalent circuit of the BLDC motor drive system is shown in Figure 1. Assuming that the three-phase stator windings are symmetrical, the phase resistance and inductance are constant, and the armature reaction is negligible. Then, the voltage equations are described as

$$
\begin{bmatrix} u_a \\ u_b \\ u_c \end{bmatrix} = \begin{bmatrix} R & 0 & 0 \\ 0 & R & 0 \\ 0 & 0 & R \end{bmatrix} \begin{bmatrix} i_a \\ i_b \\ i_c \end{bmatrix} + \begin{bmatrix} L & 0 & 0 \\ 0 & L & 0 \\ 0 & 0 & L \end{bmatrix} \frac{d}{dt} \begin{bmatrix} i_a \\ i_b \\ i_c \end{bmatrix} + \begin{bmatrix} e_a \\ e_b \\ e_c \end{bmatrix} + \begin{bmatrix} u_N \\ u_N \\ u_N \end{bmatrix} \tag{1}
$$

where u_a, u_b, and u_c are the three-phase voltages; i_a, i_b, and i_c are the three-phase currents; e_a, e_b, and e_c are the phase back-EMFs; R is the stator resistance; L is the phase inductance; and u_N is the neutral voltage.

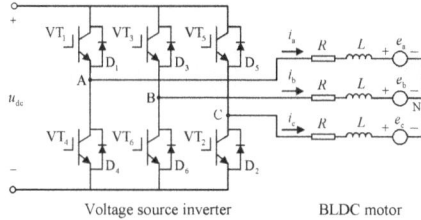

Figure 1. Equivalent circuit of a brushless DC (BLDC) motor drive system.

The line voltage equations are expressed as

$$
\begin{cases} u_{ab} = u_a - u_b = R(i_a - i_b) + L\frac{d(i_a - i_b)}{dt} + e_a - e_b \\ u_{bc} = u_b - u_c = R(i_b - i_c) + L\frac{d(i_b - i_c)}{dt} + e_b - e_c \\ u_{ca} = u_c - u_a = R(i_c - i_a) + L\frac{d(i_c - i_a)}{dt} + e_c - e_a \end{cases} \tag{2}
$$

In general, the back-EMF of BLDC motors is between those of a trapezoidal wave and a sine wave since it contains many high-order harmonics. Hence, the back-EMF of the BLDC motor can be expressed as

$$
\begin{cases} e_a = \sum_{n=0}^{\infty} E_{2n+1} \sin((2n+1)\theta) = e_{a1} + e_3 + e_{a5} + e_{a7} + e_9 + \dots \\ e_b = \sum_{n=0}^{\infty} E_{2n+1} \sin((2n+1)(\theta - 2\pi/3)) = e_{b1} + e_3 + e_{b5} + e_{b7} + e_9 + \dots \\ e_c = \sum_{n=0}^{\infty} E_{2n+1} \sin((2n+1)(\theta + 2\pi/3)) = e_{c1} + e_3 + e_{c5} + e_{c7} + e_9 + \dots \end{cases} \tag{3}
$$

where E_{2n+1} is the back-EMF coefficient; $(2n+1)$ is the order of harmonics; θ is the rotor position; e_{am}, e_{bm}, and e_{cm} represent the mth harmonics of the three-phase back-EMF and the subscript m denotes harmonic order; e_3 and e_9 represent the third harmonics of the back-EMF.

Figure 2 shows phase diagram between back-EMFs and phase current, where e_a, e_b, and e_c denote the back-EMFs, and i_a, i_b, and i_c denote phase current. α represents the commutation error, which is assumed $\alpha \in [-\pi/6, \pi/6]$. The sign of the commutation error α reflects the commutation information (i.e., delayed or advanced commutation). A negative commutation error α denotes advanced commutation, whereas a positive error represents a delayed one.

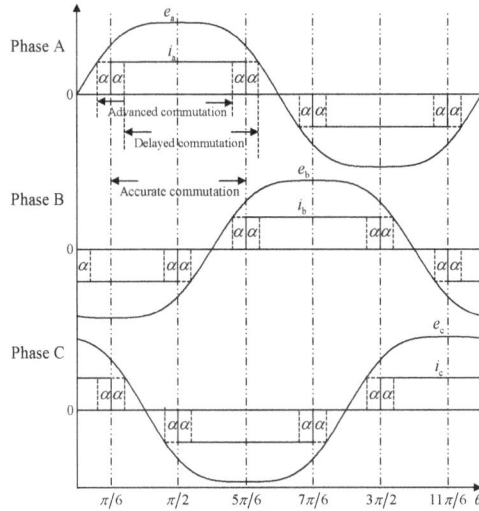

Figure 2. Phase diagram of the back electromotive forces (EMFs) and phase current.

2.2. Line Voltage Difference Integral without Consideration of Freewheeling Period

For a BLDC motor, there are six operation modes and every operation modes last 60 electrical degrees. In the 60 electrical degree conduction interval, two periods are included: the freewheeling period and the normal conduction period. In the freewheeling period t_{fw}, the incoming phase current begins to rise, the un-commutated current remains conducting, and the outgoing phase current gradually decays due to the existence of phase inductance. When the outgoing phase current drops to zero, the freewheeling period is over and the motor starts operating in the normal conduction period.

Assume at some step that VT_1 and VT_6 are conducting, while VT_5 is floating. The line voltage difference $(u_{bc} - u_{ca})$ is expressed as

$$u_{bc} - u_{ca} = R(i_a + i_b - 2i_c) + Ld(i_a + i_b - 2i_c)/dt + e_a + e_b - 2e_c \tag{4}$$

Considering a BLDC motor with a negligible freewheeling period t_{fw}, the current equation satisfies $i_a = -i_b$ and $i_c = 0$ during the entire 60-electrical degree period when $t \in [\pi/6, \pi/2]$, as shown in Figure 2. Substituting the current equation into (4), the line voltage difference $(u_{bc} - u_{ca})$ is simplified as

$$u_{bc} - u_{ca} = e_a + e_b - 2e_c \tag{5}$$

By integrating (5), the line voltage difference integral d is achieved as follows:

$$d = \int_{t_{\pi/6}}^{t_{\pi/2}} (u_{bc} - u_{ca})dt = \int_{t_{\pi/6}}^{t_{\pi/2}} (e_a + e_b - 2e_c)dt \tag{6}$$

Figure 3 shows the three-phase back-EMF under accurate, delayed, and advanced commutation; the shadowed areas in each subfigure represent the integral of each back-EMF. As seen in Figure 3a, the integral of the back-EMFs e_a and e_b yields equal amplitudes but opposite signs, i.e., $\int_{t_{\pi/6}}^{t_{\pi/2}} e_a dt = -\int_{t_{\pi/6}}^{t_{\pi/2}} e_b dt$; hence, the sum of the integrated back-EMFs e_a and e_b is zero in the interval. Similarly, the integral of the back-EMF e_c is zero, i.e., $\int_{t_{\pi/6}}^{t_{\pi/2}} e_c dt = 0$, due to the AC symmetry of the back-EMF e_c. From this analysis, (6) satisfies the condition

$$d = \int_{t_{\pi/6}}^{t_{\pi/2}} (u_{bc} - u_{ca})dt = \int_{t_{\pi/6}}^{t_{\pi/2}} (e_a + e_b - 2e_c)dt = 0 \tag{7}$$

The line voltage difference integral d under accurate commutation yields a value of zero. When the commutation instants are delayed or advanced by α, the conduction period of VT_1 and VT_6 lasts for the interval $t \in [\alpha/6 + \alpha, \alpha/2 + \alpha]$, and the line voltage difference integral d is written as

$$d = \int_{t_{\pi/6+\alpha}}^{t_{\pi/2+\alpha}} (u_{bc} - u_{ca})dt = \int_{\pi/6+\alpha}^{\pi/2+\alpha} (e_a + e_b - 2e_c)d\theta = d_{trp} + d_{oth} \tag{8}$$

where

$$d_{trp} = \int_{\pi/6+\alpha}^{\pi/2+\alpha} \left[\sum_{\substack{n=1, \\ 4,7,\ldots}} E_{2n+1}\sin((2n+1)\theta) + \sum_{\substack{n=1, \\ 4,7,\ldots}} E_{2n+1}\sin((2n+1)(\theta - 2\pi/3)) \right.$$

$$\left. -2\sum_{\substack{n=1, \\ 4,7,\ldots}} E_{2n+1}\sin((2n+1)(\theta + 2\pi/3)) \right] d\theta$$

$$d_{oth} = \int_{\pi/6+\alpha}^{\pi/2+\alpha} \left[\sum_{\substack{n=0, \\ 2,3,5,\ldots}} E_{2n+1}\sin((2n+1)\theta) + \sum_{\substack{n=0, \\ 2,3,5,\ldots}} E_{2n+1}\sin((2n+1)(\theta - 2\pi/3)) \right.$$

$$\left. -2\sum_{\substack{n=0, \\ 2,3,5,\ldots}} E_{2n+1}\sin((2n+1)(\theta + 2\pi/3)) \right] d\theta$$

Since the third harmonics of the three three-phase back-EMFs are equal, $d_{trp} = 0$. Then, (8) is simplified as

$$\int_{\pi/6+\alpha}^{\pi/2+\alpha} (e_a + e_b - 2e_c)d\theta = d_{oth} \tag{9}$$

Solving and simplifying the definite integral gives

$$\int_{\pi/6+\alpha}^{\pi/2+\alpha} (e_a + e_b - 2e_c)d\theta = 3\int_{\pi/6+\alpha}^{\pi/2+\alpha} \left[\sum_{\substack{n=0,2, \\ 3,5,\ldots}} E_{2n+1}\sin((2n+1)(\theta - \pi/3)) \right] d\theta$$

$$= -3\sum_{\substack{n=0,2, \\ 3,5,\ldots}} (E_{2n+1}/(2n+1))\cos((2n+1)(\alpha + \pi/6)) \tag{10}$$

$$+3\sum_{\substack{n=0,2, \\ 3,5,\ldots}} (E_{2n+1}/(2n+1))\cos((2n+1)(\alpha - \pi/6))$$

As the back-EMF harmonic coefficients satisfy the condition $E_1/E_7 \gg 1$, the high-order terms of the back-EMF in (10) are negligible, and (10) can be simplified as

$$\int_{\pi/6+\alpha}^{\pi/2+\alpha} (e_a + e_b - 2e_c)d\theta = -3E_1[\cos(\alpha + \pi/6) - \cos(\alpha - \pi/6)]$$

$$-\tfrac{3}{5}E_5[\cos(5\alpha + 5\pi/6) - \cos(5\alpha - 5\pi/6)] \tag{11}$$

As seen in (11), the three-phase back-EMF integral $\int_{\pi/6+\alpha}^{\pi/2+\alpha} (e_a + e_b - 2e_c)d\theta$ is a function of the commutation error α. The relationship between the three-phase back-EMF integral and the commutation error α will be discussed in the following section.

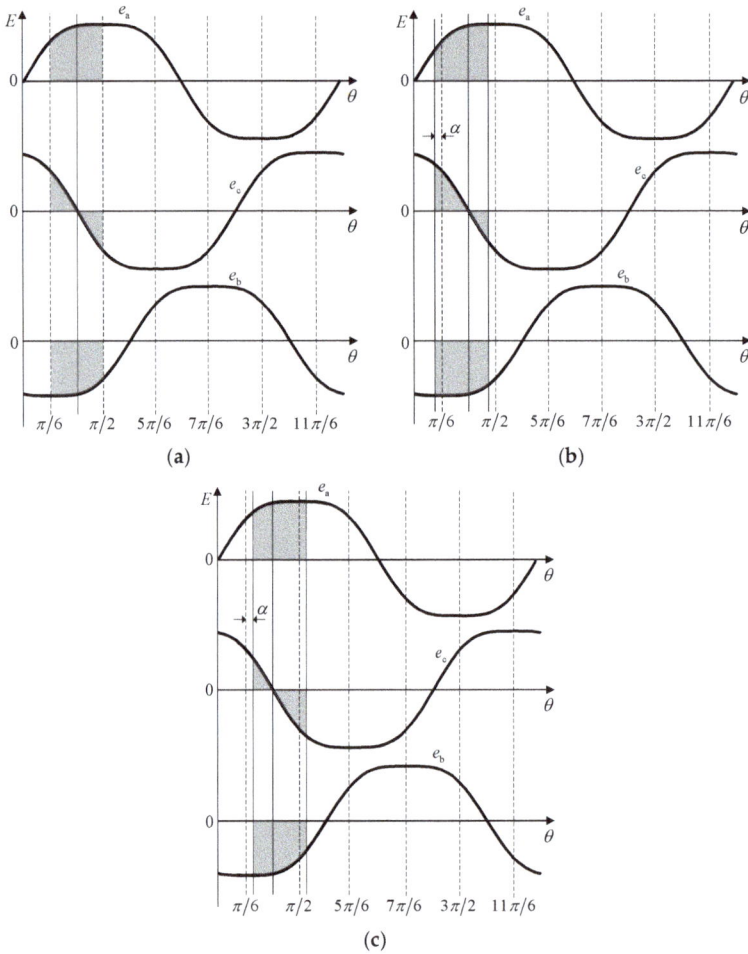

Figure 3. Three-phase back-EMF. (**a**) Accurate commutation; (**b**) Advanced commutation; (**c**) Delayed commutation.

Accurate commutation: when the commutation error $\alpha = 0$, the motor works under accurate commutation, as shown in Figure 3a, and the back-EMF integral $\int_{\pi/6}^{\pi/2}(e_a + e_b - 2e_c)d\theta = 0$.

Advanced commutation: when the commutation error $\alpha \in [-\pi/6, 0)$, the motor works under advanced commutation, as shown in Figure 3b, and the three-phase back-EMF integral $\int_{\pi/6+\alpha}^{\pi/2+\alpha}(e_a + e_b - 2e_c)d\theta < 0$.

Delayed commutation: when the commutation error $\alpha \in (0, \pi/6]$, the motor works under delayed commutation, as shown in Figure 3c, and the three-phase back-EMF integral $\int_{\pi/6+\alpha}^{\pi/2+\alpha}(e_a + e_b - 2e_c)d\theta > 0$.

Similarly, when VT_3 and VT_4 are in conduction state, the phase current flows in opposite direction compared with the direction during the conduction period of VT_1 and VT_6. The three-phase back-EMF integral $\int_{7\pi/6}^{3\pi/2}(e_a + e_b - 2e_c)d\theta = 0$ under accurate commutation, $\int_{7\pi/6+\alpha}^{3\pi/2+\alpha}(e_a + e_b - 2e_c)d\theta < 0$ under delayed commutation, and $\int_{7\pi/6+\alpha}^{3\pi/2+\alpha}(e_a + e_b - 2e_c)d\theta > 0$ under advanced commutation.

From previous research on three-phase back-EMF integral and the line voltage difference integral expression in (7), when the freewheeling period t_{fw} is negligible, the line voltage difference integral d is proportional to the commutation error α and can be used as an indicator of commutation errors.

The relationship between the conduction switch, the line voltage difference integral, and the commutation information is shown in Table 1; here, t_s and t_e are the lower and upper limits of integral, respectively.

Table 1. Relationship between the integral d, conduction switch, and commutation error.

Conduction Switch	d	Sign of d	Advanced/Delayed
VT_1-VT_6	$\int_{t_s}^{t_e}(u_{bc}-u_{ca})dt$	>0	Delayed
		<0	Advanced
VT_1-VT_2	$\int_{t_s}^{t_e}(u_{ab}-u_{bc})dt$	<0	Delayed
		>0	Advanced
VT_3-VT_2	$\int_{t_s}^{t_e}(u_{ba}-u_{ac})dt$	>0	Delayed
		<0	Advanced
VT_3-VT_4	$\int_{t_s}^{t_e}(u_{bc}-u_{ca})dt$	<0	Delayed
		>0	Advanced
VT_5-VT_4	$\int_{t_s}^{t_e}(u_{ab}-u_{bc})dt$	>0	Delayed
		<0	Advanced
VT_5-VT_6	$\int_{t_s}^{t_e}(u_{ba}-u_{ac})dt$	<0	Delayed
		>0	Advanced

When the motor is in no-load state, the effect of the freewheeling period t_{fw} is negligible. Figure 4 shows the diagram of the line voltage difference integral d in the 60-electrical degree interval of A+B− and A−B+, where T_{A+B-} and T_{A-B+} represents the conduction period of phase A+B− and A−B+, respectively. The integral d in the A+B− conduction interval is zero under accurate commutation (Figure 4a), negative under advanced commutation (Figure 4b), and positive under delayed commutation (Figure 4c). In the A−B+ conduction interval, the integral d is opposite compared with that in the A+B− conduction interval. These results agree with the theoretical analysis.

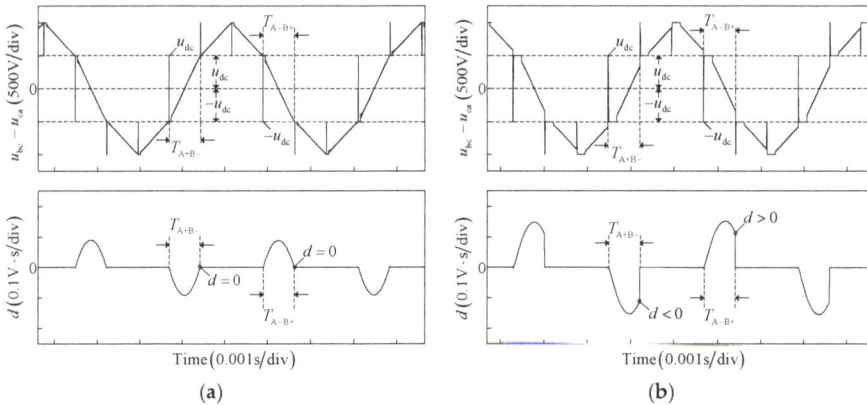

Figure 4. *Cont.*

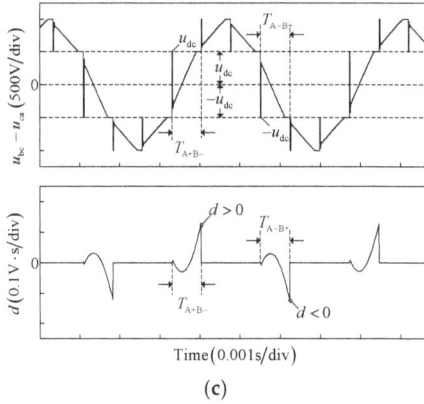

Figure 4. Line voltage difference ($u_{bc} - u_{ca}$) and integral d in the conduction period of phase A+B− and A−B+. (**a**) Accurate commutation; (**b**) Advanced commutation; (**c**) Delayed commutation.

2.3. Line Voltage Difference Integral Considering Freewheeling Period

During the freewheeling period t_{fw}, the line voltage difference is distorted by the DC-link voltage and clamped at the level of the DC-link voltage u_{dc}. When line voltage difference integral is performed, the distorted line voltage difference in the interval t_{fw} is also integrated, which will cause the line voltage difference integral to deviate from the accurate line voltage difference integral. In this section, the line voltage difference integral of a BLDC motor considering the freewheeling period will be discussed.

Assume that the commutation error is α and the conduction interval of VT$_1$ and VT$_6$ lasts for the period when $t \in [t_{\pi/6+\alpha}, t_{\pi/2+\alpha}]$, as shown in Figure 2. In the conduction period, the line voltage difference integral d^* can be expressed as

$$d^* = \int_{t_{\pi/6+\alpha}}^{t_{\pi/2+\alpha}} \left(u_{bc} - u_{ca} \right) dt = \int_{t_{\pi/6+\alpha}}^{t_{\pi/6+\alpha}+t_{fw}} \left(u_{bc} - u_{ca} \right) dt + \int_{t_{\pi/6+\alpha}+t_{fw}}^{t_{\pi/2+\alpha}} \left(u_{bc} - u_{ca} \right) dt \qquad (12)$$

Figure 5 shows the equivalent circuit when the commutation occurs from VT$_5$ to VT$_1$. In the commutation period t_{fw}, VT$_5$ is turned off, VT$_1$ is turned on, and VT$_6$ continues conducting. The three phase voltages are expressed as

$$\begin{cases} u_a = Ri_a + L\frac{di_a}{dt} + e_a + u_N = u_{dc} \\ u_b = Ri_b + L\frac{di_b}{dt} + e_b + u_N = 0 \\ u_c = Ri_c + L\frac{di_c}{dt} + e_c + u_N = 0 \end{cases} \qquad (13)$$

The line voltage difference meets the condition $u_{bc} - u_{ca} = u_{dc}$, and the line voltage difference integral during the freewheeling period t_{fw} is reformulated as

$$\int_{t_{\pi/6+\alpha}}^{t_{\pi/6+\alpha}+t_{fw}} \left(u_{bc} - u_{ca} \right) dt = \int_{t_{\pi/6+\alpha}}^{t_{\pi/6+\alpha}+t_{fw}} u_{dc} dt \qquad (14)$$

In (14), the line voltage difference ($u_{ab} - u_{bc}$) is equal to the DC-link voltage u_{dc} in the freewheeling period t_{fw}; hence, the line voltage difference integral shows a trend of linear growth in the freewheeling period t_{fw}.

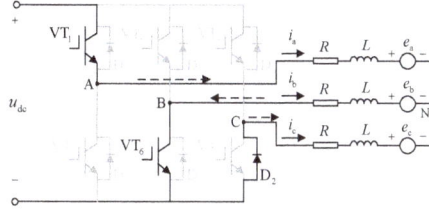

Figure 5. Equivalent circuit when commutation occurs from VT$_5$ to VT$_1$.

Figure 6 shows the diagram of the line voltage difference ($u_{bc} - u_{ca}$) and its integral d^* in the conduction period of phase A+B− and A−B+. The as seen in Figure 6, the line voltage difference ($u_{bc} - u_{ca}$) is equal to the DC-link voltage u_{dc} in the freewheeling period t_{fw} and the corresponding integral d^* linearly increases in the interval. Compared with that in Figure 4, since the disturbed voltage difference is integrated in the freewheeling period t_{fw}, the integral d^* under different commutation produces an amplitude deviation. Especially under accurate commutation, for example, the integral d^* deviates from zero, and becomes positive and negative in the conduction period of phase A+B− and A−B+, respectively.

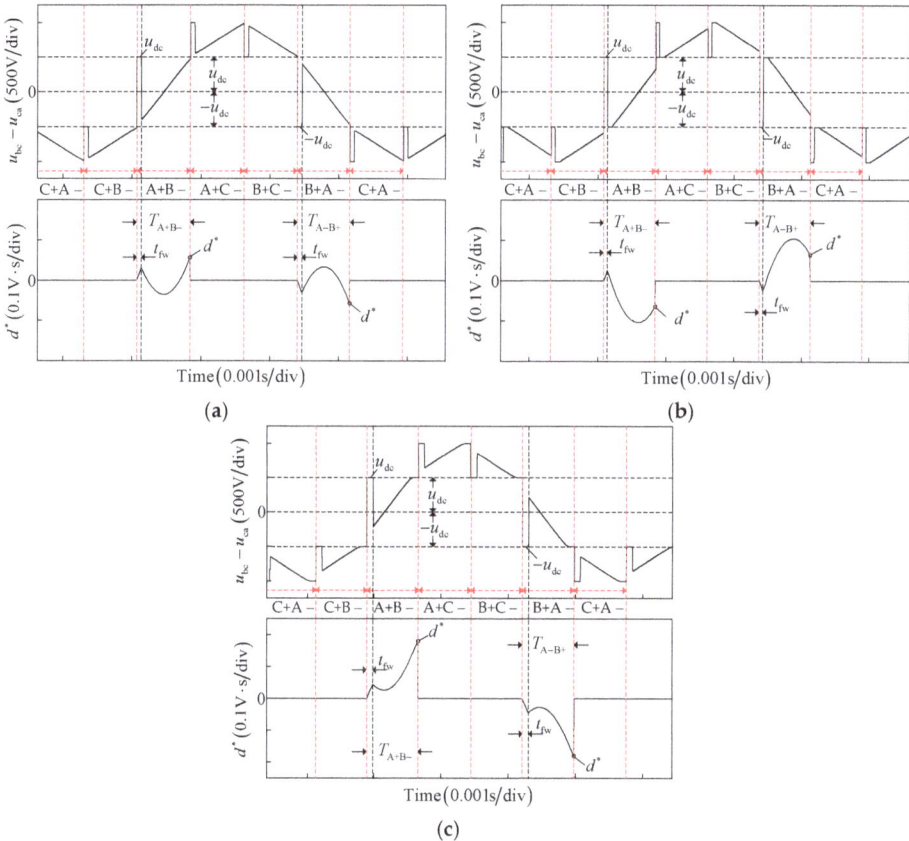

(a)

(b)

(c)

Figure 6. The line voltage difference ($u_{bc} - u_{ca}$) and corresponding integral d^* in the conduction period of phase A+B− and A−B+. (**a**) Accurate commutation; (**b**) Advanced commutation; (**c**) Delayed commutation.

The other voltage spikes in Figure 6 are analyzed as followed. When the commutation occurs, the outgoing phase current continues conducting through the anti-parallel diode, so the terminal voltage is clamped at the level of 0 V or u_{dc}. When the motor commutates between different phases, the clamped voltage value of three-phase voltage changes, hence the line voltage difference $(u_{bc} - u_{ca})$ will also change accordingly. The cause of the voltage spikes is the same under delayed, advanced, and accurate commutation. To facilitate the analysis, the voltage spike under the accurate commutation is taken as an example.

(1) When commutation occurs from A+B− to A+C−, it meets $u_a = u_{dc}$, $u_b = u_{dc}$, and $u_c = 0$ in the freewheeling period t_{fw}. Therefore, $u_{bc} - u_{ca} = 2u_{dc}$, and it corresponds to the positive spike whose magnitude exceeds u_{dc}.

(2) When commutation occurs from B+A− to C+A−, it meets $u_a = 0$, $u_b = 0$, and $u_c = u_{dc}$ in the freewheeling period t_{fw}. Therefore, $u_{bc} - u_{ca} = -2u_{dc}$, and it corresponds to the negative spike whose magnitude exceeds $-u_{dc}$.

(3) When commutation occurs from A+C− to B+C−, it meets $u_a = 0$, $u_b = u_{dc}$, and $u_c = 0$ in the freewheeling period t_{fw}. Therefore, $u_{bc} - u_{ca} = u_{dc}$, and it corresponds to the positive spike whose magnitude is equal to u_{dc}.

(4) When commutation occurs from C+A− to C+B−, it meets $u_a = u_{dc}$, $u_b = 0$, and $u_c = u_{dc}$ in the freewheeling period t_{fw}. Therefore, $u_{bc} - u_{ca} = -u_{dc}$, and it corresponds to the negative spike whose magnitude is equal to $-u_{dc}$.

The voltage spikes of the line voltage difference $(u_{bc} - u_{ca})$ in 360-degree electrical angle are listed in Table 2.

Table 2. Voltage spike values in 360-degree electrical angle.

Commutation Case	u_a	u_b	u_c	$u_{bc} - u_{ca}$
C+A− → C+B−	u_{dc}	0	u_{dc}	$-u_{dc}$
C+B− → A+B−	u_{dc}	0	0	u_{dc}
A+B− → A+C−	u_{dc}	u_{dc}	0	$2u_{dc}$
A+C− → B+C−	0	u_{dc}	0	u_{dc}
B+C− → B+A−	0	u_{dc}	u_{dc}	$-u_{dc}$
B+A− → C+A−	0	0	u_{dc}	$-2u_{dc}$

In the normal conduction period, when $t \in [t_{\pi/6+\alpha} + t_{fw}, t_{\pi/2+\alpha}]$, the phase currents i_a and i_b conduct, while the outgoing phase current i_c remains zero, i.e., $i_a = -i_b$ and $i_c = 0$. Substituting the current equation and (2) into the second item of (12), the line voltage difference integral $\int_{t_{\pi/6+\alpha}+t_{fw}}^{t_{\pi/2+\alpha}} (u_{bc} - u_{ca})dt$ is rewritten as

$$\int_{t_{\pi/6+\alpha}+t_{fw}}^{t_{\pi/2+\alpha}} (u_{bc} - u_{ca})dt = \int_{t_{\pi/6+\alpha}+t_{fw}}^{t_{\pi/2+\alpha}} [R(i_a + i_b - 2i_c) + Ld(i_a + i_b - 2i_c)/dt + e_a + e_b - 2e_c]dt$$
$$= \int_{t_{\pi/6+\alpha}+t_{fw}}^{t_{\pi/2+\alpha}} (e_a + e_b - 2e_c)dt \tag{15}$$

By decomposing (15) into a combination of two items, it is reformulated as

$$\int_{t_{\pi/6+\alpha}+t_{fw}}^{t_{\pi/2+\alpha}} (u_{bc} - u_{ca})dt = \int_{t_{\pi/6+\alpha}+t_{fw}}^{t_{\pi/2+\alpha}} (e_a + e_b - 2e_c)dt = \int_{t_{\pi/6+\alpha}}^{t_{\pi/2+\alpha}} (e_a + e_b - 2e_c)dt - \int_{t_{\pi/6+\alpha}}^{t_{\pi/6+\alpha}+t_{fw}} (e_a + e_b - 2e_c)dt \tag{16}$$

Substituting (14) and (16) into (12), and (12) is rewritten as

$$d^* = \int_{t_{\pi/6+\alpha}}^{t_{\pi/2+\alpha}} (u_{bc} - u_{ca})dt = \int_{t_{\pi/6+\alpha}}^{t_{\pi/2+\alpha}} (e_a + e_b - 2e_c)dt + \int_{t_{\pi/6+\alpha}}^{t_{\pi/6+\alpha}+t_{fw}} (u_{dc} - e_a - e_b + 2e_c)dt \tag{17}$$

The phase current satisfies $i_a + i_b + i_c = 0$ in the freewheeling period t_{fw}. Substituting the phase current relationship, (13) into the second item of (17), it is simplified as

$$\int_{t_{\pi/6+a}}^{t_{\pi/6+a}+t_{fw}}(u_{dc}-e_a-e_b+2e_c)dt = -3\int_{t_{\pi/6+a}}^{t_{\pi/6+a}+t_{fw}}[Ri_c+Ldi_c/dt]dt = -3R\int_{t_{\pi/6+a}}^{t_{\pi/6+a}+t_{fw}}i_c dt + 3LI_c \qquad (18)$$

where I_c is the final value of the outgoing phase current before commutation.

During the commutation period t_{fw}, the outgoing phase current drops from the final value I_c to zero, the integral item $\int_{t_{\pi/6+a}}^{t_{\pi/6+a}+t_{fw}} i_c dt$ can be approximated by $-I_c t_{fw}/2$, and (18) is rewritten as

$$\int_{t_{\pi/6+a}}^{t_{\pi/6+a}+t_{fw}}(u_{dc}-e_a-e_b+2e_c)dt = 3RI_c t_{fw}/2 + 3LI_c \qquad (19)$$

From (13), the phase current i_c in the freewheeling period t_{fw} is solved as

$$i_c(t) = \frac{-\left[\beta - (RI_c + \beta)e^{-Rt/L}\right]}{R} \qquad (20)$$

where $\beta = \frac{u_{dc}+2e_c-e_a-e_b}{3}$.

From (20), the freewheeling period t_{fw} for the outgoing phase current i_c dropping from final value I_c to zero is solved as

$$t_{fw} = -\frac{L}{R}\ln\frac{\beta}{\beta + RI_c} \qquad (21)$$

Substituting (21) into (19), and (19) is shown as

$$\int_{t_{\pi/6+a}}^{t_{\pi/6+a}+t_{fw}}(u_{dc}-e_a-e_b+2e_c)dt = 3LI_c\left(1 - \sqrt{\ln\frac{\beta}{\beta + RI_c}}\right) \qquad (22)$$

In (22), $\int_{t_{\pi/6+a}}^{t_{\pi/6+a}+t_{fw}}(u_{dc}-e_a-e_b+2e_c)dt$ is related to β. Next, the effect of β will be discussed. In the freewheeling period when $t \in [t_{\pi/6+a}, t_{\pi/6+a}+t_{fw}]$, $(2e_c - e_a - e_b)$ is expressed as

$$
\begin{aligned}
2e_c - e_a - e_b = \quad & 2\sum_{n=0}^{\infty}E_{2n+1}\sin((2n+1)(\theta+2\pi/3+\alpha)) \\
& -\sum_{n=0}^{\infty}E_{2n+1}\sin((2n+1)(\theta-2\pi/3+\alpha)) - \sum_{n=0}^{\infty}E_{2n+1}\sin((2n+1)\theta+\alpha)
\end{aligned} \qquad (23)
$$

Considering $E_1 \gg E_7$, the high-order terms of the back-EMF in (23) are negligible, and (23) can be simplified as

$$
\begin{aligned}
2e_c - e_a - e_b &= 2(e_{c1}+e_3+e_{c5}) - (e_{a1}+e_3+e_{a5}) - (e_{b1}+e_3+e_{b5}) \\
&= 2e_{c1}+2e_{c5}-e_{a1}-e_{a5}-e_{b1}-e_{b5} \\
&= 2E_1\sin(5\pi/6+\alpha) + 2E_5\sin(\pi/6+5\alpha) - E_1\sin(\pi/6+\alpha) \\
&\quad -E_5\sin(5\pi/6+5\alpha) - E_1\sin(\alpha-\pi/2) - E_5\sin(5\alpha-\pi/2)
\end{aligned} \qquad (24)
$$

By calculating the trigonometric functions, (24) is simplified as

$$2e_c - e_a - e_b = \frac{3}{2}E_1\cos\alpha - \frac{3\sqrt{3}}{2}E_1\sin\alpha - \frac{3}{2}E_5\cos(5\alpha) - \frac{3\sqrt{3}}{2}E_5\sin(5\alpha) \qquad (25)$$

Both sides of (25) are multiplied by $1/E_5$, and refined as

$$f(\alpha) = \frac{2e_c - e_a - e_b}{E_5} = \frac{3}{2}\cdot\frac{E_1}{E_5}\cdot\cos\alpha - \frac{3\sqrt{3}}{2}\cdot\frac{E_1}{E_5}\cdot\sin\alpha - \frac{3}{2}\cos(5\alpha) - \frac{3\sqrt{3}}{2}\sin(5\alpha) \qquad (26)$$

In general, the EMF harmonic coefficients of a BLDC motor with a trapezoidal back EMF [25] $E_1/E_5 \in [5, 125]$. The relationship between $f(\alpha)$, E_1/E_5, and the commutation error α during the freewheeling period t_{fw} can be obtained in (26). When $E_5 > 0$, $E_1/E_5 \in [5, 125]$, and the corresponding 3-D graph is shown in Figure 7a. It is observed that with respect to $E_1/E_5 \in [5, 125]$, $f(\alpha)$ is greater than zero when the commutation error $\alpha \in [-\pi/6, \pi/6]$, so $2e_c - e_a - e_b = f(\alpha) \cdot E_5 > 0$. When $E_5 < 0$, $E_1/E_5 \in [-125, -5]$, and the corresponding 3-D graph is shown in Figure 7b. $f(\alpha)$ is less than zero when the commutation error $\alpha \in [-\pi/6, \pi/6]$, and $2e_c - e_a - e_b = f(\alpha) \cdot E_5 > 0$. Based on the relationship, it is derived that $\beta = \frac{u_{dc} + 2e_c - e_a - e_b}{3} > \frac{u_{dc}}{3}$ and $\frac{RI_c}{\beta} < \frac{3RI_c}{u_{dc}}$. Both sides of $\frac{3RI_c}{u_{dc}}$ are multiplied by I_c, and the item $\frac{3RI_c}{u_{dc}}$ is represented as $\frac{3RI_c^2}{u_{dc}I_c} = \frac{1.5P_{cu}}{P_{in}}$, where P_{cu} denotes copper loss and $P_{cu} = 2I_c^2 R$. P_{in} denotes the input power of the motor and $P_{in} = u_{dc}I_c$. In general, the copper loss P_{cu} is negligible, hence it satisfies $\frac{3RI_c}{u_{dc}} = \frac{3RI_c^2}{u_{dc}I_c} = \frac{1.5P_{cu}}{P_{in}} \ll 1$, which implies $\ln\sqrt{\frac{\beta}{\beta + RI_c}} = \ln\sqrt{\frac{1}{1 + RI_c/\beta}} \approx 0$ Therefore, (19) is simplified as

$$\int_{t_{\pi/6+\alpha}}^{t_{\pi/6+\alpha}+t_{fw}} (u_{dc} - e_a - e_b + 2e_c)dt \approx 3LI_c \tag{27}$$

Combining (27) and (17), the latter is reformulated as

$$d^* = \int_{t_{\pi/6+\alpha}}^{t_{\pi/2+\alpha}} (u_{bc} - u_{ca})dt = \int_{t_{\pi/6+\alpha}}^{t_{\pi/2+\alpha}} (e_a + e_b - 2e_c)dt + 3LI_c \tag{28}$$

According to the previous analysis, the general function of the line voltage difference integral d^* is given as

$$d^* = \int_{t_s}^{t_e} (u_{xz} - u_{zy})dt = \int_{t_s}^{t_e} (e_x + e_y - 2e_z)dt + 3LI_z \tag{29}$$

where the subscripts x and y represent the active phase, and z represents the inactive phase. t_s and t_e are the lower and upper limits of integral, respectively.

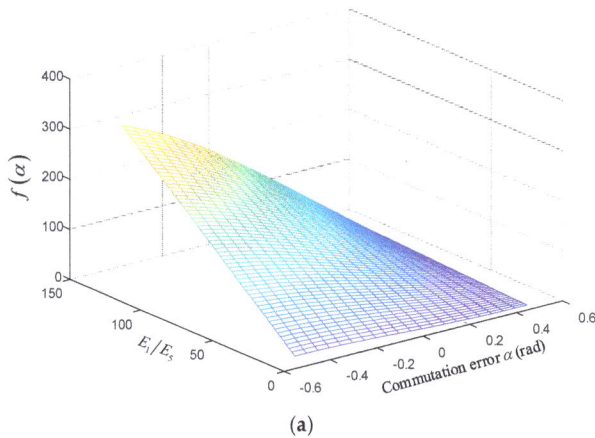

(a)

Figure 7. *Cont.*

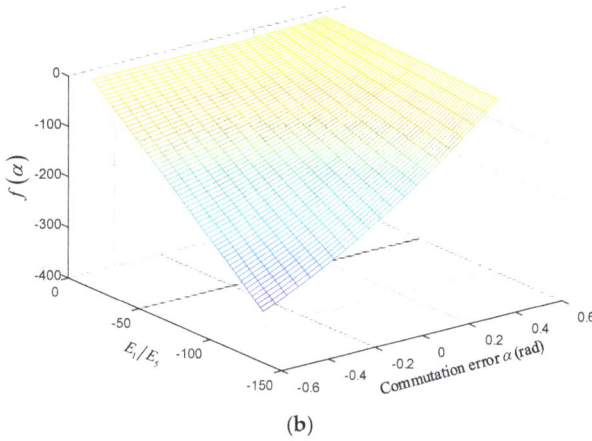

(b)

Figure 7. Relationships between $f(\alpha)$, E_1/E_5, and the commutation error α during the freewheeling period t_{fw}. (**a**) $E_5 > 0$; (**b**) $E_5 < 0$.

2.4. Commutation Error Compensation Strategy

In Sections 2.2 and 2.3, the respective line voltage difference integrals d and d^* without and with consideration of the freewheeling period t_{fw} are discussed. Next, the commutation error compensation strategy is described based on the analysis of the line voltage difference integral.

From (29), the three-phase back-EMF integral is rewritten as

$$\int_{t_s}^{t_e} \left(e_x + e_y - 2e_z\right)dt = d^* - 3LI_z \tag{30}$$

According to the research in Section 2.2, three-phase back-EMF integral can reflect commutation errors. In (30), three-phase back-EMF integral can be obtained from the combination of the integral d^* and $-3LI_z$ to determine commutation errors when the freewheeling period t_{fw} is considered.

In general, the commutation instants are obtained by delaying ZCPs by 30 electrical degrees in the sensorless technique and reactivating the motor windings according to the predefined commutation sequence. The predefined commutation sequences are shown in Table 3; here, H_{va}, H_{vb}, and H_{vc} are virtual hall signals.

Table 3. Predefined commutation sequence.

H_{va}, H_{vb}, H_{vc}	001	101	100	110	010	011
Direction			Clockwise			
Conduction switch	VT_3-VT_2	VT_3-VT_4	VT_5-VT_4	VT_5-VT_6	VT_1-VT_6	VT_1-VT_2
Direction			Anti-clockwise			
Conduction switch	VT_5 VT_6	VT_1-VT_6	VT_1-VT_2	VT_3-VT_2	VT_3-VT_1	VT_5-VT_4

In Table 1, the sign of the three-phase back-EMF integral changes according to the direction of the phase current. The three-phase back-EMF integral d_c can be updated as

$$d_c = (-1)^{(H_{va}+H_{vb}+H_{vc}+\rho)} \cdot \int_{t_s}^{t_e} \left(e_x + e_y - 2e_z\right)dt \tag{31}$$

where ρ denotes the rotation direction of the motor rotor. When $\rho = 0$, the motor runs in the anti-clockwise rotation, and when $\rho = 1$, the motor rotates in the opposite one. When the sensorless

BLDC motor runs, the windings are excited in according with the predefined commutation sequence listed in Table 3, and the conduction switches are obtained. Then the line voltage difference integral d^* is calculated with Table 1, and the outgoing phase current I_z before commutation is sampled. By using (30), the three-phase back-EMF integral $\int_{t_s}^{t_e}(e_x + e_y - 2e_z)dt$ can be given. Considering the rotation direction ρ of the motor rotor is known, the three-phase back-EMF integral d_c is converted to a positive value under delayed commutation and to a negative one under advanced commutation with (31), which also makes PI controller available.

Here, a PI regulator is also designed to compensate commutation errors, and the PI regulator is defined as

$$u(k) = k_p(e(k) - e(k-1)) + k_iT_ie(k) + u(k-1) \tag{32}$$

where three-phase back-EMF integral error $e(k) = d_{ref} - d_c(k)$, and the reference value $d_{ref} = 0$. k_p is the proportional coefficient, and k_i is the integral coefficient. T_i is the integral period and $T_i = 0.0001$ s. $u(k)$ is the kth output value of the PI regulator, $u(k-1)$ is the $(k-1)$th output value of the PI regulator. Here, k represents the number of iterations.

For the sensorless BLDC motor based on ZCP detection of the back-EMF, the accurate commutation instants are obtained by 30 electrical degrees delayed shift of ZCPs. Therefore, after the commutation error is determined, the commutation shift is updated as

$$\theta(k) = \theta(k-1) - u(k) \tag{33}$$

where $\theta(0) = \pi/6$.

Figure 8 shows the block diagram of a sensorless BLDC motor drive system and the proposed commutation error compensation strategy is shown in the shadowed area. The system contains a speed-loop controller, a current-loop controller, and a commutation error compensation link. In the commutation error compensation link, when commutation signals are detected, the outgoing phase current final value I_z is sampled and the line voltage difference is integrated until the commutation occurs again. Then, the line voltage difference integral d^* is recorded, and the three-phase back-EMF integral d_c is calculated from the sampled current value I_z and the line voltage difference integral d^*. The entire compensation process is then resumed. A flowchart of the proposed method is shown in Figure 9.

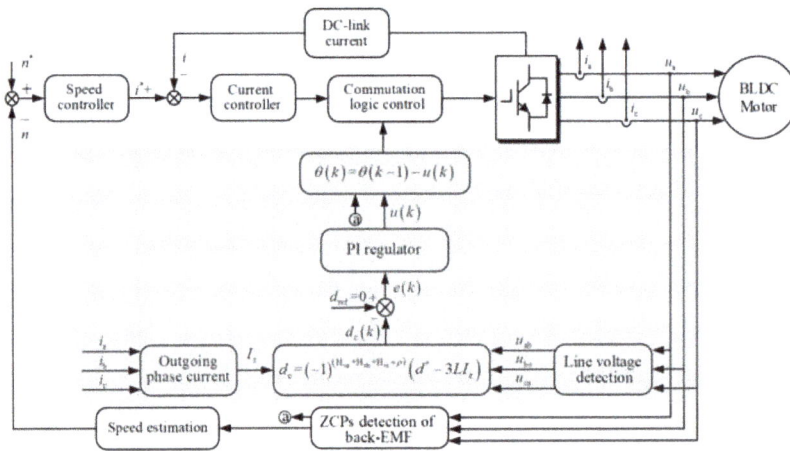

Figure 8. Sensorless BLDC motor drive system with the proposed strategy.

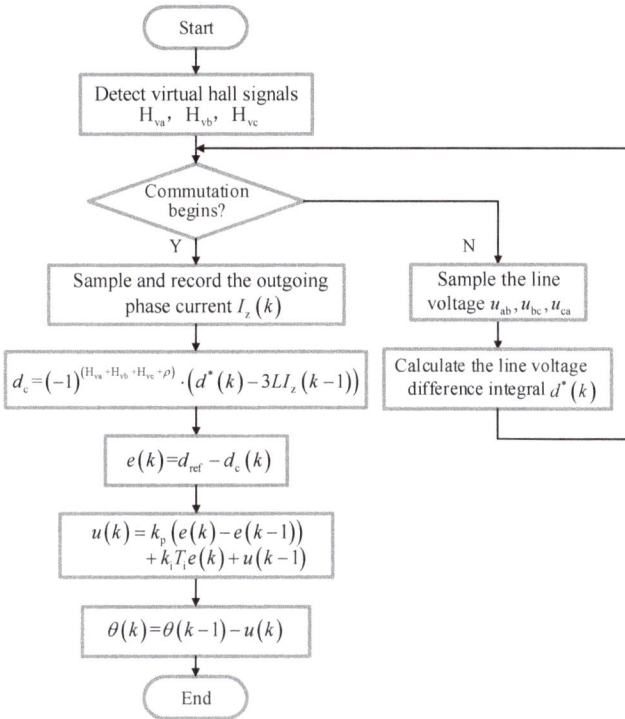

Figure 9. Flowchart of the commutation error compensation procedure.

Figure 10 shows the back-EMF waveforms by the constant speed test of 1200 rpm.

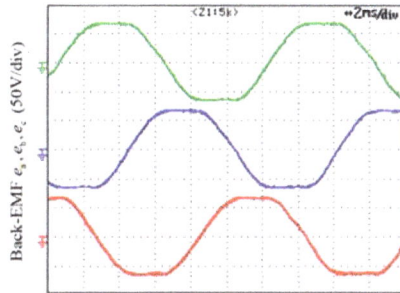

Figure 10. Measured back-EMF waveforms of the BLDC motor at 1200 rpm.

3. Experiment Results

An experimental prototype is set up to verify the feasibility and effectiveness of the proposed strategy. The experimental platform, which consists of the control system and the experimental motor, is shown in Figure 11. The BLDC motor is connected to the generator by a flexible coupling, and its rated parameters are shown in Table 4. The core processor is DSP (TMS320F28335) with a 150 MHz clock frequency, and the three-phase inverter is IPM PM50RL1A060 (Mitsubishi). The line voltage and phase current are sampled by three voltage sensors (LV25-P) and three current sensors (LTS15-NP), respectively. To remove the effect of switching noise, a low-pass FIR filter with the Hamming window function was designed. The cutoff frequency and the order of the filter were selected as 3 kHz and

50, respectively. The filter provides 65 dB of signal attenuation at the switching frequency. Many experiments were conducted to obtain the proportional coefficient k_p and the integral coefficient k_i of the PI regulator. Based on the test results, the proportional coefficient and integral coefficient are designed as $k_p = 8.16$ and $k_i = 0.023$, respectively. The modulation method of the three-phase inverter is PWM-ON, and the switching frequency of three-phase inverter and the sampling frequency of the line voltage are 10 and 200 kHz, respectively. The experimental results were monitored and recorded by a digital oscilloscope DL750 (YOKOGAWA), and the line voltage difference integral was observed through a DAC.

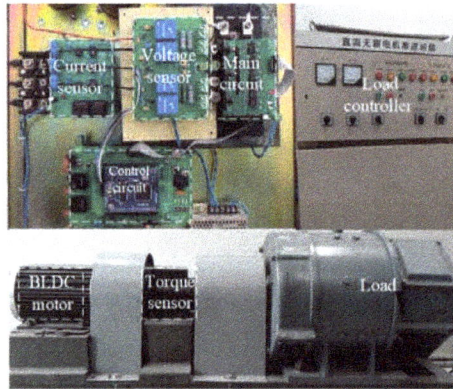

Figure 11. Experimental setup of the BLDC motor drive system.

Table 4. Motor parameters.

Motor Parameter	Value	Unit
Rated voltage	200	V
Rated power	3.15	kW
Rated speed	1500	rpm
Rated torque	20	Nm
Pole pairs	4	
Phase inductance	1.234	mH
Phase resistance	0.0654	Ω
Back-EMF coefficient	0.528	V/(rad/s)

Figure 12 shows the phase current i_c and the line voltage difference integral d^* under advanced, delayed, and accurate commutation. In the experiment, the motor runs at a speed of 500 rpm, the load torque is 7.5 Nm, and the commutation instants are set to be advanced and delayed by $10°$. The conduction periods of A+B− and A−B+ are taken as examples, and the outgoing phase current final value I_c before commutation and the line voltage difference integral $\int_{t_s}^{t_e} (u_{ca} - u_{ab}) dt$ are marked in solid circles. In Figure 12a,b, there is an obvious current ripple in phase current i_c and the line voltage difference integral d^* is different. Figure 12c shows that the current ripple is attenuated obviously and the integral d^* also changes correspondingly when the commutation instants are accurate. In addition, as shown in the dashed circle, the line voltage difference integral d^* takes on an approximately linear variation trend during the freewheeling period t_{fw}. The experimental results are consistent with the theoretical analysis.

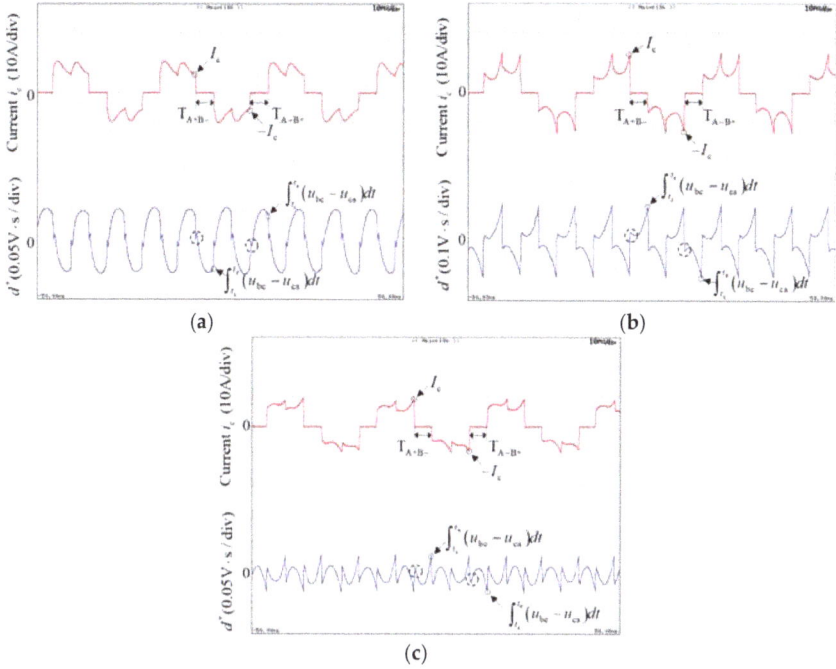

Figure 12. Phase current and the line voltage difference integral. (**a**) Advanced commutation; (**b**) Delayed commutation; (**c**) Exact commutation.

The performance of the proposed strategy is tested in the steady and transient states. In the steady-state experiment, the motor runs at a fixed speed and load. In the transient experiment, the motor works at varying speeds and a fixed load. Figure 13 shows the phase currents and integral d^* before and after compensation under different test conditions. It is observed that the phase current i_c and the integral d^* fluctuates obviously under delayed and advanced commutation, yet the current ripple is effectively reduced after the proposed compensation strategy is used, and the integral d^* also changes correspondingly when the commutation instants are accurate. The experimental results show that the proposed strategy can accurately compensate commutation errors at different speeds and load torques.

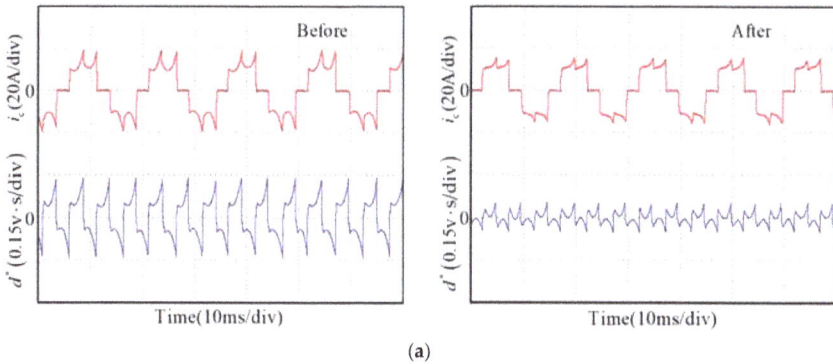

Figure 13. *Cont.*

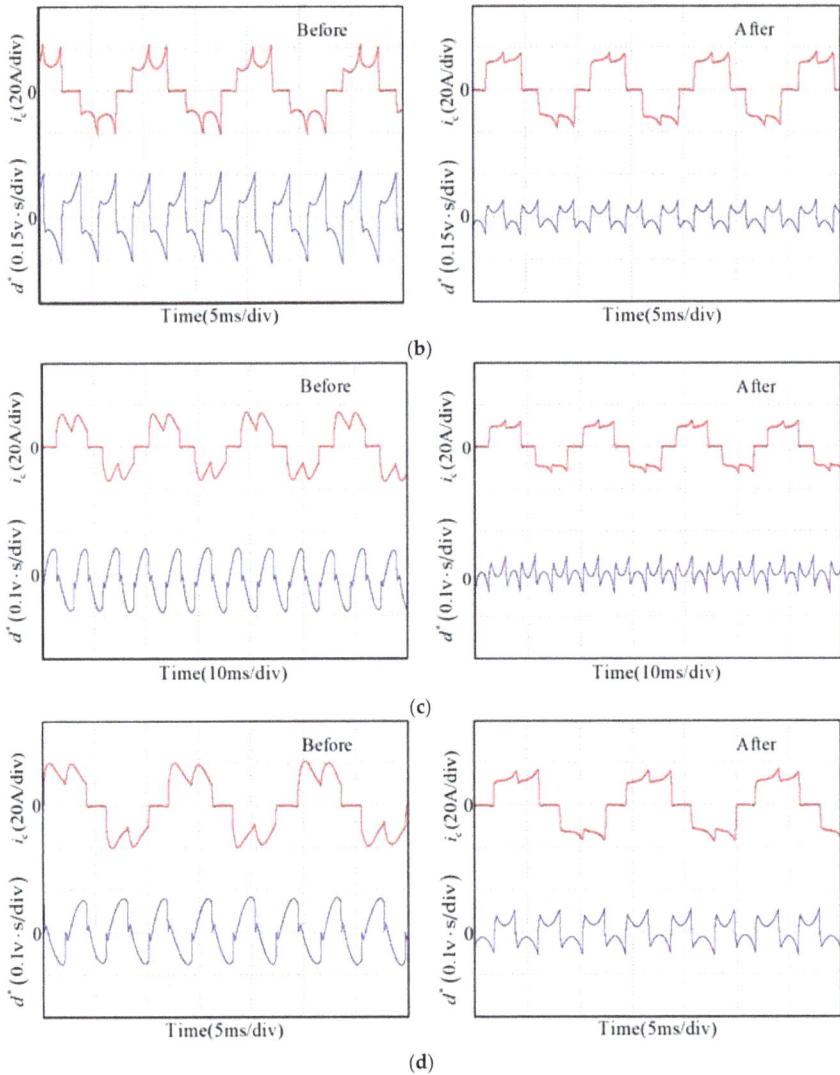

Figure 13. Current waveform of phase C before and after compensation. (**a**) Commutation instants are delayed by 10° at a speed of 1000 rpm and load torque of 12 Nm; (**b**) Commutation instants are delayed by 12° at a speed of 1500 rpm and load torque of 16 Nm; (**c**) Commutation instants are advanced by 12° at a speed of 850 rpm and load torque of 10 Nm; (**d**) Commutation instants are advanced by 14° at a speed of 1200 rpm and load torque of 14 Nm.

Subsequently, the performance of the proposed strategy during the transient process is tested. In the experiment, the commutation instants are delayed by about 10°, and the motor is accelerated from 300 rpm to 1500 rpm at a load torque of 12 Nm. Figure 14 shows the phase current, the line voltage difference integral, and their enlarged vision with the proposed strategy. When the motor speed ramps up, the compensation strategy is initiated at the points S_1 and S_2, and the commutation error is almost removed at the points T_1 and T_2, respectively. It is seen that, after the compensation strategy is activated, the current fluctuation is effectively reduced. Therefore, the proposed compensation strategy exhibits satisfactory dynamic performance during variable-speed operation.

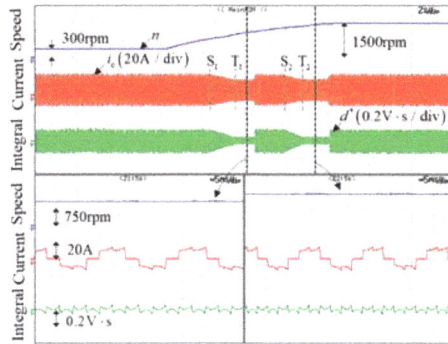

Figure 14. Transient performance and enlarged view with the proposed strategy when the motor accelerates from 300 to 1500 rpm.

To test the convergence rate of the proposed strategy under the steady speed operation, the commutation error is set to about $10°$, and the motor runs at a speed of 800 rpm and load torque of 12 Nm. Figure 15 shows the phase current, the line voltage difference integral d^*, and their enlarged vision with PI controller. It is found that, before the point of S_a, the phase current i_c curls up and the integral d^* is high, while the compensation strategy with PI controller is activated at the point S_a, the current ripple and the integral d^* gradually decreases. After 1.05 s, the current fluctuation is effectively suppressed, and the commutation error is almost eliminated. In addition, the convergence rate of the proposed method is tested at different speeds, and the test results are shown in Table 5, and the results show the proposed strategy converges quickly in a wide speed range.

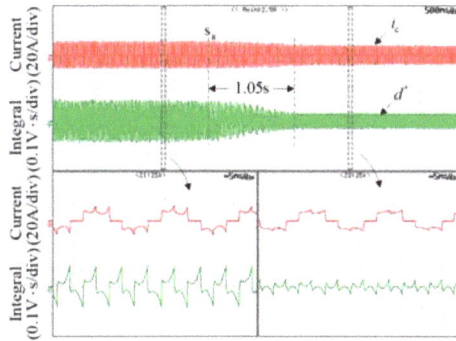

Figure 15. Comparative results of the phase current, the integral of the line voltage difference, and their enlarged view at a speed of 800 rpm and load torque of 12 Nm.

Table 5. Convergence rate of proposed strategy.

Speed (rpm)	300	500	800	1200	1500
Convergence rate (s)	2.52	1.59	1.05	0.713	0.565

Moreover, the power consumption is reduced after the commutation instants are compensated with the proposed strategy. In these power consumption tests, the commutation instants are set to be delayed and advanced by $10°$ and $12°$, respectively, the speed range is 300–1500 rpm and the load torques T_L are about 11, 12, 13 and 15 Nm. Figure 16 shows the power consumption of the BLDC motor before and after compensation after steady operation for 40 min. The experimental results in Figure 16 show that the proposed strategy can reduce power consumption over wide speed ranges, especially in the high-speed ranges.

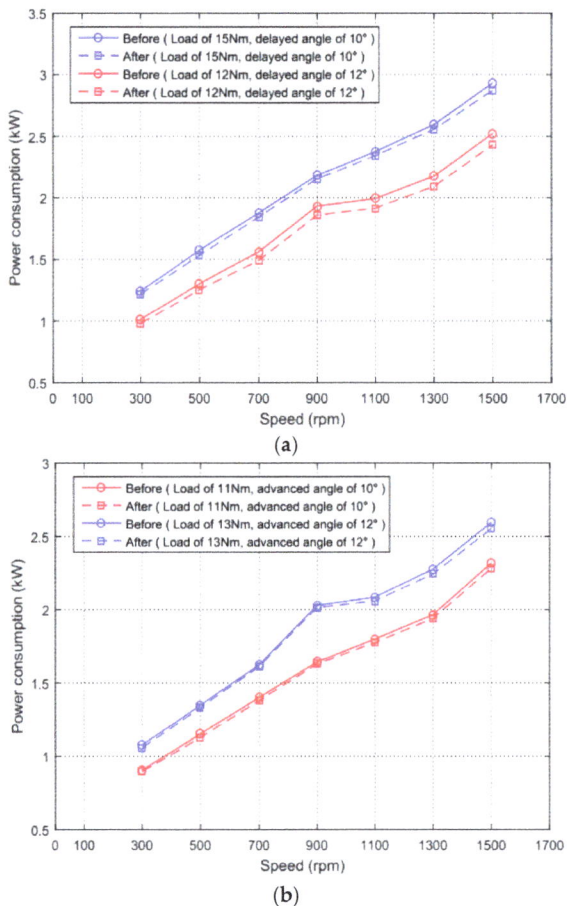

Figure 16. Power consumption comparison before and after compensation. (**a**) Under delayed commutation; (**b**) Under advanced commutation.

4. Conclusions

This paper presents a novel compensation strategy for commutation error of sensorless BLDC motors. The method is suitable in both the steady-speed and variable-speed range. The relationship between the line voltage difference integral in 60 electrical degrees interval and the commutation error is established. A commutation error commutation strategy with respect of the line voltage difference integral is proposed to correct the inaccurate commutation instants by regulating the three-phase back-EMF integral to zero. In this method, the influence of the voltage pulses caused by the diode-freewheeling conduction on the line voltage difference integral is analyzed, making the proposed strategy applicable to BLDC motors considering the freewheeling period. Furthermore, a PI controller is also designed to realize closed-loop compensation for commutation errors. Finally, experimental results show the effectiveness and feasibility of the proposed strategy under different test conditions.

Author Contributions: This paper is the result of the hard work of all of the authors. X.Y. and J.Z. conceived and designed the proposed method. X.Y. and J.Z. conceived and analyzed the data. X.Y. provided funding support. J.Z. and G.L. conducted the experiment and verified the proposed method. J.Z., G.L., and H.L. wrote the paper, and H.L. and J.W. contributed to the review of the document. All authors gave advice for the manuscript.

Funding: This work was supported by the High-tech Ship Project of China under Grant No. GJYF-043/6.

Conflicts of Interest: The authors declare no conflict of interest.

References

1. Capolino, G.A.; Cavagnino, A. New trends in electrical machines technology—Part I. *IEEE Trans. Ind. Electron.* **2014**, *61*, 4281–4285. [CrossRef]
2. Acarnley, P.P.; Watson, J.F. Review of position-sensorless operation of brushless permanent-magnet machines. *IEEE Trans. Ind. Electron.* **2006**, *53*, 352–362. [CrossRef]
3. Samoylenko, N.; Han, Q.; Jatskevich, J. Dynamic performance of brushless DC motors with unbalanced hall sensors. *IEEE Trans. Energy Convers.* **2008**, *23*, 752–763. [CrossRef]
4. Hwang, C.; Li, P.; Chuang, F.; Liu, C.; Huang, K. Optimization for reduction of torque ripple in an axial flux permanent magnet machine. *IEEE Trans. Magn.* **2009**, *45*, 1760–1763. [CrossRef]
5. Yang, Y.; Ting, Y. Improved angular displacement estimation based on Hall-effect sensors for driving a brushless permanent-magnet motor. *IEEE Trans. Ind. Electron.* **2014**, *61*, 504–511. [CrossRef]
6. Cheng, C.; Cheng, M. Implementation of a highly reliable hybrid electric scooter drive. *IEEE Trans. Ind. Electron.* **2007**, *54*, 2462–2473. [CrossRef]
7. Huang, X.; Goodman, A.; Gerada, C.; Fang, Y.; Lu, Q. A single sided matrix converter drive for a brushless DC motor in aerospace applications. *IEEE Trans. Ind. Electron.* **2012**, *59*, 3542–3552. [CrossRef]
8. Chun, T.; Tran, Q.; Lee, H.; Kim, H. Sensorless control of BLDC motor drive for an automotive fuel pump using a hysteresis comparator. *IEEE Trans. Power Electron.* **2014**, *29*, 1382–1391. [CrossRef]
9. Bhogineni, S.; Rajagopal, K.R. Position error in sensorless control of brushless dc motor based on average line to line voltages. In Proceedings of the IEEE International Conference on Power Electronics, Drives and Energy Systems (PEDES), Karnataka, India, 16–19 December 2012; pp. 1–6.
10. Shen, J.; Tseng, K. Analyses and compensation of rotor position detection error in sensorless PM brushless DC motor drives. *IEEE Trans. Energy Convers.* **2003**, *18*, 87–93. [CrossRef]
11. Cheng, K.; Tzou, Y. Design of a sensorless commutation IC for BLDC motors. *IEEE Trans. Power Electron.* **2003**, *18*, 1365–1375. [CrossRef]
12. Park, J.; Hwang, S.; Kim, J. Sensorless control of brushless DC motors with torque constant estimation for home appliances. *IEEE Trans. Ind. Appl.* **2012**, *48*, 677–683. [CrossRef]
13. Wang, S.; Lee, A. A 12-step sensorless drive for brushless DC motors based on back-EMF differences. *IEEE Trans. Energy Convers.* **2015**, *30*, 646–654. [CrossRef]
14. Bernardes, T.; Montagner, V.F.; Grundling, H.A.; Pinheiro, H. Discrete-time sliding mode observer for sensorless vector control of permanent magnet synchronous machine. *IEEE Trans. Ind. Electron.* **2014**, *61*, 1679–1691. [CrossRef]
15. Lee, H.; Lee, J. Design of iterative sliding mode observer for sensorless PMSM control. *IEEE Trans. Control. Syst. Technol.* **2013**, *21*, 1394–1399. [CrossRef]
16. Alonge, F.; D'Ippolito, F.; Sferlazza, A. Sensorless control of induction-motor drive based on robust Kalman filter and adaptive speed estimation. *IEEE Trans. Ind. Electron.* **2014**, *61*, 1444–1453. [CrossRef]
17. Quang, N.K.; Hieu, N.T.; Ha, Q.P. FPGA-based sensorless PMSM speed control using reduced-order extended kalman filters. *IEEE Trans. Ind. Electron.* **2014**, *61*, 4574–4582. [CrossRef]
18. Idkhajine, L.; Monmasson, E.; Maalouf, A. Fully FPGA-based sensorless control for synchronous AC drive using an extended Kalman filter. *IEEE Trans. Ind. Electron.* **2012**, *59*, 3908–3918. [CrossRef]
19. Chen, H.C.; Liaw, C.M. Sensorless control via intelligent commutation tuning for brushless dc motor. *Inst. Electron. Eng. Electron. Power Appl.* **2002**, *46*, 678–684. [CrossRef]

20. Fang, J.; Li, W.; Li, H. Self-compensation of the commutation angle based on dc-link current for high-speed brushless DC motors with low inductance. *IEEE Trans. Power Electron.* **2014**, *29*, 428–439. [CrossRef]

21. Wu, X.; Zhou, B.; Song, F.; Chen, F.; Wei, J. A closed loop control method to correct position phase for sensorless Brushless DC motor. In Proceedings of the International Conference on Electrical Machines and Systems, Wuhan, China, 17–20 October 2008; pp. 1460–1464.

22. Krishnan, G.; Ajmal, K.T. A neoteric method based on PWM ON PWM scheme with buck converter for torque ripple minimization in BLDC drive. In Proceedings of the Annual International Conference on Emerging Research Areas: Magnetics, Machines and Drives (AICERA/iCMMD), Kottayam, India, 24–26 July 2014; pp. 1–6.

23. Lee, A.; Wang, S.; Fan, C. A current index approach to compensate commutation phase error for sensorless brushless dc motor with non-ideal back EMF. *IEEE Trans. Power Electron.* **2016**, *31*, 4389–4399. [CrossRef]

24. Jang, G.; Kim, M. Optimal commutation of a BLDC motor by utilizing the symmetric terminal voltage. *IEEE Trans. Magn.* **2006**, *42*, 3473–3475. [CrossRef]

25. Zhou, X.; Chen, X.; Lu, M.; Zeng, F. Rapid self-compensation method of commutation phase error for low inductance BLDC motor. *IEEE Trans. Ind. Inform.* **2017**, *13*, 1833–1842. [CrossRef]

energies

MDPI

Article

On Speed Control of a Permanent Magnet Synchronous Motor with Current Predictive Compensation

Meiling Tang [1,2,*] and Shengxian Zhuang [1]

[1] School of Electrical Engineering, Southwest Jiaotong University, Chendu 610031, Sichuan, China; sxzhuang@home.swjtu.edu.cn
[2] Department of Manufacturing and Transportation, Hangzhou Wanxiang Polytechnic College, Hangzhou 310023, Zhejiang, China
* Correspondence: Ips2032@sina.com; Tel.: +86-15868871989

Received: 4 November 2018; Accepted: 10 December 2018; Published: 26 December 2018

Abstract: In this study, a current model predictive controller (MPC) is designed for a permanent magnet synchronous motor (PMSM) where the speed of the motor can be regulated precisely. First, the mathematical model, the specifications, and the drive topology of the PMSM are introduced, followed by an elaboration of the design of the MPC. The MPC is then used to predict the current in a discrete-time calculation. The phase current at the next sampling step can be estimated to compensate the current errors, thereby modifying the three-phase currents of the motor. Next, Simulink modeling of the MPC algorithm is given, with three-phase current waveforms compared when the motor is operated under the designed MPC and a traditional vector control for PMSM. Finally, the speed responses are measured when the motor is controlled by traditional control methods and the MPC approach under varied speed references and loads. In comparison with traditional controllers, both the simulation and the experimental results suggest that the MPC for the PMSM can improve the speed-tracking performance of the motor and that this motor has a fast speed response and small steady-state errors under the rated load.

Keywords: MPC; PMSM; vector control; speed tracking

1. Introduction

Permanent magnet synchronous motor (PMSM) technology is widely explored and employed in industrial equipment, aerospace aircraft, domestic appliances, and electric vehicles. The motor takes advantage of high working efficiency, high power density, highly accurate position tracking, and low power factor, as compared to induction motors [1]. One popular application of PMSM is the surface-mounted permanent magnet (PM) motor, with embedded PM motors already widely used in industry. The direct-axis inductance L_d and quadrature-axis inductance L_q of the surface-mounted PM motor are equal because this motor has identical air gaps around the stator [2]. The identical inductance values enhance the performance of the power converter for the motor as a load. Consequently, control of the motor becomes easier, since the motor is at a constant load. As of now, field-oriented control (FOC) is widely employed for PMSM [3]. After sampling the phase currents of the motor, the controller of the motor converts three-phase current signals to two orthogonal values via Clark and Parker transformations. The alternating-current motor can be controlled like a direct-current motor because the amplitude of phase voltage or current can be directly regulated to control the motor, thus improving its performance and simplifying the control [4]. Generally, the current loop regulation for PMSM includes hysteresis control and pulse width modulation (PWM) [5]. Vector control further simplifies the control by using seven voltage vectors to directly drive the motor without current modulation, improving its dynamic response [6]. Space vector pulse width modulation (SVPWM) is becoming increasingly popular for the motor. Moreover, the mechanical performance of the motor can be enhanced by SVPWM in combination with flux-weakening modulation when the motor is operated in high-speed applications. Also, some papers have investigated direct torque control (DTC) to improve the performance of the motor. DTC can directly govern the torque of the motor by observing flux linkage and speed. As reported in the literature [7], DTC is employed to reduce the torque ripples of the motor by an optimized duty cycle. This control method has also been investigated to address the fault tolerance, maximum torque outputs, and evaluation of power factors for PMSM [8–10]. In one study in particular, a predictive DTC has been developed to propose a fault-tolerance function for PMSM [10]. The performance and dynamic responses of the motor can be further modified.

Recently, model predictive control (MPC) as an advanced control method has become more attractive for the control of PMSM and has been strongly developed in the last three decades. For power electronics development, MPC can merge cascaded control loops into one loop for power converters in the initial exploration algorithm, especially for multivariable systems, to tackle parameter varieties [11]. An MPC consists mainly of three parts: a cost function, a predictive model, and a model of the load. By minimizing the cost function, the motor will reach the desired behavior defined by the function that compares the output of the predictive model with a reference [5]. Weighting factors play an important role for fast responses and the stability of the control system [12]. Two main branches of MPC are finite control set MPC (FCS-MPC) and continuous control set MPC (CCS-MPC) [13,14]; CCS-MPC calculates a continuous control solution and outputs a desired voltage in the power converter through a modulator with a fixed switching frequency. In addition, modified generalized predictive control (GPC) and explicit MPC (EMPC) are popular and increase some scholars' interest [15,16]; FCS-MPC discretizes the signal to formulate the MPC algorithm without external modulation and can be sorted in two types, namely, optimal switching vector MPC (OSV-MPC) and optimal switching sequence MPC (OSS-MPC) [17,18]. OSV-MPC computes predictive values for the power converter only via an enumerated searching algorithm, making this MPC very intuitive. Only one voltage vector is employed in the entire switching period. This drawback of this MPC is avoided by OSS-MPC, which produces a limited number of possible switching sequences in a working period. In general, all MPC methods heretofore mentioned are time-consuming for microprocessors. CCS-MPC outweighs FCS-MPC in the aspect of computational cost, as the former predicts the output value and optimizes it offline. Under this condition, CCS-MPC can be used on a long predictive horizon, which can stabilize the whole system because more steps need to be predicted by the controller [19]. By contrast, FCS-MPC methods are usually adopted in short predictive horizons. Therefore, the key factors of MPC are the selection of a cost function, the design of the weighting factor, the reduction of the computational cost, and the predictive horizon extension. In recent studies, some practical MPCs of PMSM have been developed to simplify the control, to save computing time, to eliminate harmonic currents, and to reduce torque ripples [20–23]. Designing a practical, stable MPC quickly and applying the controller on a PMSM to replace traditional methods is meaningful for industrial applications.

In this study, an MPC is introduced and designed for a PMSM, with phase currents regulated by a simple predictive control, and a computation period reduced for microchips. Speed control of the motor by using the MPC is carried out under varied loads. The performance of MPC is compared with those of FOC, SVPWM, and DTC. In Section 2 of this paper, the mathematical model of the PMSM is elaborated. In Section 3, a Simulink model is built. Experiments are carried out in Section 4. Conclusions are presented in Section 5. First, the model of a surface-mounted PMSM is introduced as well as the basic control method mentioned, particularly the influence of the calculation step time for three-phase bridge topology. Then, the principles of MPC are given, followed by a predictive model of the inverter and the model of the PMSM. Finally, an MPC for the PMSM is designed, built, and programmed, with simulations and experimental results showing the improvement of the performance of the motor. Phase currents of the motor can be regulated accurately under limited calculation steps. The whole performance of the motor is improved by MPC with a fast response and smaller response errors.

2. Mathematical Model of the PMSM

2.1. PMSM Model

The voltage equation for the PMSM under Cartesian coordinates can be expressed by Equation (1), neglecting local magnetic circuit saturation, eddy losses, and hysteresis losses.

$$u_{3s} = R_{3s}i_{3s} + \frac{d\lambda_{3s}}{dt} \tag{1}$$

The equation for flux linkage of the motor can be expressed by

$$\lambda_{3s} = L_{3s}i_{3s} + \varphi_f \cdot F_{3s}(\theta) \tag{2}$$

where λ_{3s} is the flux linkage of the three windings. u_{3s}, R_{3s} and i_{3s} are voltages, resistances, and currents of the motor, respectively. L_{3s} is the three-phase inductance, and $F_{3s}(\theta)$ is the three-phase angle.

$$i_{3s} = \begin{bmatrix} i_A \\ i_B \\ i_C \end{bmatrix} \tag{3}$$

$$R_{3s} = \begin{bmatrix} R & & \\ & R & \\ & & R \end{bmatrix} \tag{4}$$

$$\lambda_{3s} = \begin{bmatrix} \lambda_A \\ \lambda_B \\ \lambda_C \end{bmatrix} \tag{5}$$

$$u_{3s} = \begin{bmatrix} u_A \\ u_B \\ u_C \end{bmatrix} \tag{6}$$

$$F_{3s}(\theta) = \begin{bmatrix} \sin\theta \\ \sin(\theta - 2\pi/3) \\ \sin(\theta + 2\pi/3) \end{bmatrix} \tag{7}$$

$$L_{3s} = L_{m3} \begin{bmatrix} 1 & \cos(2\pi/3) & \cos(4\pi/3) \\ \cos(2\pi/3) & 1 & \cos(2\pi/3) \\ \cos(4\pi/3) & \cos(2\pi/3) & 1 \end{bmatrix} + L_{l3} \begin{bmatrix} 1 & & \\ & 1 & \\ & & 1 \end{bmatrix} \tag{8}$$

where L_{m3} are mutual inductances, and L_{l3} are inductance leakages. The torque of the motor can be calculated as

$$T_e = \frac{1}{2}P_n \frac{\partial}{\partial \theta_m}\left(i_{3s}^T \cdot \lambda_{3s}\right) \tag{9}$$

P_n are pole pairs of the motor. The dynamic behavior of the motor can be expressed by

$$J\frac{d\omega_m}{dt} = T_e - T_L - B\omega_m \tag{10}$$

where J is the rotating inertia of the motor, B is the damping coefficient, and T_L is the loading torque. ω_m is the angular speed of the motor. Two classical current regulations, hysteresis comparator and

PWM regulation, are popular for the current loop of PMSM and they are able to regulate the phase current of PMSMs. The main specifications of the motor are listed in Table 1.

Table 1. Main specifications of the motor.

Specifications	Quantity (SI)
Rated power	500 W
Rated current	8 A
Pole number (p)	8
Length of the stator (l)	350 mm
Number of turns of each coil	60
Width of the mover plate (w)	16 mm
Height of the mover plate (h)	16 mm
Width of the coil area (c)	10 mm

2.2. Drive Topology

The three-phase inverter topology for PMSM is shown in Figure 1. Six metal-oxide semiconductor field-effect transistors (MOSFETs) and their states are given and indicated as S_1–S_6. For one bridge, S_1 and S_4 cannot switch on simultaneously in case of a short circuit between Vdc and the ground. Therefore, when S_1 switches on, S_4 will switch off and vice versa. For three-phase alternative current motors, six MOSFETs construct eight switching modes, i.e., V_1–V_6, as shown in Figure 2. The eight vectors divide the vector plane into six sections for the motor.

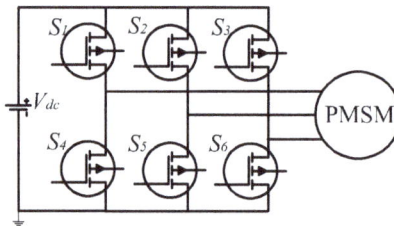

Figure 1. Inverter topology for the permanent magnet synchronous motor (PMSM).

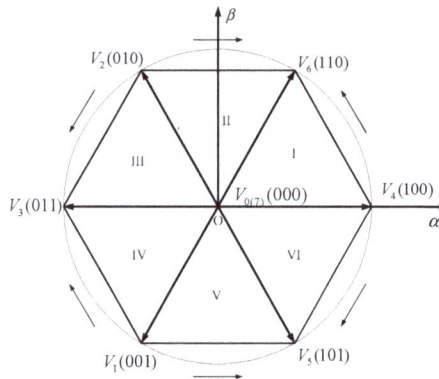

Figure 2. Voltage vectors for the PMSM.

2.3. Model Predictive Control

An MPC consists mainly of three parts: (1) a cost function, (2) a predictive model of the inverter, and (3) a model of the load. For the control of the PMSM, the load is the PMSM. The entire control scheme is shown in Figure 3. A predictive current is obtained by a predictive model that includes parts 2 and 3. The predictive current and the current reference are the inputs of part 1. If a static power converter can be controlled by a set of switching states within a finite number, these states can be predicted by an MPC according to criteria that can estimate the predictive values. This criterion is called the cost function. This function is used to predict the future variables for the system or the switching states of the inverter. The state that minimizes the cost function is usually the criterion.

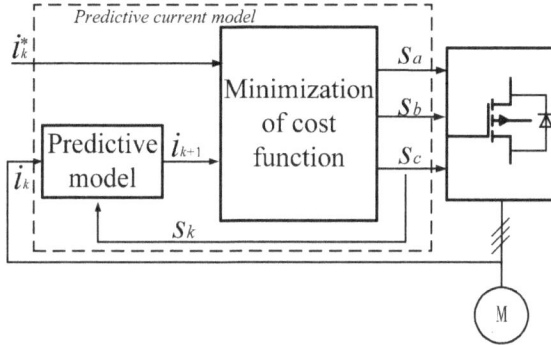

Figure 3. Scheme of current model predictive controller (MPC) for the PMSM.

In this study, a current MPC is employed to control a PMSM. As a load, the model of the PMSM has been given. However, the discrete-time model of the motor is needed, and this model will be an important part of the predictive model. In addition, this predictive model should take the model of the inverter into consideration as well. The states of the switches can be expressed by Equation (11):

$$s = \frac{2}{3}\left(s_a + as_b + a^2 s_c\right) \tag{11}$$

where $a = e^{j2\pi/3}$ and the voltage vector generated by the inverter is formulated as Equation (12).

$$s = \frac{2}{3}\left(v_{aN} + av_{aN} + a^2 v_{aN}\right) \tag{12}$$

Consequently, the load voltage vector can be calculated by the switching states.

$$v = V_{dc}S \tag{13}$$

According to the voltage vector of Equation (12), the current vectors for the PMSM can be expressed as Equation (14), and the back electromagnetic force s Equation (15).

$$i = \frac{2}{3}\left(i_a + ai_b + a^2 i_c\right) \tag{14}$$

$$e = \frac{2}{3}\left(e_a + ae_b + a^2 e_c\right) \tag{15}$$

The load current dynamic is calculated in Equation (16):

$$v = Ri + L\frac{di}{dt} + e. \tag{16}$$

This function can be expressed by two equations after Clark and Parker transformation:

$$v_q = Ri_q + L\frac{di_q}{dt} + L\omega i_d + \lambda_m\omega \tag{17}$$

$$v_d = Ri_d + L\frac{di_d}{dt} - L\omega i_q \tag{18}$$

The discretized dynamic functions of the predictive model from Equations (17) and (18) can be expressed as

$$v_q(k) = Ri_q(k) + \frac{L}{T_s}\left[i_q(k+1) - i_q(k)\right] + L\omega i_d(k) + \lambda_m \omega \tag{19}$$

$$v_d(k) = Ri_d(k) + \frac{L}{T_s}\left[i_d(k+1) - i_d(k)\right] - L\omega i_q(k) \tag{20}$$

The discrete currents derived from Equations (19) and (20) according to the dynamic functions are

$$i_q(k+1) = i_q(k) + \frac{T_s}{L}\left[v_q(k) - Ri_q(k) - L\omega i_d(k) - \lambda_m \omega\right] \tag{21}$$

$$i_d(k+1) = i_d(k) + \frac{T_s}{L}\left[v_d(k) - Ri_d(k) + L\omega i_q(k)\right]. \tag{22}$$

The cost function is used to select the voltage vector for the PMSM. The voltage vectors will be applied to the inverter so that the phase currents of the PMSM will reach the expected current references in the next sampling time after minimizing the cost function. The consequence is that the predictive current will be applied to the motor, and it will be operated according to the references. For current prediction of the PMSM, the cost function is given as Equation (23). The flow chart of the current predictive control is shown in Figure 4. Assuming that the initial value of the cost function is infinite, when the temporary variable *j* changes from 0 to 7, the states of the MOSFETs will be determined. First, the voltage vector can be obtained according to Equation (13). Then, the predicted current in the *d-q* axis can be calculated from Equations (22) and (23), respectively. Finally, according to the cost function, as given in Equation (23), the states of the MOSFETs will be obtained if the temporary variable *j* = 7. Otherwise, the program will jump to Equation (13). The corresponding control algorithm code is shown in the Appendix A.

$$g = \left|i_{q_{ref}} - i_q(k+1)\right| + \left|i_{d_{ref}} - i_d(k+1)\right| \tag{23}$$

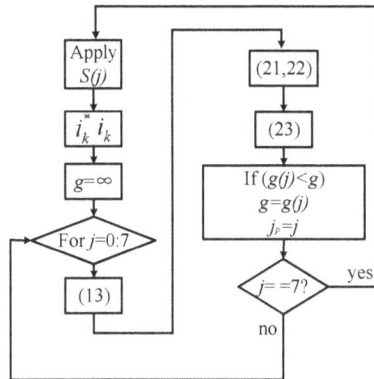

Figure 4. Main flow chart of the controller for the MPC.

3. Simulink Modeling of the MPC Algorithm

Simulations for an MPC of a PMSM are designed as shown in Figure 5. The whole operation time is 0.1 s. The reference speed changes from 500 rpm to 1000 rpm at 0.03 s, and the load varies from 5 Nm to 8 Nm at 0.07 s.

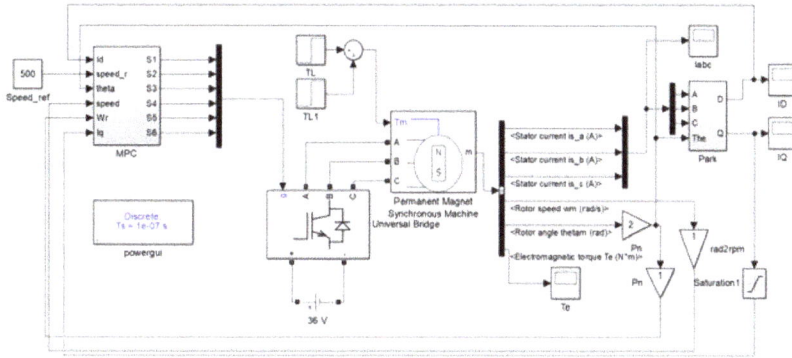

Figure 5. MPC blocks for the PMSM.

Three-phase currents are shown in Figure 6. Also, simulations are carried out when the motor is controlled by an SVPWM controller. Three-phase currents are obtained in Figure 7. I_d and I_q are also obtained. Compared with the three-phase currents under the two controllers, the inverter outputs nonsinusoidal currents for the MPC but totally sinusoidal waves for the SVPWM. The amplitudes of the currents from the SVPWM method change dramatically when the speed reference changes and the load varies.

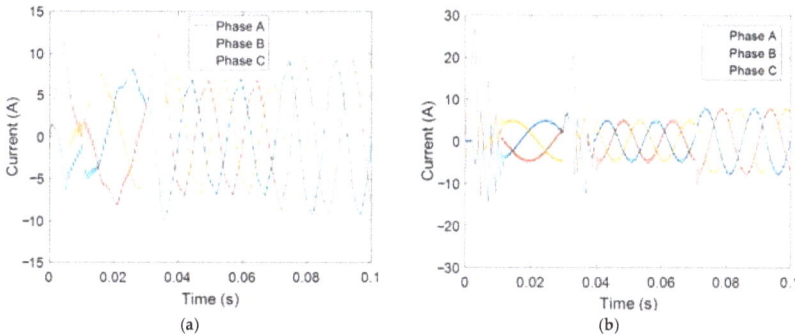

Figure 6. Three-phase currents of the motor. (**a**) The currents under MPC, (**b**) the currents under vector control.

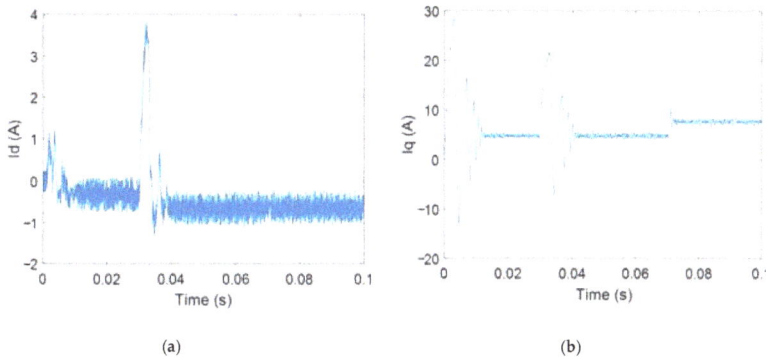

Figure 7. The currents of the motor after coordination transformation from three phases to the *d-q* axis. (**a**) I_d of the motor and (**b**) I_q of the motor.

4. Experimental Verification

Experimental Setup

The experimental setup is established by employing a dSPACE DS1104 card, which can be regarded as a microprocessor. The microprocessor is the controller of the whole system after the proposed algorithm is programmed and downloaded. The experimental hardware is shown in Figure 8, including power suppliers, a computer, a dSPACE control card, three highly accurate current sensors, a driver of the three-phase PMSM, and the motor. The entire control part of the PMSM is built using MATLAB/Simulink software, and the designed program can be debugged into the dSPACE card. An encoder interface of the card is employed to obtain the angles and the speeds of the rotor, and six analog-to-digital converters are used to sample the phase currents and voltages of the motor for the controller.

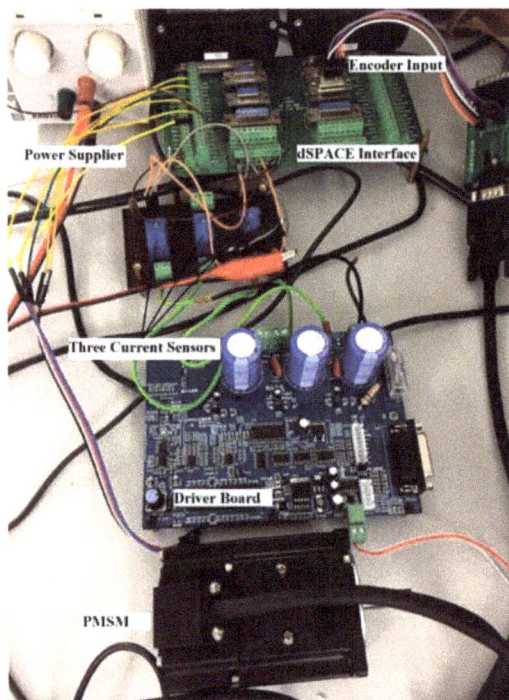

Figure 8. Experimental setup of the PMSM control system.

Three-phase currents of the motor under steady state with no load and in loaded state at the speed of 500 rpm are shown in Figure 9a,b. When the motor works with no load, the amplitude of the three-phase current is 0.3 A, and the currents are nearly sinusoidal. When a rated load of 8 Nm is produced for the motor at this speed, it can be seen from Figure 9b that the three-phase currents of the motor with the MPC controller are increased accordingly, with slight current deformation. This deformation begins mainly at the interchange section between the top half of the current and the bottom half; the dead time of MOSFETs could be the cause. In Figure 10, the reference speed increases to 1000 rpm, the initial currents are shown when the motor starts, and the load of 5 Nm is carried on at 0.05 s. The motor performs well at the beginning and in the loaded states.

(a)

(b)

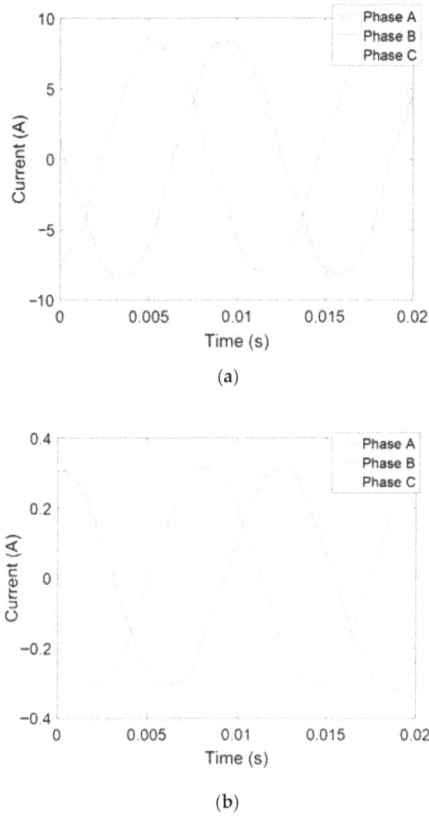

Figure 9. Three-phase currents of the motor. (**a**) The currents when the motor is operated with a rated load and (**b**) with no load.

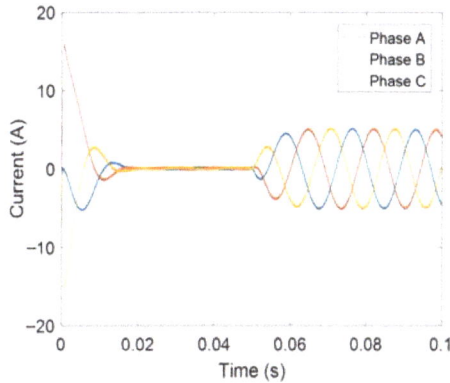

Figure 10. Sstart of three-phase currents and the currents with load of 5 Nm.

The speed control of the motor with the current MPC is shown in Figure 11, compared with classic control methods. The speed reference is 500 rpm under the load of 5 Nm, increasing to 1000 rpm at 0.03 s. The load increases from 5 Nm to 8 Nm at 0.07 s. There is no overshoot if the motor is governed by the MPC. The speed cannot fully reach the reference when the motor enters a steady state via the controllers of FOC, SVPWM, and DTC. Three details are shown in Figure 12, including the motor start (zoom 1), the loading variation (zoom 2), and the steady-state errors (zoom 3). The three-phase measured currents of the motor are shown in Figure 13. These details show that the dynamic responses under MPC outweigh those of other controllers, with fewer overshoots and small steady-state errors. The MPC system is not vulnerable to varied loads or to the parameter change of the control system. In addition to the MPC, other controllers cannot avoid overshoot and vibration when the motor starts. Although the DTC has less rising time, the steady-state error from this controller is obvious. In Figure 12b, the controllers of both the FOC and the MPC are stable, with the loading torque varying from 1 Nm to 2 Nm. The steady-state errors are shown in Figure 12c. The FOC can curb the steady error at the lowest scope. However, this method has an obvious overshoot. The steady errors can be limited at 5 rpm by using MPC and other control methods, suggesting higher steady errors that exceed 5 rpm. This shows that the performance of the designed MPC outweighs that of other controllers under identical conditions.

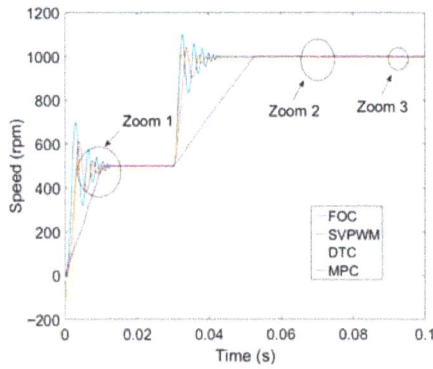

Figure 11. Speed responses of the motor under different controllers. FOC: field-oriented control, SVPWM: space vector pulse width modulation, DTC: direct torque control.

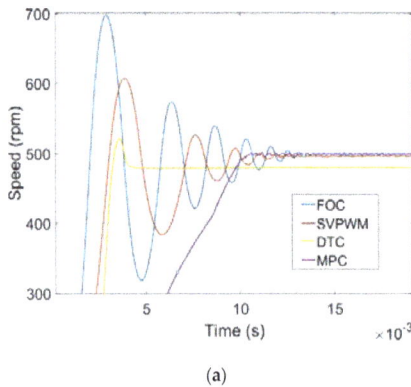

(a)

Figure 12. *Cont.*

(b)

(c)

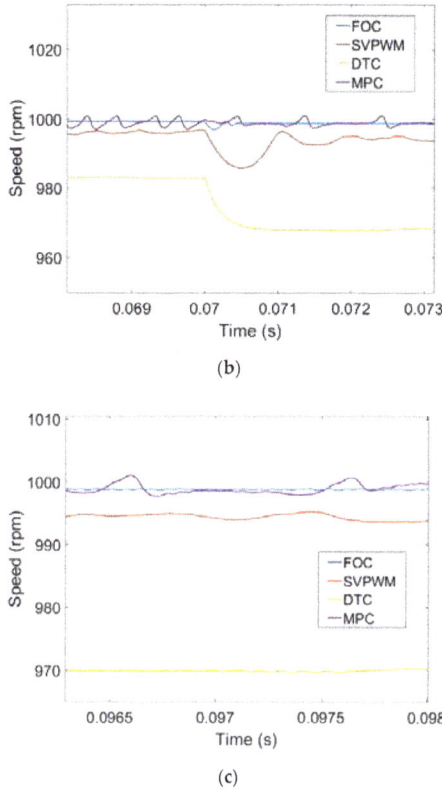

Figure 12. Details of the speed responses with zoomed parts: (**a**) zoom 1, (**b**) zoom 2, and (**c**) zoom 3.

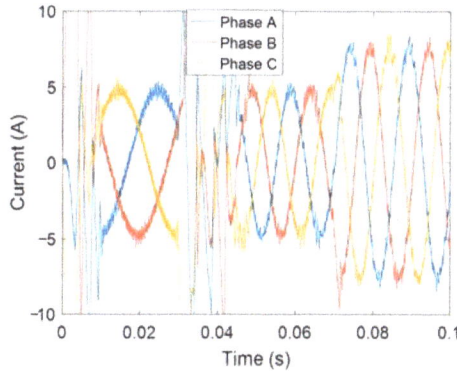

Figure 13. Three-phase measured currents of the motor.

Figure 14 shows the speed response of the motor under different speed references. The speed reference starts at 100 rpm in the beginning and increases to 500 rpm at 0.02 s and 1000 rpm at 0.05 s. The entire speed response is shown in Figure 14a. The details of the speed variations at each step are shown in Figure 14b–d. There are a few overshoots within 1% at the stepping times. Speed steady-state errors could be on the rise with the increase of the speed reference. However, the speed steady-state errors can be limited to 0.5 rpm according to the experimental results.

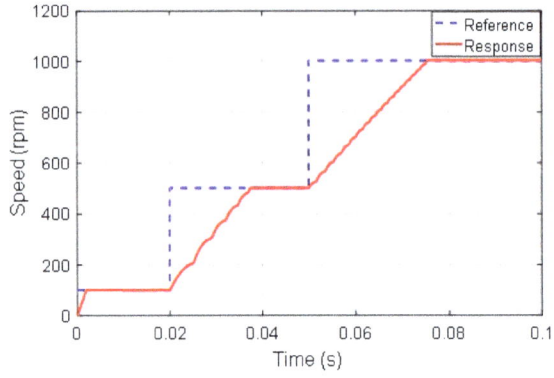

(a)

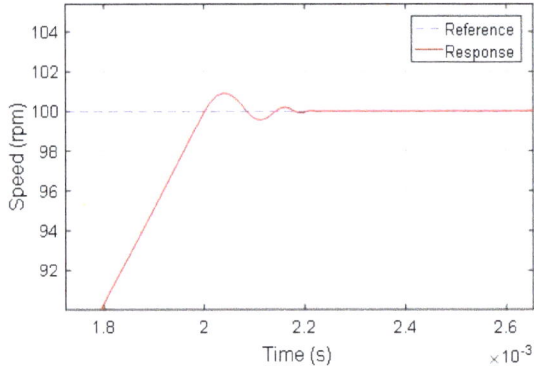

(b)

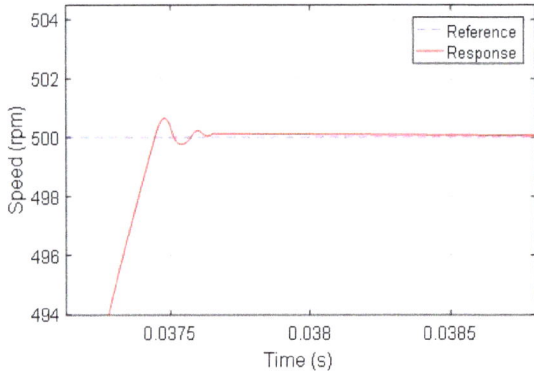

(c)

Figure 14. *Cont.*

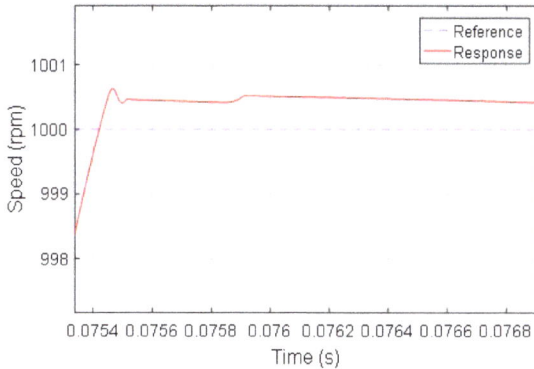

(d)

Figure 14. Speed responses under different speed references. (**a**) Complete graph of the speeds changing from 100 rpm to 500 rpm and finally to 1000 rpm. (**b**) Speed changes from 0 rpm to 100 rpm. (**c**) Speed changes from 100 rpm to 500 rpm. (**d**) Speed changes from 500 rpm to 1000 rpm.

5. Conclusions

In this study, a current MPC of a PMSM was investigated. The mathematical model of the motor and the Simulink modeling of the MPC method are given. Both the simulation and experimental results suggest that the designed MPC is feasible. For practical application, the main programmed algorithm is also given. The speed responses show that the MPC performs better than traditional controllers, as supported by the simulation and experimental results. MPC is shown to possess fast-tracking capability and small steady errors in speed regulation, and the speed reference has a lower overshoot under varied loads. The speed responses could be limited to 0.5 rpm, with experimental results validated under different speed references. The experimental results proved the feasibility and effectiveness of the designed MPC. This control method could be a promising candidate to control the PMSM in the future.

Author Contributions: M.T. developed the analysis, hardware design, and measurement. He also conducted the simulation. S.Z. was responsible for the background theory. He also provided guidance and supervision.

Funding: The authors gratefully acknowledge the support by the government of Hang Zhou city (grant code 2017JD60) and Zhejiang Provincial Education Department funding with code YB201705.

Conflicts of Interest: The authors declare no conflict of interest.

Appendix A

The main function of the code is to determine the states of MOSFET according to the cost function *g*. According to the topology in Figure 1, the entire states of all MOSFETs can be determined if states S1, S2, and S3 are obtained. States S4, S5, and S6 are correspondingly opposite to the three former states. The control program corresponding to Figure 4 is shown below.

```
function Smin = MPC[wr,id,iq,theta]
s = [0 0 0; 0 0 1; 0 1 0; 0 1 1;1 0 0; 1 0 1; 1 1 0; 1 1 1]
gm = zeros(8,1);
s1 = s2 = s3 = s4 = s5 = s6 = s6 = s7 = 0;
for i = 1:8
sa = s(i,1); sb = s(i,2); sc = s(i,3);
Vinva = (Vdc*(2*Sa − b − Sc))/3;
Vinvb = (Vdc*(2*Sb − Sa − Sc))/3;
Vinvc = (Vdc*(2*Sc − Sb − Sa))/3;
```

vsq = (2/3)*((Vinva*cos((Wr*t + theta))) + (Vinvb*cos((Wr*t) + theta + (4*pi/3))) + (Vinvc*cos((Wr*t) + theta + (2*pi/3))));
vsd = (2/3)*((Vinva*sin((Wr*t + teta))) + (Vinvb*sin((Wr*t) + theta + (4*pi/3))) + (Vinvc*sin((Wr*t) + theta + (2*pi/3))));
iq1 = iq + Ts*[vsq − R*iq − L*wr*id − phi*wr]/L;
id1 = id + Ts*[vsd − R*id + L*wr*Iq]/L
g = abs(idr − id1) + abs(iqr − iq1);
if(g < gm)
i_min = i;
g_min = g;
end
end
v = v(i_min);
iq = iq1;
id = id1;
s = s(:,i_min);

References

1. Krishnan, R. *Permanent Magnet Synchronous and Brushless DC Motor Drives*; CRC Press/Taylor & Francis: Boca Raton, FL, USA, 2010.
2. Hanselman, D.C. *Brushless Permanent-Magnet Motor Design*; McGraw-Hill: New York, NY, USA, 1994.
3. Abu-Rub, H.; Iqbal, A.; Guzinski, J. *High Performance Control of AC Drives with MATLAB/Simulink Models*; John Wiley & Sons: Hoboken, NJ, USA, 2012.
4. Quang, N.P.; Dittrich, J. *Vector Control of Three-Phase AC Machines System Development in the Practice*; Springer: Berlin/Heidelberg, Germany, 2015.
5. Rodriguez, J.; Pontt, J.; Silva, C.A.; Correa, P.; Lezana, P.; Cortes, P.; Ammann, U. Predictive Current Control of a Voltage Source Inverter. *IEEE Trans. Ind. Electron.* **2007**, *54*, 495–503. [CrossRef]
6. Bida, V.M.; Samokhvalov, D.V.; Al-Mahturi, F.S. PMSM vector control techniques—A survey. In Proceedings of the IEEE Conference of Russian Young Researchers in Electrical and Electronic Engineering (EIConRus), Moscow, Russia, 29 January–1 February 2018; pp. 577–581. [CrossRef]
7. Mohan, D.; Zhang, X.; Foo, G.H.B. A Simple Duty Cycle Control Strategy to Reduce Torque Ripples and Improve Low-Speed Performance of a Three-Level Inverter Fed DTC IPMSM Drive. *IEEE Trans. Ind. Electron.* **2017**, *64*, 2709–2721. [CrossRef]
8. Shinohara, A.; Inoue, Y.; Morimoto, S.; Sanada, M. Maximum Torque Per Ampere Control in Stator Flux Linkage Synchronous Frame for DTC-Based PMSM Drives Without Using q-Axis Inductance. *IEEE Trans. Ind. Appl.* **2017**, *53*, 3663–3672. [CrossRef]
9. Xia, C.; Wang, S.; Gu, X.; Yan, Y.; Shi, T. Direct Torque Control for VSI-PMSM Using Vector Evaluation Factor Table. *IEEE Trans. Ind. Electron.* **2016**, *63*, 4571–4583. [CrossRef]
10. Zhou, Y.; Chen, G. Predictive DTC Strategy with Fault-Tolerant Function for Six-Phase and Three-Phase PMSM Series-Connected Drive System. *IEEE Trans. Ind. Electron.* **2018**, *65*, 9101–9112. [CrossRef]
11. Wang, S.; Xu, D.D.; Li, C. Dynamic control set-model predictive control for field-oriented control of VSI-PMSM. *IEEE Appl. Power Electron. Conf. Expos.* **2018**, 2630–2636. [CrossRef]
12. Caseiro, L.M.A.; Mendes, A.M.S.; Cruz, S.M.A. Dynamically Weighted Optimal Switching Vector Model Predictive Control of Power Converters. *IEEE Trans. Ind. Electron.* **2018**. [CrossRef]
13. Siami, M.; Khaburi, D.A.; Rivera, M.; Rodríguez, J. A Computationally Efficient Lookup Table Based FCS-MPC for PMSM Drives Fed by Matrix Converters. *IEEE Trans. Ind. Electron.* **2017**, *64*, 7645–7654. [CrossRef]
14. Ahmed, A.A.; Koh, B.K.; Lee, Y.I. A Comparison of Finite Control Set and Continuous Control Set Model Predictive Control Schemes for Speed Control of Induction Motors. *IEEE Trans. Ind. Inform.* **2018**, *14*, 1334–1346. [CrossRef]

15. Shoukry, Y.; El-Kharashi, M.W.; Hammad, S. MPC-On-Chip: An Embedded GPC Coprocessor for Automotive Active Suspension Systems. *IEEE Embed. Syst. Lett.* **2010**, *2*, 31–34. [CrossRef]

16. Pejcic, I.; Korda, M.; Jones, C.N. Control of nonlinear systems with explicit-MPC-like controllers. In Proceedings of the IEEE 56th Annual Conference on Decision and Control (CDC), Melbourne, VIC, Australia, 12–15 December 2017; pp. 4970–4975. [CrossRef]

17. Vazquez, S.; Aguilera, R.P.; Pablo Acuna, J.P.; Agelidis, V.G. Model Predictive Control for Single-Phase NPC Converters Based on Optimal Switching Sequences. *IEEE Trans. Ind. Electron.* **2016**, *63*, 7533–7543. [CrossRef]

18. Kakosimos, P.; Abu-Rub, H. Predictive Speed Control with Short Prediction Horizon for Permanent Magnet Synchronous Motor Drives. *IEEE Trans. Power Electron.* **2018**, *33*, 2740–2750. [CrossRef]

19. Rubino, S.; Bojoi, R.; Odhano, S.A.; Zanchetta, P. Model predictive direct flux vector control of multi three-phase induction motor drives. *IEEE Trans. Ind. Appl.* **2018**. [CrossRef]

20. Zhang, Y.; Huang, L.; Xu, D.; Liu, J.; Jin, J. Performance evaluation of two-vector-based model predictive current control of PMSM drives. *Chin. J. Electr. Eng.* **2018**, *4*, 65–81. [CrossRef]

21. Zhang, X.; He, Y. Direct Voltage-Selection Based Model Predictive Direct Speed Control for PMSM Drives without Weighting Factor. *IEEE Trans. Power Electron.* **2018**. [CrossRef]

22. Luo, Y.; Liu, C. Elimination of Harmonic Currents Using a Reference Voltage Vector Based-Model Predictive Control for a Six-Phase PMSM Motor. *IEEE Trans. Power Electron.* **2018**. [CrossRef]

23. Zhou, Z.; Xia, C.; Yan, Y.; Wang, Z.; Shi, T. Torque Ripple Minimization of Predictive Torque Control for PMSM With Extended Control Set. *IEEE Trans. Ind. Electron.* **2017**, *24*, 6930–6939. [CrossRef]

energies

MDPI

Article

Finite Element Method Investigation and Loss Estimation of a Permanent Magnet Synchronous Generator Feeding a Non-Linear Load

Alexandra C. Barmpatza * and **Joya C. Kappatou**

Department of Electrical and Computer Engineering, University of Patras, Patras 26504, Greece;
joya@ece.upatras.gr
* Correspondence: abarmpatza@upatras.gr; Tel.: +30-694-24779943

Received: 24 October 2018; Accepted: 2 December 2018; Published: 4 December 2018

Abstract: The purpose of this paper is the performance investigation of a Permanent Magnet Synchronous Generator (PMSG) system, suitable for wind power applications and the comparison of the machine electromagnetic characteristics under open and closed control loop implementations. The copper and iron losses are estimated and compared for the above control systems with the use of the Steinmetz-Bertotti loss separation equation. In addition, the effect of the rotating magnetic field on the total losses is studied. The generator is simulated using Finite Element Analysis (FEA), while the rest of the components are connected to the machine model using a drawing window of the FEA software and suitable command files. The close loop control used in the present study results to less losses and greater machine efficiency. The main novelty of the paper is the simulation of the PMSG coupled with a converter and control schemes using FEA, which ensures more accurate results of the whole system and allows the detailed machine electromagnetic study, while the majority of existing papers on this topic uses simulation tools that usually simulate in detail the electric circuits but not the machine. The FEM model is validated by experimental results.

Keywords: finite element analysis; pulse width modulation; permanent magnet synchronous generator; wind generator

---

1. Introduction

The utilization of wind energy for electrical energy production is becoming more and more attractive nowadays. In [1,2], various wind turbine concepts have been studied in order to achieve the most efficient system. As the cost of rare-earth materials has come down, PMSGs have become more and more popular due to their tolerable air-gap depth, high efficiency, high power factors, stability and the absence of any additional DC supply and slip rings.

The majority of existing papers referring to the analysis of wind energy systems [3–8], use simulation tools that are more focused on the detailed simulation of the control system rather than the machine. Consequently, in most of the cases Matlab/Simulink is used as a simulation tool. In [3], a wind generator system, simulated in a general-purpose circuit simulator, is presented. For the simulation, a magnetic circuit model of a permanent magnet reluctance generator (PMRG) is introduced and the simulation results are compared with experimental data. In [4,5], a multi-physics system for wind turbines with permanent magnets and full conversion power electronics is presented. The study uses two models, an analytical model and a FEA model, in order to simulate the generator while both the power electronics and the DC link voltage control are simulated by Simplorer. However, as stated in [5], although this co-simulation enables the researchers to investigate the full wind system, there is a waste of simulation speed due to the communication and synchronization subroutines, additional effort required to couple the simulation tools that come from different vendors and additional effort needed to develop models for different simulation and user interface environments. In [6], a variable speed wind generator system using a PI controller has been investigated too. The system is simulated in Matlab software which gives no opportunity for a detailed analysis of the system. In [7], a variable speed wind energy conversion system, which uses a PMRG is investigated. The whole system is simulated via a Simplorer-Simulink interface. In [8], a SRG coupled with an OULTON converter for a WECS is studied, using experimental results and Matlab/Simulink. In [9] a DC-bus grid connected hybrid wind/photovoltaic/fuel cell for supplying power to a low voltage distribution system is presented and simulated using PSIM software. In [10] a PMSG that feeds a diode rectifier load is simulated using FEM and the influence of gap consideration on load identification under various Halbach-array-based topologies is investigated.

In this study the performance and the loss estimation of a PMSG system [11,12], suitable for wind power applications, consisting of a PMSG, a three-phase single switch boost rectifier and the controller are investigated. The generator is simulated using FEA, while the rest of the components are connected to the machine model using a drawing window (circuit editor) of the FEA software and suitable command files. The FEA approach uses a RM analysis solver. A time-stepping solution to the transient electromagnetic equation is obtained allowing the rotor to rotate by the appropriate angle at each time-step. In the present study, the non-linear BH characteristic of ferromagnetic material is taken into account, which is time consuming, but it can provide more accurate results. Furthermore, the machine and the winding configuration can be captured with more detail using FEA. Therefore, the detailed simulation adopted to the present paper ensures better accuracy and reliability of the results concerning all the system components and the ability to study the machine electromagnetic behavior. In other words, a more detailed analysis has been conducted, which enables to obtain more accurate waveforms and more precise information about the harmonic components that stress the machine and distort the output voltage and current waveforms producing overheating. To the authors knowledge, the existence of similar system studies, in which a detailed simulation of all components is conducted, using modern simulation software tools, like in the present paper, is very limited in the international literature and this fact can be considered as the originality of this particular article. In addition, the FEM model is validated by experimental results which ensure its accuracy.

The simulated system is investigated for three different cases. At first, the generator feeds a resistive load without a converter connection. Two generator models are used and compared. The first one has magnets mounted on the rotor in an asymmetrical way, while the second has a symmetric magnet topology. The real machine used in the present study has an asymmetrical topology, but the symmetrical topology is simulated too, because the authors want to emphasize the accuracy of FEM which gives the opportunity to simulate the machines in detail. A comparison in the cogging torque and the stator current harmonic components of the two models has been made, that could not be achieved using machine models from a Simulink software library. Secondly the generator, with the asymmetric magnet topology, coupled with the converter and an open loop control, where the duty cycle remains constant, has been investigated. Comparisons of the electromagnetic characteristics and generator losses for various duty cycles and rotor speeds have been made. Then a close loop control, that varies the duty cycle with the use of a PI controller, has been applied and the simulation results are compared with those of the previous case. Finally, the losses due to the rotating magnetic field have been investigated and estimated also.

2. The Wind Power System

Generally, wind power generation systems consist of a generator, a rectifier, a boost DC-DC converter and an inverter. The terminal voltage of the machine is rectified and then boosted by the DC-DC converter. The inverter converts the DC voltage to the appropriate voltage for the AC grid connection [3–8]. The simulated system consists of the PMSG, the three-phase single switch boost rectifier and the control system. Figure 1 depicts the schematic diagram of the system. In other words, this paper focuses mainly on the first stage of the conversion, i.e., in the AC-DC converter, while the connection of the inverter, as well as the synchronization to the grid will be studied in the future.

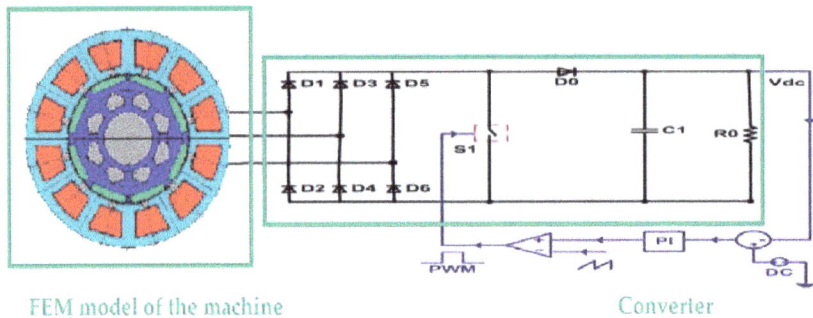

FEM model of the machine   Converter

Figure 1. The complete interconnected close loop system.

The investigated generator, that is a real commercial machine [11,12] with 12 slots, four pole pairs, 200 Hz nominal frequency, 660 W rated power, 48 V nominal voltage and 3000 rpm nominal speed, has been simulated. The stator winding is star connected, single layer, concentrated and non-overlapping, while the magnets are mounted on the rotor in an asymmetrical way, that is different distances between the magnets, in order a more sinusoidal voltage waveform to be induced [13]. For comparison purposes both the asymmetric and the symmetric model were simulated. Figure 2 presents the meshed cross sections of the FEM machine models and a photo of the rotor of the machine with the asymmetrically placed magnets with three different distances (a, b, c) between them. The machine has been modeled using Opera mesher and the model consists of approximately 45,000 linear elements. On a modern PC (Intel i7-4770 with 8 GB RAM) the finite element analysis required 4 min in order to reach the steady state condition when the generator feeds a resistive load without a converter connection, 48 min when the generator is coupled with a rectifier and an open loop voltage control is implemented, while 140 min for the close loop system.

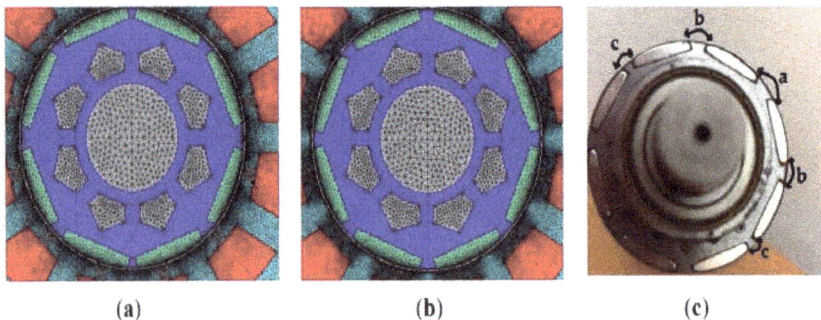

(a)   (b)   (c)

Figure 2. (a) The meshed cross section of the PMSG with asymmetrically placed rotor magnets, (b) the meshed cross section of the PMSG with symmetrically placed rotor magnets, (c) a photo of the rotor of the machine with the asymmetrically placed magnets (a > b > c).

The three-phase single switch boost rectifier, whose principle of operation is analyzed in [14,15], has been chosen to be connected to the generator terminals. By using the circuit editor of the FEA software the model of the converter was connected to the FEA model of the PMSG. However, in this particular study, we have omitted the boost inductors, in order to investigate the system dynamic behavior exclusively due to the inductances of the machine [16]. The machine inductances act as the boost inductors for the voltage source rectifier. Consequently, the FEA model of the PMSG can provide not only an electromagnetic insight of the machine but also a more precise sizing of the boost inductors, if there are essential for the operation of the converter.

Open and close loop control systems are implemented (Figure 1) by inserting suitable command files. The simulation results correspond to a specific operation at 3000 rpm, with the output load (R_0) 10 Ω and the DC-link capacitor (C_1) 2 mF. In the open loop control the duty cycle remains invariable, while in the close loop control the duty cycle changes, with the use a PI controller. The PI compensator input is the difference of the DC reference voltage and the output voltage V_{dc}, Figure 1.

3. Loss Terms

3.1. Copper Losses

The investigated machine is a PMSM and therefore only the stator has copper losses. It is known that the stator copper losses consist of two components: a DC and an AC owing to skin and proximity effects. In this study, for the copper losses per phase calculation, only the DC component is taken into consideration, as the main interest is on the iron losses:

$$P_{cu} = I^2 R, \tag{1}$$

where I is the stator current rms value and R the stator winding resistance that is considered constant.

3.2. Iron Losses

The iron losses have an important effect on the machine efficiency and performance and could provoke machine overheating, reduction of the rated torque and the efficiency. One of the most widespread of the literature methods for iron loss estimation is based on the Steinmetz-Bertotti equation. According to this equation the total iron losses can be separated into hysteresis and eddy current losses. The generator eddy-current losses are created due to three main reasons. The first reason is the MMF distribution, especially when the machine has fractional slot concentrated windings. In this case the amount of magneto-motive force (MMF) harmonics is large, which results in increased eddy-current losses. Secondly the eddy current losses could be created due to the permeance variation, while the third reason is the existence of PWM harmonics. The converter provokes time harmonics in the stator current waveform and consequently losses are created on the generator [14,15,17–21]. Moreover, the eddy current losses are separated into classical losses, that are the eddy currents induced in materials by an external alternating magnetic field, and excess losses that are result of the internal movement of domain walls between different magnetic domains [22–25]. Taking into account the above consideration, the total iron losses are the sum of hysteresis, classical eddy current and excess losses [22,26,27]:

$$P_{fe} = P_h + P_e + P_a = \varrho_{fe} k_h f B^\alpha + \varrho_{fe} k_e f^2 B^2 + \varrho_{fe} k_a f^{1.5} B^{1.5}, \tag{2}$$

where ϱ_{fe} is the iron density in Kg/m^3, α is the constant of Steinmetz, k_h is the hysteresis loss coefficient, k_e is the eddy current loss coefficient, k_a is the excess loss coefficient, f is the electrical frequency in Hz and B is the peak value of the magnetic flux density in T. The coefficients α, k_h, k_e, and k_a, are determined by fitting based on Epstein frame measurements and for the specific analysis they have the values α = 2, k_h = 0.0061 W/kg/Hz/T$^\alpha$, k_e = 0.00013334 W/kg/Hz^2/T^2, and k_a = 0.00027221 W/kg/Hz$^{1.5}$/T$^{1.5}$. For the FEM simulations, the right choice of the material parameters and the power supply conditions is the most critical factor.

Equation (2) is sufficiently accurate to predict the iron losses for an alternating magnetic field. Nevertheless, for a rotating magnetic field, Equation (2) is not appropriate for the iron losses estimation. In that case the magnetic field vector, except from the space rotation, creates rotating magnetic fields in some regions of the electrical machine core which results in differentiated losses. The above losses due to the rotating magnetic field are calculated in Section 6.

4. FEM Model Validation

In order to validate the FEM model of the machine, essential experiments have been conducted in the laboratory and the experimental results are compared with the corresponding simulation results. The experiments were performed for the no load condition and when the generator is feeding a 10 ohm resistive load through a three-phase diode bridge rectifier. The experimental system is depicted in Figure 3. It consists of the PMSG coupled with a permanent magnet DC motor, the DC power supply that feeds the permanent magnet DC motor, the three-phase diode bridge and a variable resistive load. The current, voltage probes and the oscilloscope are used for the experimental procedure.

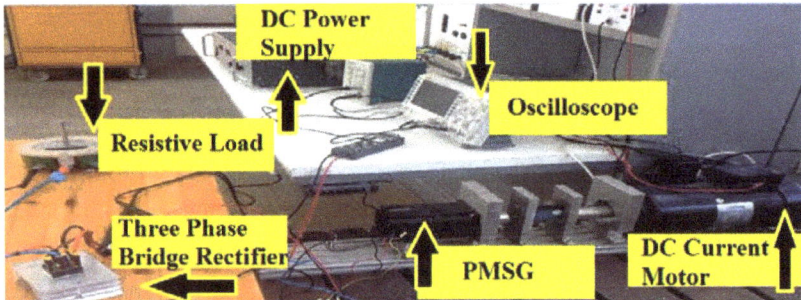

Figure 3. The experimental system in the Electromechanical Energy Conversion Laboratory.

At first experiments were conducted for the no load condition at 2000 rpm and the results are compared with the corresponding FEA simulation results. In Figure 4a the line voltage waveform is depicted, while in Figure 4b it can be seen the line voltage waveform that was exported for the simulation. The rms value of the line voltage waveform resulted from experiment is 18.8 V, while the same rms value resulted from simulation is 18.6 V. The difference is approximately 1%.

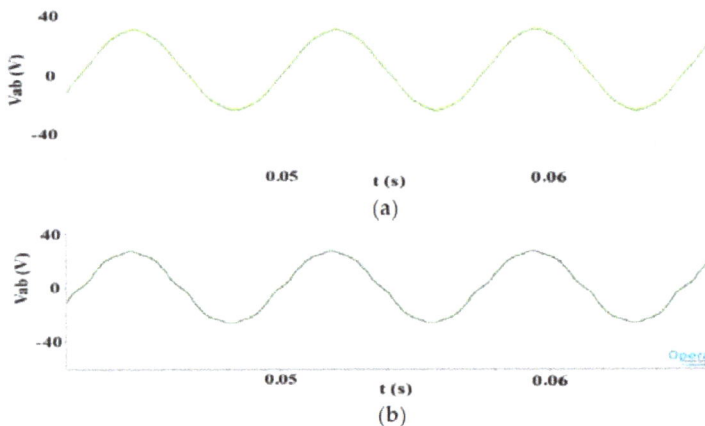

Figure 4. The waveforms of line voltage for the no load condition and 2000 rpm: (**a**) experimental results, (**b**) FEA simulation results.

Experimental results are also exported when the generator is coupled with a three-phase diode bridge rectifier and a resistive load. In Figure 5 the experimental and simulation results for 1500 rpm are presented, while in Figure 6 the same results are depicted for 2000 rpm. The experimental results agree very well qualitative as well as quantitative with the simulation results. When the generator speed is 1500 rpm, the experimental rms line voltage is 13.5 V and the rms phase current 1.49 A, while the corresponding simulation values are 13.9 V and 1.32 A, respectively. For 2000 rpm, the rms line voltage is 18.9 V and 18.6 V considering the experimental and simulation results, respectively, while the rms phase current is 2.11 A for the experimental and 1.80 A for the simulation case. The differences between the experimental and simulation results are very small and thus the FEM model is considered valid.

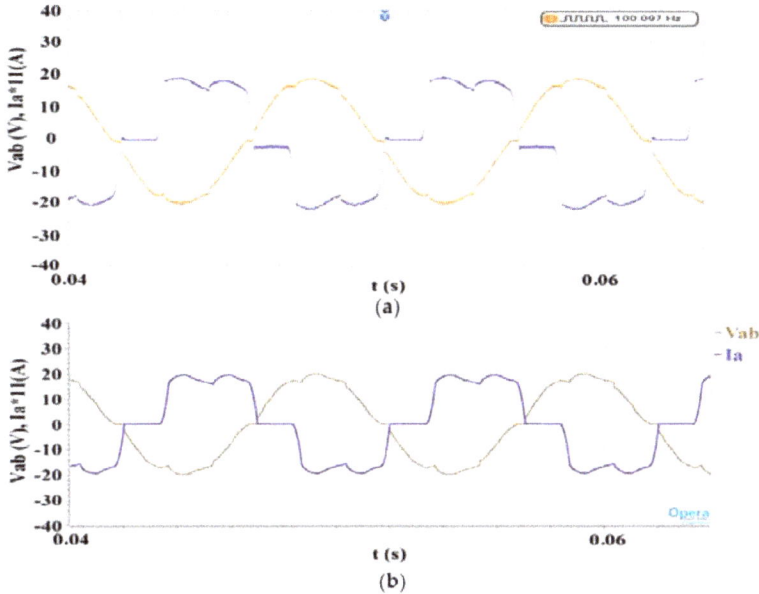

Figure 5. The waveforms of line voltage and phase current when the generator is coupled with a three-phase diode bridge and a resistive load of 10 ohm for 1500 rpm: (**a**) experimental results, (**b**) FEA simulation results.

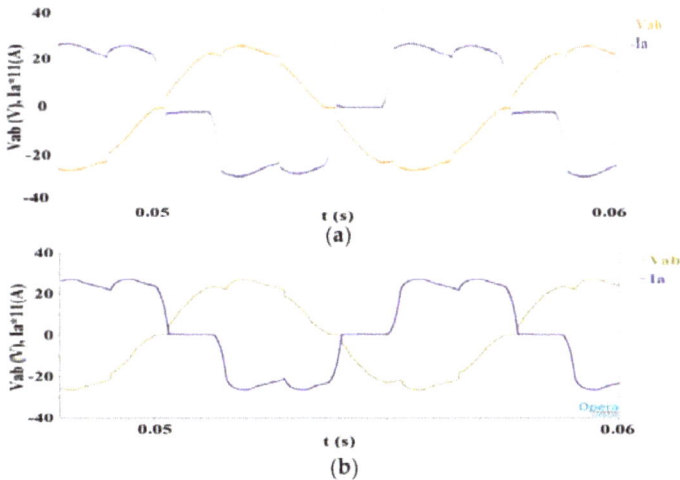

Figure 6. The waveforms of line voltage and phase current when the generator is coupled with a three-phase diode bridge and a resistive load of 10 ohm for 2000 rpm: (**a**) experimental results, (**b**) FEA simulation results.

5. Simulation Results

5.1. Symmetric and Asymmetric Model of the Generator with a Resistive Load

At first, the generator feeding directly a resistive load of 10 Ω, without the power electronic converter, is investigated. In Figure 7a the phase current spectra for the two models, the asymmetric and the symmetric, are compared. In both cases, the stator current contains higher harmonics caused by the configuration of the stator winding and the rotor topology. In other words, the stator current does not form a pure sinusoidal waveform, due to higher harmonics caused by the spatial non-sinusoidal magnetic field distribution of the machine. The configuration of the stator winding and the rotor topology strongly influences the harmonic content of the air gap magnetic field and thus the voltage and the stator current. From the simulation, it can be verified that the generator with the asymmetric configuration induces a more sinusoidal current. Indeed, the amplitude of the higher harmonic components of the current is lower when magnets are placed asymmetrically, Figure 7a. According to [13,28,29], when the magnets are placed asymmetrically on the rotor there is a reduction in the cogging torque and also an EMF with lower harmonic distortion is induced. The cogging torque is the result of the interaction between the stator slot openings with the PMs on the rotor, so the period of the cogging torque is linked with the number of slots, Q, and poles, 2p. The period of the cogging torque, T_c, can be expressed by the following relationship [29]:

$$T_c = \frac{2\pi}{LCM\{Q, 2p\}},$$ (3)

where the LCM is the least common multiple. As far as the studied generator concerned, the period of the cogging torque is $T_c = \pi/12$ s which can be verified by Figure 7b where the cogging torque appearing in the two models is compared. From the same Figure it can be seen also that the amplitude of the cogging toque is almost doubled in the case of the symmetric model of the machine.

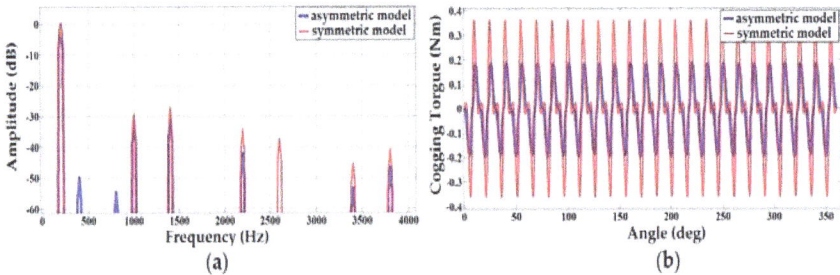

Figure 7. (**a**) The phase current spectrum for the model for 3000 rpm rotor speed and a resistive load 10 Ω, (**b**) The machine's cogging torque versus angle. (asymmetric model: blue line, symmetric model: red line).

As the machine used for the experiment has asymmetrically placed magnets, the simulated machine should have also the same rotor topology in order the results to be accurate. Therefore, in all the simulations bellow the FEM model of the machine with asymmetrically placed magnets is used.

5.2. Open Loop Control of the System

The three-phase single switch boost rectifier is connected to the generator terminals and an open loop PWM control is applied to the switch. In Figure 8a the output open loop DC voltages for three different duty cycles when the machine supplies a resistive load of 10 Ω, are depicted and compared. When the duty cycle increases, the output open loop DC voltage increases too. In Figure 8b the distribution of the absolute value of the magnetic flux density (B_{mod}) is depicted for the case of the open loop control with 50% duty cycle. The switching period is equal to $T_s = 10^{-4}$ s. The duty cycle does not change and the conducting time interval of the switch is equal to $t_{on} = 5 \times 10^{-5}$ s. In Figure 9a the waveforms of the phase current for three different duty cycles are presented, while in Figure 9b the FFT analysis in the machine phase current is shown. The ripples appearing in the waveforms of Figure 9a result from the machine inductances and the switching frequency of the PWM technique [30,31]. As the machine operates at 200 Hz and the switching frequency equals to 10 kHz, it can be seen 50 ripples in the current waveform for one electrical period, Figure 9a. The aforementioned analysis reveals that, this converter increases the amplitude of already existing harmonic components with order numbers μ $= 5, 7, 11, 13, \ldots$ in the machine phase currents. Especially the fifth and the seventh harmonic of the stator current are increased, Figures 7a and 9b [14,15,19,20]. The relation between the boost rectifier duty cycle and the voltage ratio is given by the equation below:

$$\frac{V_i}{V_o} = 1 - d, \tag{4}$$

where, V_i is the input voltage, V_{dc} is the DC output voltage and d the duty cycle. By (4), it is revealed that the voltage ratio and the duty cycle are inversely proportional variables. Thus, when the voltage ratio presents a reduction, the duty cycle presents increment. As is it referred in [21], when the ratio of V_i/V_{dc} is small, the stator current waveform has a shape closer to sinusoidal which means smaller harmonic content. Figure 9b depicts that the PWM current harmonics amplitude raises with the decrease of PWM duty cycle [32]. The increased harmonic amplitudes increase the stress of the machine and the deformation of the output voltage and current waveforms. So, a precise estimation, using proper simulation tools, is needed.

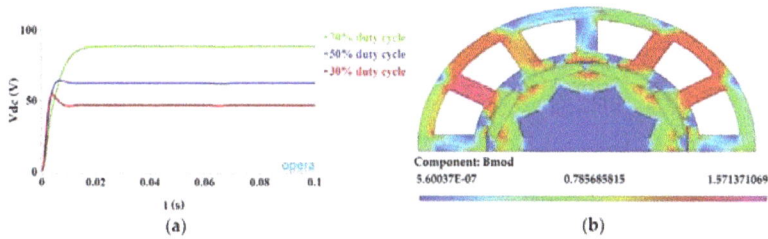

Figure 8. Waveforms of: (**a**) the output open loop DC voltage for a resistive load equal to 10 Ω and various duty cycles: (green line—70%, blue line—50%, red line—30%), (**b**) The distribution of B_{mod} for the nominal speed and 50% duty cycle.

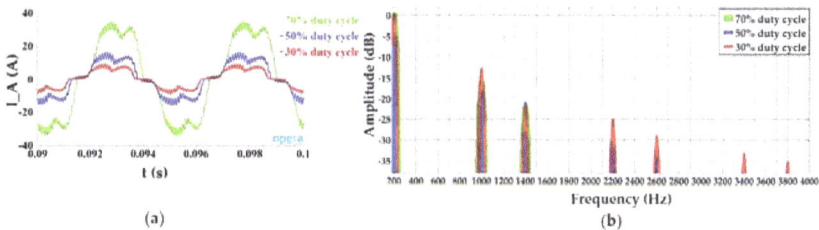

Figure 9. Waveforms for various duty cycles of: (**a**) the stator current of phase A versus time for the open loop system, (**b**) the phase current spectrum of the phase currents of the machine. (green line—70%, blue line—50%, red line—30%).

As it is seen above, when a PMSG supplies a rectifier, like in wind generation systems, the generator operation is influenced by the induced in the stator windings voltage and current harmonics [19]. The time harmonics, which are introduced, not only distort the current waveform [14,15,19–21], but also lead to the increment of the losses [19,33,34]. The losses of the machine due to the converter operation overheat the PMSM and affect its operation degrading the generator reliability and efficiency. In Table 1 the copper losses, the iron losses and the generator efficiency are quantified when the duty cycle changes. In all cases both the rotor speed and the switching frequency, f_s, remain invariable with values 3000 rpm and 10 kHz respectively. Observing the Table 1 values, it emerges that the increment of the duty cycle provokes the increment of the copper losses. On the contrary the iron losses present decrement [11,32]. If a comparison between the iron losses reduction and the corresponding increment of the copper losses is made, it will be observed that the first variation is much smaller. Actually the change of the duty cycle has greater influence on the copper losses. Finally, it can be observed that the generator has the best efficiency for duty cycle equal to 50%.

Table 1. Copper, Iron Losses and Efficiency for the Open Loop System for Different Duty Cycles and Constant Rotor Speed.

Duty Cycle (%)	Copper Losses (W)	Iron Losses (W)	η (%)
70	143.77	4.90	70.14
50	23.73	5.17	86.23
30	8.89	5.23	77.42

Figure 10a presents the iron losses versus harmonic order for the three selected duty cycles, while in Figure 10b the basic harmonic is omitted in order to focus on the behavior of the higher harmonic components when the duty cycle changes. It can be observed that the basic harmonic reduces when the duty cycle increases, while the higher order harmonics increase.

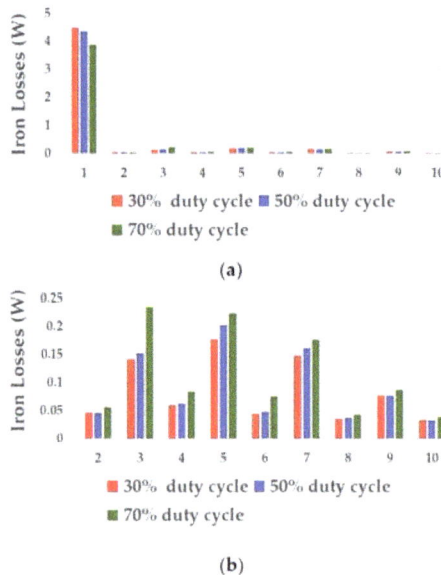

(a)

(b)

Figure 10. Total iron losses versus harmonic order for different duty cycles (green line—70%, blue line—50%, red line—30%): (a) the full spectrum from the first to the tenth harmonic, (b) the spectrum without the basic harmonic for better clarity, focusing on higher order harmonics.

Figure 11 depicts the values of the three iron loss components resulting by using the loss separation equation of Steinmetz-Bertotti. The implementation of the above theorem in the FEM software is made by inserting the essential command files. All the iron loss components present reduction with the increment of the duty cycle. If the three iron loss components are compared, it can be observed that the eddy current losses have the greatest value, while the excess losses have the minimum value.

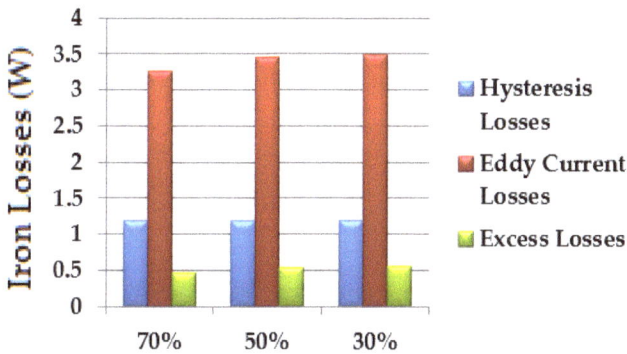

Figure 11. Total iron losses versus duty cycle for 3000 rpm rotor speed (blue line—hysteresis losses, red line—eddy current losses, green line—excess losses).

In Table 2 the copper losses, the iron losses and the efficiency are presented for three different rotor speeds when the duty cycle is 50% and the output load is invariable. The increment of the rotor speed leads to increment on both loss terms [35]. The best efficiency is achieved when the rotor rotates with the nominal speed.

Table 2. Copper, Iron Losses and Efficiency for the Open Loop System for Different Rotor Speeds and Constant Duty Cycle.

n (rpm)	Copper Losses (W)	Iron Losses (W)	η (%)
2000	10.27	2.57	83.25
2500	15.58	3.77	85.70
3000	23.73	5.17	86.23

Figure 12a presents the iron losses versus harmonic order for the three above rotor speeds, while in Figure 12b the basic harmonic is omitted in order to focus on the behavior of the higher harmonic components when the rotor speed changes. It can be seen that when the speed increases all the harmonic iron loss components increase also.

Observing Figure 13 it can be seen that the eddy current losses have the greatest contribution to the total iron losses, while the excess losses have the minimum contribution. When the speed rises, the eddy current losses present greatest variation compared to the hysteresis losses. This is logical because according to Equation (2) the hysteresis losses have proportional relationship with the frequency, while the eddy current losses have proportional relationship with the frequency square. Consequently, when the operating frequency increases, the eddy current losses increase more than the hysteresis losses [36].

(a)

(b)

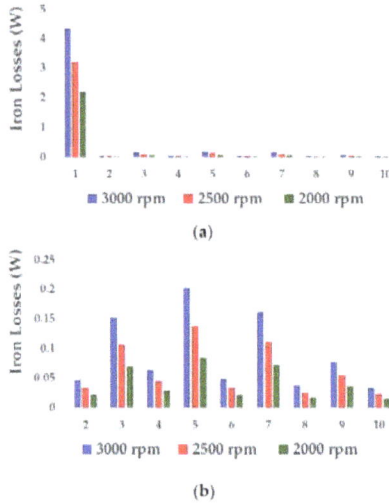

Figure 12. Total iron losses versus harmonic order for different rotor speeds (blue line—3000 rpm, red line—2500 rpm, green line—2000 rpm): (**a**) the full spectrum from the first to the tenth harmonic; (**b**) the spectrum without the basic harmonic for better clarity, focusing on higher order harmonics.

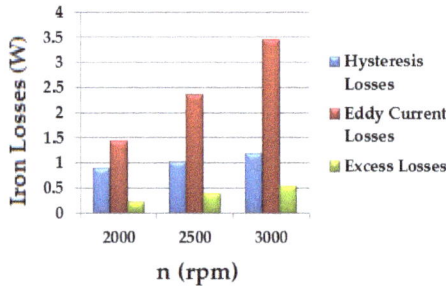

Figure 13. Total iron losses versus rotor speed for 50% duty cycle (blue line—hysteresis losses, red line—eddy current losses, green line—excess losses).

5.3. Close Loop Control of the System

In this section a close loop control has been applied, as it is seen in Figure 1. In this control scheme the duty cycle is not constant, but it is modulated by the PI controller, as described previously. The switching period T_s remains constant, while the conducting time interval of the switch, t_{on}, changes as it is determined by the close loop control. The electromagnetic variables of the machine in the steady state of the close loop operation and the waveform of the output close loop DC voltage, when the load varies, are studied. In order to study the operation of the control scheme a variation of the load was imposed. At time t_1 the load decreases from 10 Ω to 5 Ω and at time t_2 the load is reestablished in 10 Ω. Using the aforementioned close loop control, the DC output voltage of 50 V with small fluctuations, independent from the load changes, is achieved as Figure 14a presents. The rise time is small enough, so that the system is driven fairly quickly in steady state. Figure 14b presents the distribution of the absolute value of the magnetic flux density (B_{mod}) for the close loop system. The comparison between Figures 8b and 14b, reveals that in the case of the open loop control the machine is more saturated.

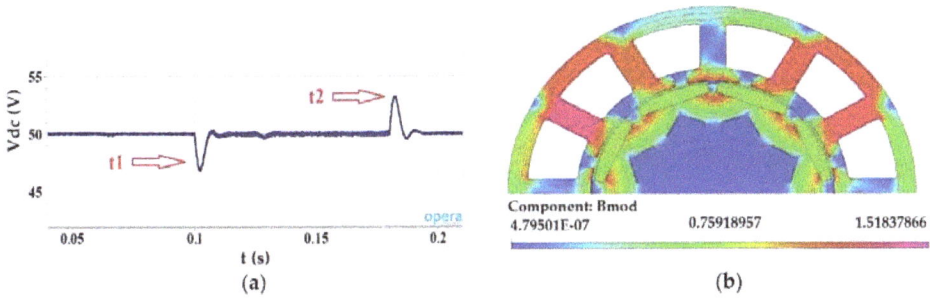

Figure 14. Waveforms of: (**a**) the output closed loop voltage for a resistive load equal to 10 Ω and nominal speed where there is a change in load from 10 Ω to 5 Ω during the time period 0.1–0.18 s, (**b**) The distribution of B_{mod} for the close loop system.

In Figure 15a the machine phase current is depicted, while in Figure 15b the corresponding phase current spectrum is shown. Both stator current waveform and spectrum for the close loop system compared with the ones of open loop system with 50% duty cycle. In both simulations the speed is constant at 3000 rpm. The rectifier input current exhibits a large fifth-order harmonic, although the seventh-order harmonic presents a slight reduction compared with the open loop control system, as shown in Figure 15b. Additionally, the second order harmonic component is missing when open loop voltage control is used, while it is significantly increased in the case of the close loop control. Table 3 compares the efficiency for the different output resistive loads when an open and a close loop is implemented respectively. From the simulation results it emerges that the efficiency is increased with the close loop control. This is an advantage of the close loop control implementation, as the losses that cause fatigue and overheating to the machine reducing its lifetime are reduced. For all the resistive load values, the increment of efficiency, in close loop control, is in the range of 3 to 5% comparing with the corresponding ones in the open loop control.

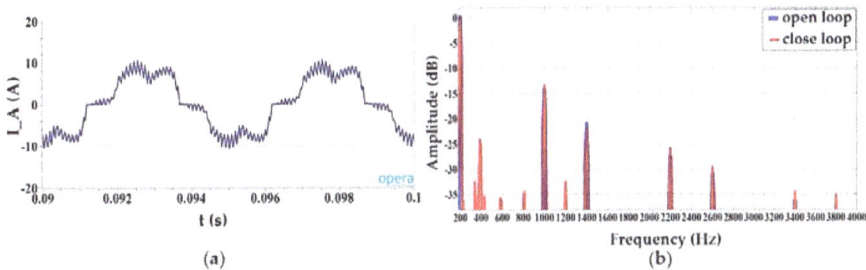

Figure 15. Waveforms of: (**a**) the stator current of phase A versus time for the close loop system, (**b**) the phase current spectrum (blue line—open loop system, red line—close loop system).

Table 3. Efficiency Comparison for Open and Close Loop System for Different Loads.

R_0 (Ω)	η_{OL} (%)	η_{CL} (%)
4	80.33	83.24
6	83.61	86.45
8	85.61	87.98
10	86.23	90.94

In Figure 16a the waveforms of the radial component of the magnetic field flux density in the middle of the airgap for both open and close loop control schemes are compared. In Figure 16b the corresponding harmonic spectra are depicted. The first harmonic is neglected in Figure 16b for clarity purposes. Comparing the spectra one concludes that the amplitude of most magnetic field harmonics is minimally reduced when the close loop control is applied.

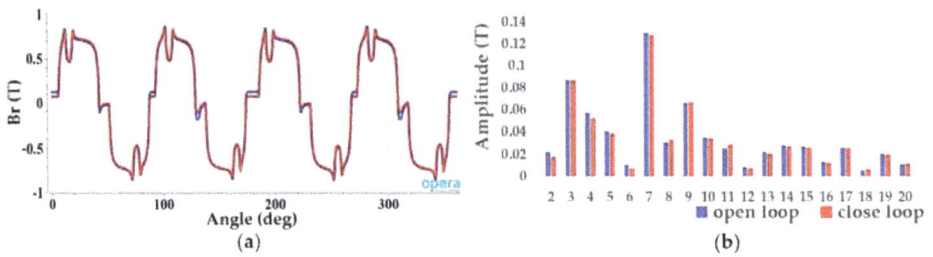

Figure 16. Comparison of the open and close loop system: (blue line—open loop system, red line—close loop system) (a) The radial component of the magnetic field flux density in the middle of the airgap, (b) The corresponding harmonic spectrums.

The copper, iron losses and the efficiency are presented in Table 4 for the close loop system. These values compared with the ones of the open loop system (Table 2—3000 rpm), reveal that in the case of closed loop the copper losses are reduced significantly, the iron losses seem to remain almost invariable, while the generator efficiency increases.

Table 4. Copper, Iron Losses and Efficiency for Close Loop System.

Copper Losses (W)	Iron Losses (W)	η (%)
8.75	5.24	90.94

Figure 17 compares the iron losses harmonic spectra for the cases of open loop system with 50% duty cycle and the close loop system. In both simulations the speed is constant at 3000 rpm. The first harmonic is missed out in Figure 17 for better clarity. It can be observed that although the iron losses are almost invariable when the close loop control is implemented, in comparison with the open loop control, the amplitude of some harmonic components changes. More specifically, the variation is evident in odd harmonics (3th, 5th, 7th) that present reduction on the case of the close loop control.

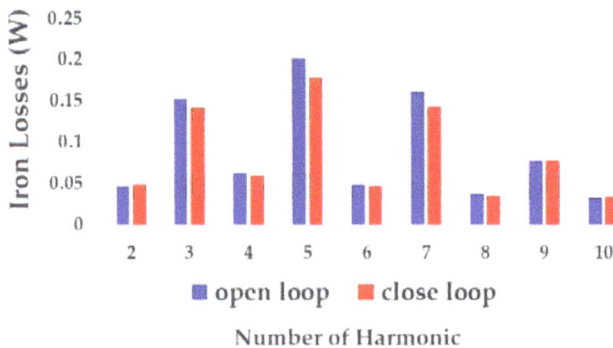

Figure 17. Total iron losses versus harmonic order (blue line—open loop, red line—close loop).

In Figure 18 the total iron losses divided into the three components are presented. Likewise with the results in the open loop control section, the eddy current losses have the major contribution on the total losses, the excess losses have the smallest part and the hysteresis losses are located in the middle.

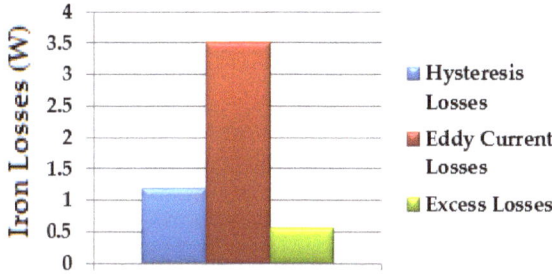

Figure 18. The three losses terms that create the total iron losses for the close loop system.

6. Losses Due to the Rotating Magnetic Field

In order the issue of the rotating magnetic field to be addressed, two alternating fields have to be considered. One field with maximum value the maximum value of the rotational field (B_{max}) and a second field with maximum value the minimum value (B_{min}) of the rotational field. Figure 19 depicts the expected form of the geometrical locus in the machine. In this section, two regions of the stator core are selected, as it can be seen from Figure 20 and both the magnetic induction waveforms and the geometrical locus of the vector B_t versus B_r are investigated for these regions. The rotating magnetic field appears as the geometrical locus of the B_t versus B_r and it will have the shape of a circle or an ellipsis.

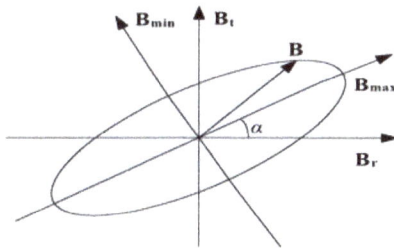

Figure 19. Elliptical locus in the rotating machine [25].

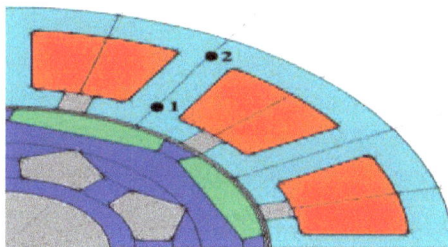

Figure 20. The two regions (1, 2) of the stator in which the magnetic induction and the geometrical locus are computed.

When the B_{min} has zero value, it means that in this region the magnetic field is not rotating but only alternating and the geometrical locus is a straight line. The locus becomes a circle, in the case that the two fields are equal. The changes of the shape of the geometrical locus are quite important, since additional core losses are created by the rotational magnetic field which cannot be estimated using Equation (2). Indeed, keeping B_{max} unchanged, B_{min} can vary on the range [0, B_{max}], the locus changes from a line to a circle, and the iron losses increase in relation with the change of B_{min} [25].

In Figure 21 the absolute value of the magnetic induction versus time is depicted for the two different regions, while in Figure 22 the corresponding variation of B_t versus B_r over one electrical period is presented. The study is referred to open loop system with 50% duty cycle and nominal rotor speed 3000 rpm. By Figure 22, it can be observed that the rotational vector of the magnetic induction creates extra iron losses in the generator. Indeed, the losses in the machine region where the machine induction value falls to zero periodically are not the same with the ones of a region where the machine induction has the same maximum value but it doesn't fall to zero. Consequently in the region 1 the geometrical locus is an ellipsis and the magnetic induction presents space rotation. In the case of the region 2 the ellipsis increases and the locus shape looks like to a circle. Obviously, in both cases the vector of the magnetic induction has no zero value, while concurrently it presents space rotation.

Therefore, in order to compute the total iron losses, one major direction in the geometrical locus that will be the dominant should be taken into account and one minor direction also. The total iron losses are the sum of the iron losses in the major and the minor direction. Thus, the classical equation of Steinmetz-Bertotti should be rewritten by taking into account the two fields:

$$P_{fe} = P_h + P_e + P_a = \varrho_{fe} k_h f B_{max}{}^\alpha + \varrho_{fe} k_h f B_{min}{}^\alpha + \varrho_{fe} k_e f^2 B_{max}{}^2 + \varrho_{fe} k_e f^2 B_{min}{}^2 + \varrho_{fe} k_a f^{1.5} B_{max}{}^{1.5} + \varrho_{fe} k_a f^{1.5} B_{min}{}^{1.5}, \quad (5)$$

The total iron losses resulting from the addition of the rotating magnetic field losses are presented in Table 5. The results of Table 5 have been exported using Equation (5) in which the existence of one major and one minor field is considered. The losses resulting from the rotating magnetic field are small, as it was expected.

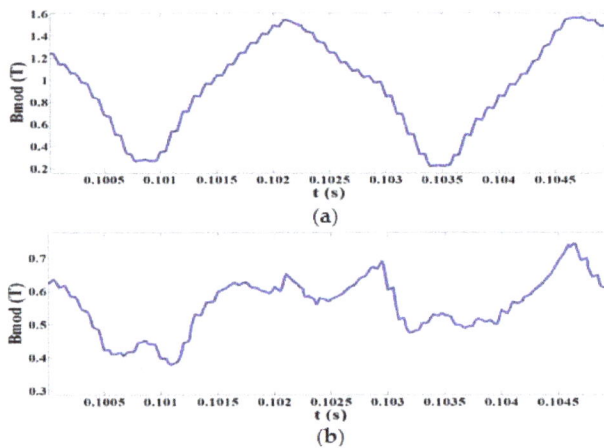

Figure 21. The absolute value of the magnetic induction versus time for an open loop control with 50% duty cycle and rotor speed 3000 rpm: (**a**) Region 1, (**b**) Region 2.

Table 5. Total Iron Losses When the Losses Due to Rotating Magnetic Field Taken into Account.

	Hysteresis Losses (W)	Eddy Current Losses (W)	Excess Losses (W)
Without Rotating Magnetic Field	1.18	3.45	0.54
Rotating Magnetic Field	0.09	0.40	0.06
Total Losses	1.27	3.85	0.60

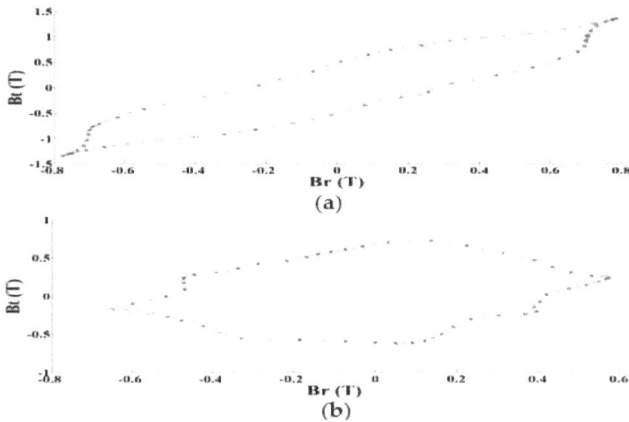

Figure 22. Magnetic induction geometrical locus for an open loop control with 50% duty cycle and rotor speed 3000 rpm: (**a**) Region 1, (**b**) Region 2.

7. Conclusions

In this paper, the performance of a PMSG system, appropriate for wind power applications is studied. Furthermore, the copper and iron losses are estimated. The iron losses are estimated with the use of the Steimetz-Bertotti loss separation equation. Moreover, the iron losses due to the rotating magnetic field are also taken in to account. The PMSG is simulated using FEA, thus the results are more accurate and an electromagnetic insight of the machine can be determined. The validation of this FEM model has been done by setting up in the laboratory an experimental system and comparing the simulation with experimental results. Firstly, the PMSG without a converter connection is investigated. Two models, with asymmetric and symmetric magnets mounted on rotor, are compared. The emphasis is given on the ability of the FEM analysis to obtain qualitatively and quantitatively more accurate results. These results cannot be obtained using Simulink machine models, because this kind of software cannot simulate the machine in detail. From the comparison between the asymmetric and the symmetric model of the generator it is shown that the asymmetric model has more sinusoidal input currents with smaller harmonic component and smaller cogging torque. Secondly, the PMSG coupled with the three-phase single switch boost rectifier is investigated. An open loop control has been implemented to the converter switch which keeps the duty cycle constant. The rectifier increases the harmonics in the machine phase currents which have been estimated and discussed. Especially the fifth and the seventh harmonic are increased when the generator is connected to the rectifier and an open loop control scheme is applied. Moreover, the reduction of the PWM duty cycle leads to current harmonics increment and copper losses reduction. In opposition the iron losses are increased when the duty cycle reduces, but the increment is small compared to the copper losses reduction. Regarding the generator efficiency the greatest value is achieved for 50% duty cycle and rotor speed at its nominal value. The increment of the speed leads to both copper and iron losses increment. Then, a close loop control has been applied to the converter switch. The close loop control scheme leads to a slight reduction of the seventh-order harmonic, although the fifth harmonic of the machine phase current remains high. Furthermore, the efficiency presents increment in the case of close loop control. The waveforms of the radial component of the magnetic field flux density in the middle of the airgap and the corresponding harmonic spectrums for both open and close loop control schemes are compared. The comparison shows that the amplitude of most magnetic field harmonics is slightly reduced when the close loop control is applied. The copper losses present remarkably reduction. The iron losses maintained almost invariable, but the amplitude of the iron losses associated with odd harmonics (3th, 5th, 7th) present reduction in the case of the close loop control. Moreover, the existence of the rotating magnetic fields leads to extra machine losses. To conclude, in both open and close control simulations, the eddy current losses have the major contribution on the total iron losses, and the excess losses have the minor contribution.

The significance of this work lies in the fact that by using a detailed analysis of the machine, using FEM, at the design stage, the electromagnetic results can be accurately obtained in order to choose the most essential configuration for the whole system. An important outcome is that by using the close loop control it is possible to achieve less losses and greater efficiency of the machine, while the research in the field of control techniques can be expanded. In a next step the losses on the magnets can be added in order to have the full impact of the losses in machines efficiency, as the early estimation of the losses can prevent generator overheating and damaging. In addition, this study can be expanded using the full grid connection system. Other future investigation can be the optimization of the control of the boost rectifier in order to increase further the efficiency or the application of other control methods and the comparison of these methods and their influence on the machine electromagnetic behavior. The full grid system will be investigated exclusively using the FEM software for the machine, unlike most papers on this topic which use simulation tools that simulate in detail the electric circuits but not the machine.

Author Contributions: The presented work was carried out through the cooperation of all authors. A.C.B. conducted the research and wrote the paper and J.C.K. supervised the whole study and edited the manuscript.

Funding: This research received no external funding.

Conflicts of Interest: The authors declare no conflict of interest.

References

1. Boldea, I.; Tutelea, L.; Blaabjerg, F. High power wind generator designs with less or no PMs: An overview. In Proceedings of the 2014 17th International Conference on Electrical Machines and Systems (ICEMS), Hangzhou, China, 22–25 October 2014.

2. Polinder, H.; Ferreira, J.A.; Jensen, B.B.; Abrahamsen, A.B.; Atallah, K.; McMahon, R.A. Trends in wind turbine generator systems. *IEEE J. Emerg. Sel. Top. Power Electron.* **2013**, *1*, 174–185. [CrossRef]

3. Goto, H.; Guo, H.J.; Inchinokura, O. A micro wind power generation system using permanent magnet reluctance generator. In Proceedings of the 2009 13th European Conference on Power Electronics and Applications (ECPEA), Barcelona, Spain, 8–10 September 2009.

4. Novakovic, B.; Duan, Y.; Solveson, M.; Nasiri, A.; Ionel, D.M. Multi-physics system simulation for wind turbines with permanent magnet generator and full conversion power electronics. In Proceedings of the 2013 13th IEEE International Conference on Electrical Machines and Drives (IEMDC), Chicago, IL, USA, 12–15 May 2013.

5. Novakovic, B.; Duan, Y.; Solveson, M.; Nasiri, A.; Ionel, D.M. Comprehensive modeling of turbine systems from wind to electric grid. In Proceedings of the 2013 5th Annual IEEE Energy Conversion Congress and Exposition (ECCE), Denver, CO, USA, 15–19 September 2013.

6. Uma, S.P.; Manikandan, S. Control technique for variable speed wind turbine using PI controller. In Proceedings of the 2013 1th IEEE International Conference on Emerging Trends in Computing, Communications and Nanotechnology (ICETCCN), Tirunelveli, India, 25–26 March 2013.

7. Raza, K.S.M.; Goto, H.; Guo, H.J.; Inchinokura, O. Maximum power point tracking control and voltage regulation of a DC grid-tied wind energy conversion system based on a novel permanent magnet reluctance generator. In Proceedings of the 2007 10th International Conference on Electrical Machines and Systems (ICEMS), Seoul, Korea, 8–10 October 2007.

8. Koreboina, V.B.; Venkatesha, L. Modelling and simulation of switched reluctance generator control for variable speed wind energy conversion systems. In Proceedings of the 2012 4th IEEE International Conference on Power Electronics, Drives and Energy Systems (ICPEDES), Bengaluru, India, 16–19 December 2012.

9. Ahmed, N.A.; Al-Othman, A.K.; AlRashidi, M.R. Development of an efficient utility interactive combined wind/photovoltaic/fuel cell power system with MPPT and DC bus voltage regulation. *Electr. Power Syst. Res.* **2011**, *81*, 1096–1106. [CrossRef]

10. Asefa, P.; Bargallo Perpina, R.; Barzegaran, M.R.; Lapthorn, A.; Mewes, D. Load identification of different Halbach-array topologies on permanent magnet synchronous generators using the coupled field-circuit FE methodology. *Electr. Power Syst. Res.* **2018**, *154*, 484–492. [CrossRef]

11. Barmpatza, A.; Pallis, I.K.; Kappatou, J. FEM modeling and study of a permanent magnet synchronous generator with DC-link voltage control for wind power systems. In Proceedings of the 2015 17th International Symposium on Electromagnetic Fields in Mechatronics, Electrical and Electronic Engineering (ISEF), Valencia, Spain, 10–12 September 2015.

12. Barmpatza, A.C.; Kappatou, J.C. PWM influence on the losses of a PMSG supplying a boost rectifier. In Proceedings of the 2016 22th International Conference on Electrical Machines (ICEM), Lausanne, Switzerland, 4–7 September 2016.

13. Tudorache, T.; Melcescu, L.; Popescu, M. Methods for cogging torque reduction of directly driven PM Wind generators. In Proceedings of the 2010 12th International Conference on Optimization of Electrical and Electronic Equipment (ICOEEE), Basov, Romania, 20–22 May 2010.

14. Kolar, J.W.; Ertl, H.; Zach, F.C. Space vector-based analytical analysis of the input current distortion of a three-phase discontinuous-mode boost rectifier system. *IEEE Trans. Power Electron.* **1995**, *10*, 733–745. [CrossRef]

15. Jangand, Y.; Jovanonic, M. A novel robust harmonic injection method for single-switch three phase discontinuous-conduction-mode boost rectifiers. *IEEE Trans. Power Electron.* **1998**, *5*, 824–834. [CrossRef]

16. Xu, Z.; Zhang, D.; Wang, F.; Boroyevich, D. Unified control for the permanent magnet generator and rectifier system. In Proceedings of the 2011 26th Annual IEEE Applied Power Electronic Conference and Exposition (APEC), Fort Worth, TX, USA, 6–11 March 2011.

17. Li, J.; Choi, D.W.; Son, D.-H.; Cho, Y.-H. Effects of MMF harmonics on rotor eddy-current losses for inner-rotor fractional slot axial flux permanent magnet synchronous machines. *IEEE Trans. Magn.* **2012**, *48*, 839–842. [CrossRef]

18. Pfingsten, G.; Steentjes, S.; Hombitzer, M.; Franck, D.; Hameyer, K. Influence of winding scheme on the iron-loss distribution in permanent magnet synchronous machines. *IEEE Trans. Magn.* **2014**, *50*. [CrossRef]

19. Stiebler, M. Sub-transients in PM synchronous generator with diode rectifier load. In Proceedings of the 2014 21th International Conference on Electrical Machines (ICEM), Berlin, Germany, 2–5 September 2014.

20. Jang, Y.; Jovanonic, M. A Comparative study of single-switch, three-phase, high-power-factor rectifiers. *IEEE Trans. Ind. Appl.* **1998**, *34*, 1327–1334. [CrossRef]

21. Yao, K.; Ruan, X.; Mao, X.; Ye, Z. Variable-duty-cycle control to achieve high input power factor for DCM boost PFC converter. *IEEE Trans. Ind. Appl.* **2010**, *58*, 1856–1865. [CrossRef]

22. Krings, A.; Soulard, J.; Wallmark, O. pwm influence on the iron losses and characteristics of a slotless permanent-magnet motor with SiFe and NiFe stator cores. *IEEE Trans. Ind. Appl.* **2015**, *51*, 1457–1484. [CrossRef]

23. Juergens, J.; Ponick, B.; Winter, O.; Fricassè, A. Influences of iron loss coefficients estimation on the prediction of iron losses for variable speed motors. In Proceedings of the 2015 14th IEEE International Conference on Electrical Machines and Drives (IEMDC), Coeur d' Alene, ID, USA, 10–13 May 2015.

24. Ibrahim, M.; Pillay, P. Core loss prediction in electrical machine laminations considering skin effect and minor hysteresis loops. *IEEE Trans. Ind. Appl.* **2014**, *49*, 2061–2068. [CrossRef]

25. Huang, Y.; Dong, J.; Zhu, J.; Guo, Y. Core loss modeling for permanent-magnet motor based on flux variation locus and finite-element method. *IEEE Trans. Magn.* **2012**, *48*, 1023–1026. [CrossRef]

26. Kim, W.; Kim, J.M.; Seo, S.W.; Ahn, J.H.; Hong, K.; Choi, J.Y. core loss analysis of permanent magnet linear synchronous generator considering the 3-D flux path. *IEEE Trans. Magn.* **2018**, *54*. [CrossRef]

27. Hwang, S.W.; Lim, M.S.; Hong, J.P. Hysteresis torque estimation method based on iron-loss analysis for permanent magnet synchronous motor. *IEEE Trans. Magn.* **2016**, *52*. [CrossRef]

28. Tudorache, T.; Trifu, I. Permanent-magnet synchronous machine cogging torque reduction using a hybrid model. *IEEE Trans. Magn.* **2012**, *48*, 2627–2632. [CrossRef]

29. Bianchini, C.; Immovilli, F.; Lorenzani, E.; Bellini, A.; Davoli, M. Review of design solutions for internal permanent-magnet machines cogging torque reduction. *IEEE Trans. Magn.* **2012**, *48*, 2685–2693. [CrossRef]

30. Tsotoulidis, S.; Safacas, A. A sensorless commutation technique of a brushless DC motor drive system using two terminal voltages in respect to a virtual neutral potential. In Proceedings of the 2012 20th International Conference on Electrical Machines (ICEM), Marseille, France, 2–5 September 2012.

31. Tsotoulidis, S.; Safacas, A.; Mitronikas, E. A Sensorless Commutation strategy for a brushless DC motor drive system based on detection of back electromagnetic force. In Proceedings of the 2011 International Aegean Conference on Electrical Machines and Power Electronics and Electromotion, Joint Conference (ACEMP), Istanbul, Turkey, 8–10 September 2011.

32. Zhao, N.; Zhu, Z.Q.; Liu, W. Rotor eddy current loss calculation and thermal analysis of permanent magnet motor and generator. *IEEE Trans. Magn.* **2011**, *47*, 4199–4202. [CrossRef]

33. Ruderman, A.; Reznikov, B.; Busquets-Monge, S. Asymptotic time domain evaluation of a multilevel multiphase PWM converter voltage quality. *IEEE Trans. Ind. Electron.* **2013**, *60*, 1999–2009. [CrossRef]

34. Ruderman, A. Discussion on effect of multilevel inverter supply on core losses in magnetic materials and electrical machines. *IEEE Trans. Energy Convers.* **2015**, *30*, 1604. [CrossRef]

35. Boubakera, N.; Matta, D.; Enricia, P.; Nierlichb, F.; Durandb, G.; Orlandinic, F.; Longèrec, X.; Aïgba, J.S. Study of eddy-current loss in the sleeves and Sm–Co magnets of a high-performance SMPM synchronous machine (10 kRPM, 60 kW). *Electr. Power Syst. Res.* **2017**, *142*, 20–28. [CrossRef]

36. Chen, P.; Tang, R.; Tong, W.; Han, X.; Jia, J.; Zhu, X. Analysis of losses of permanent magnet synchronous motor with PWM supply. In Proceedings of the 2014 17th International Conference on Electrical Machines and Systems (ICEMS), Hangzhou, China, 22–25 October 2014.

energies

MDPI

Article

Cogging Torque Reduction Based on a New Pre-Slot Technique for a Small Wind Generator

Miguel García-Gracia [1,*], Ángel Jiménez Romero [1], Jorge Herrero Ciudad [2] and Susana Martín Arroyo [1]

1 Department of Electrical Engineering, University of Zaragoza, 50018 Zaragoza, Spain;
anjirom@unizar.es (Á.J.R.); smartin@unizar.es (S.M.A.)
2 For Optimal Renewable Energy Systems, S.L. (4fores), 50197 Zaragoza, Spain; herrero@4fores.es
* Correspondence: mggracia@unizar.es; Tel.: +34-976-761923

Received: 18 October 2018; Accepted: 15 November 2018; Published: 20 November 2018

Abstract: Cogging torque is a pulsating, parasitic, and undesired torque ripple intrinsic of the design of a permanent magnet synchronous generator (PMSG), which should be minimized due to its adverse effects: vibration and noise. In addition, as aerodynamic power is low during start-up at low wind speeds in small wind energy systems, the cogging torque must be as low as possible to achieve a low cut-in speed. A novel mitigation technique using compound pre-slotting, based on a combination of magnetic and non-magnetic materials, is investigated. The finite element technique is used to calculate the cogging torque of a real PMSG design for a small wind turbine, with and without using compound pre-slotting. The results show that cogging torque can be reduced by a factor of 48% with this technique, while avoiding the main drawback of the conventional closed slot technique: the reduction of induced voltage due to leakage flux between stator teeth. Furthermore, through a combination of pre-slotting and other cogging torque optimization techniques, cogging torque can be reduced by 84% for a given design.

Keywords: cogging torque; permanent magnet synchronous generator; small wind turbines; finite element method; renewable energy; energy conversion

---

1. Introduction

Increasing interest in the efficiency of electric machinery and reducing maintenance costs is making the use of permanent magnet synchronous generators (PMSGs) more common. PMSGs combine high efficiency with low maintenance and a high power density [1], factors that make them extremely attractive for use in renewable energy applications are, such as wind [2], wave power [3], and tidal power [4], or electrical mobility applications [5] and, in general, in uses where they must act as a motor or generator. Furthermore, in renewable energy applications, PMSGs allow direct-drive configurations, making the use of a gearbox unnecessary or reducing the number of gearbox stages, which decreases the overall generator volume and improves its efficiency [6].

However, machines based on permanent magnets (PMs) also have some drawbacks, and the cogging torque is one of the main ones. The magnetic interaction between the flux generated by the rotor PMs and the stator geometry results in a pulsating torque called cogging torque, which, depending on the PM machine design, can cause an undesired ripple in both the machine's induced voltage (EMF) and its mechanical torque [7,8]. Other problems with PMSGs are the vibrations and noise they make. Since this type of machine has high magnetic flux density values in the air gap, the electromagnetic forces between the PMs and the stator teeth are high [9]. These electromagnetic forces are divided into two components, one radial and the other tangential. The tangential component of the electromagnetic force contributes to the torque in the stator teeth, while the radial component causes vibrations and even deformations in the machine [10]. These radial forces act on the stator producing vibrations and noise, especially when their frequency coincides with the natural frequency of the machine's mechanical structure [11].

The cogging torque is especially important in wind energy applications as it establishes in which conditions the system will begin generating. The mechanical torque captured by the generation system must be larger than the cogging torque starting the rotation, which is why achieving a reduced cogging torque is one of the objectives for this type of machine.

There are several methods to reduce cogging torque in the PMSG design phase. The most used is skewing, which consists of preventing the stator teeth and the magnets from becoming aligned by either turning the stator teeth [6,12] or the rotor's permanent magnets [1,2,13]. The required skew angle to largely cancel out the effect of the interactions between the PMs and the slots depends on how many slots and poles the machine has. Other methods study the use of notches in stator teeth [7,14]. These notches produce the same effect in the magnetic interaction as the slots and increase the effective number of slots, which impacts on the cogging torque as it depends on how many poles and slots the machine has. Therefore, this method's effectiveness is conditioned by the number of poles and slots selected in the design.

A study is presented in [15,16] in which slot openings in one half of the stator, shift in one direction with respect to the tooth and in the other half, shift in the other direction. This means the cogging torque waveform moves in opposite directions in each machine half and the cogging caused by each machine half may be cancelled out depending on the shifted angle. Other studies focus on the shape of PM edges, concluding that their size can be reduced on the magnet sides to lessen air gap reluctance variation, which reduces the magnetic energy variation in the machine and, therefore, mitigates the cogging torque [17,18].

Several authors have conducted studies of PMSG with closed slots and their effects. Leakage fluxes caused as a result of closing stator slots are analyzed in [4], concluding that the size of PMs should be increased to compensate flux loss through closed slots. The increase in iron losses caused by tooth-tip saturation, distortion in the induced voltage this saturation causes and how the use of closed slots influences this, are studied in [19,20]. The study by [21] focuses on average torque and its ripple in machines with closed slots for several stator types. The work by [22] analyzes the use of magnetic composite wedges to close stator slots in terms of stator flux linkage, average and ripple torque, and magnetic losses.

Unlike the above-mentioned methods, which focus on minimizing the cogging torque in the machine design stage, this article proposes a cogging torque reduction method that is easy to implement without the need for any changes to the original design of the machine, a 6.3-kW generator for a small wind turbine. The suggested solution comprises sliding a metal part (a pre-slot wedge) into the slots after completing the machine winding. This technique minimizes slot openings so that induced voltage remains unaltered and the mounting of machine windings is not hampered. The results of the proposed method are analyzed using FEMM 2D (Finite Element Method Magnetics) software on an original PMSG design and compared with the results obtained experimentally. Additionally, constructive improvements are suggested to reduce cogging torque. Finally, the article shows how the proposed technique can also be combined with the skewing technique, thus significantly reducing the cogging torque to 0.03 Nm in the ideal case and 0.51 Nm when imperfections in the manufacturing process are considered.

2. Machine Type and Main Parameters

The machine involved in this study is a double layer fractional winding PMSG with an interior rotor and surface-mounted magnets comprising 36 slots and 20 poles. Figure 1a is a cross section of the generator showing the slots forming the stator, the rotor in the internal part, and the surface-mounted magnets above it in the central part. The details of the millimeter measurements of the machine's slots and PMs are shown in Figure 2. The characteristic parameters of the studied PMSG are shown in Table 1.

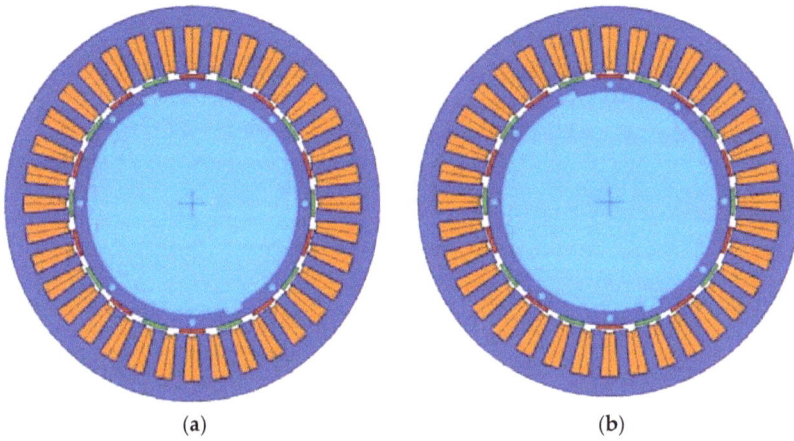

(a)   (b)

Figure 1. Cross section area of the permanent magnet synchronous generator (PMSG): (**a**) Original model; (**b**) Model with centered holes.

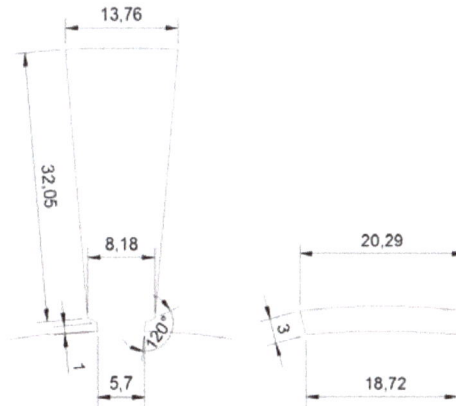

Figure 2. Slot and magnet dimensions (mm).

Table 1. Parameters of the permanent magnet synchronous generator (PMSG) machine.

Parameter	Value
Phase	3
Pole number	20
Slot number	36
Rated speed	232 rpm
Rated power	6300 W
Rated voltage	256.4 V
Rated torque	102 Nm
Air gap	1 mm
Thickness of PM	3 mm
Rotor diameter	180 mm
Material of steel	M330-50A
Material of PM	NdFeB

The analysis of the PMSG and the cogging torque reduction methods proposed in this study was performed using FEMM 2D finite element software. To validate the FEMM model used in the cogging torque reduction analysis a comparison is made with the experimental values of the original machine.

3. Cogging Torque

Cogging torque is a parasitic torque resulting from interactions between the rotor's permanent magnets and the stator slots. Air gap reluctance differs depending on the rotor's angular position to the slots. Rotor magnets tend to align with the stator in the position in which air gap permeance is larger [23], so when they are shifted from this position during rotation, they generate a torque, the cogging torque.

Electromagnetic torque can be obtained from the variation in the total energy of the magnetic field compared with the angular position of the rotor θ when excitation current is constant [14]

$$T = -\frac{\partial W_c}{\partial \theta} \tag{1}$$

The total energy stored in the magnetic field or coenergy W_c in a PMSG is given by [7]

$$W_c = \frac{1}{2} L\, i^2 + \frac{1}{2}(R + R_m)\, \varnothing_m^2 + N\, i\, \varnothing_m \tag{2}$$

where L is the inductance of the windings, i the excitation current, R and R_m are, respectively, the reluctances viewed by the magnetomotive force and by the magnetic field, $\varnothing_m$ the flux due to the magnets crossing the air gap, and N the number of winding turns.

Therefore, substituting in Equation (2) results in

$$T = \frac{1}{2} i^2 \frac{dL}{d\theta} - \frac{1}{2} \varnothing_m^2 \frac{dR}{d\theta} + N\, i\, \frac{d\varnothing_m}{d\theta} \tag{3}$$

The second term of Equation (3) corresponds to magnet reluctance torque and it is known as cogging torque [17], T_{cog}

$$T_{cog} = -\frac{1}{2} \varnothing_m{}^2 \frac{dR}{d\theta} \tag{4}$$

As observed in Equation (4), cogging torque is independent of the current and corresponds to the result of analyzing Equation (3). when the machine is in open circuit. Cogging torque depends on magnetic flux and on the rate of change of air-gap reluctance. From Equation (4), to minimize T_{cog}, reluctance R should be independent of the rotor position. Therefore, a very low cogging torque design requires an almost constant value of R for any rotor position.

4. Cogging Torque Measurement

Cogging torque is calculated in FEMM for every angular rotor position, making the machine operate off-load. The torque is calculated by integrating the Maxwell stress tensor throughout the air gap

$$T_{cog} = \frac{L}{g\, u_o} \int_S r\, B_n\, B_t\, dS \tag{5}$$

where μ_o is the air gap permeability, L is the rotor depth, g is the air gap length, B_n the normal flux density, B_t the tangential flux density and r the radius from the center of the rotor to the center of the air gap [7].

Compared with the results obtained when the calculation is based on the magnetic energy variation with respect to the angular rotor position given by Equation (1), in [18] it is shown that both methods obtain almost identical results.

The simulation in FEMM of the PMSG in Figure 1a obtained the cogging torque shown in Figure 3 ("original" curve), whose maximum value is 2.32 Nm, while the experimental results of the machine show maximum values of 3.70 Nm. The main reason for this deviation from experimental values is due to component manufacturing tolerance. Consequently, if a tolerance of ±0.1 mm is included in the 20 PMs of the PMSG model and this error is distributed randomly at the height of the PMs, the result shown in Figure 3 (curve "with manufacturing errors") is obtained. Having magnets that are not the same, impacts the cogging torque significantly, mainly because differently sized PMs cause higher magnetic flux variations in the air gap. The maximum cogging torque value obtained in the simulation is 3.90 Nm (Figure 3). This value is slightly higher than the experimental PMSG results, making it possible to validate the developed FEMM model with respect to the cogging torque analysis.

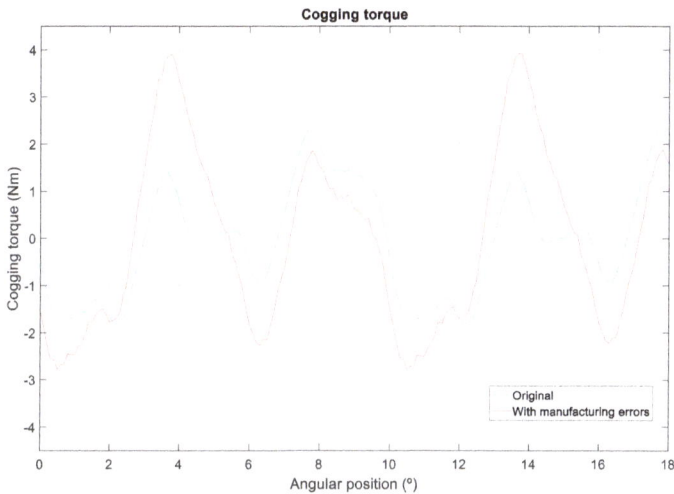

Figure 3. Simulation results of cogging torque of the original model considering manufacturing errors.

5. Cogging Torque Reduction Methods

5.1. Pre-Slot Method

The main objective is to reduce the cogging torque without affecting the machine's construction characteristics and, therefore, without making any changes in the generator's geometry.

A closed-slot stator topology reduces reluctance variation in the air gap and, therefore, the machine's maximum cogging torque value. Furthermore, the minimum dimensions of PMSG slot openings are conditioned by winding mounting factors. Their minimum size depends on the cross section of the winding conductors so that they can be inserted in the slot.

Furthermore, the slot closing method has the drawback of generating a leakage flux through the slots due to the high permeability of the magnetic core connecting the teeth, Figure 4a. These leakages reduce the flux linked by the machine windings, thus producing a drop in induced voltage. This drop in induced voltage can be seen in Figure 5, showing the induced voltage of the PMSG with open slots (the "original" curve) and with closed slots. The effective induced voltage value is 256.4 V in the original generator model with open slots, and it decreases to 221.6 V when the slots are closed.

To avoid the above-mentioned drawbacks, the proposed cogging torque reduction method consists of closing the slots by sliding in a pre-slot part made of the same ferromagnetic material as the stator, as observed in Figure 6. The pre-slot is placed between the teeth longitudinally after machine winding; this does not alter the winding or the slot fill factor. Figure 6a provides details of the space between two of the machine's stator teeth showing where the ferromagnetic part is slid into the start of each slot. The pre-slot considered in this study is 1.5 mm high; its dimensions are adjusted to the available space to render changing the machine winding unnecessary.

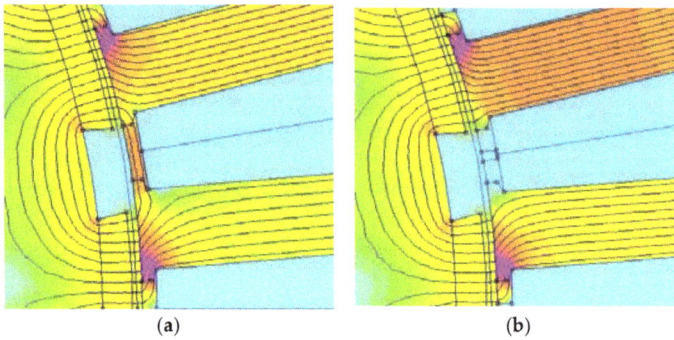

(a) **(b)**

Figure 4. Magnetic field lines ($\theta = 0°$): (**a**) Model with closed slots; (**b**) Model with pre-slots.

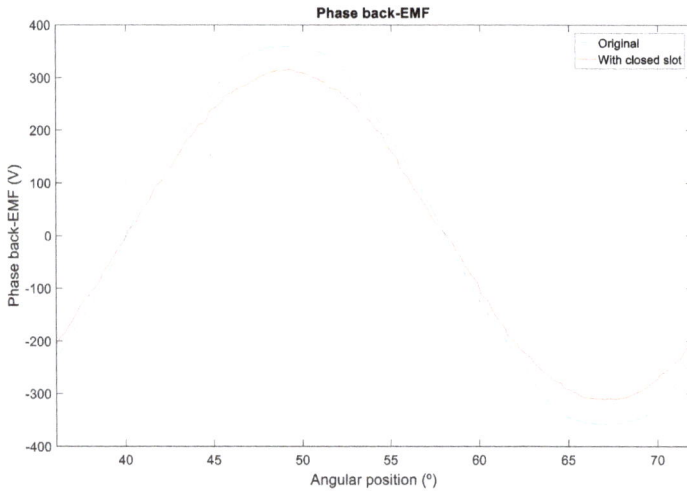

Figure 5. Effect of closed slots in back electromotive force (EMF).

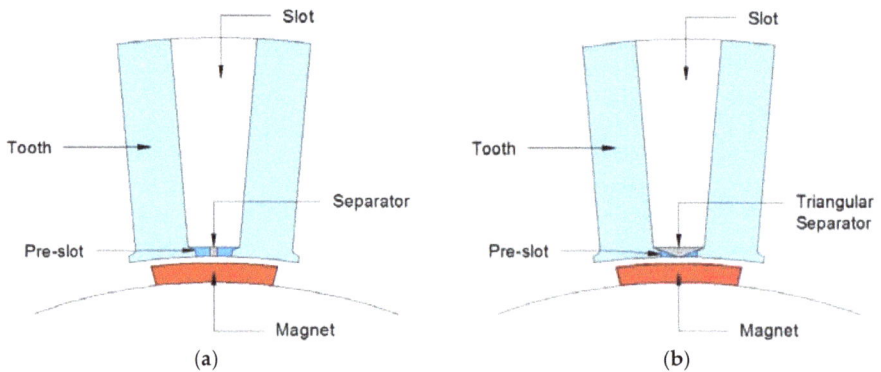

(a) **(b)**

Figure 6. Proposed cogging-torque reduction method: (**a**) Pre-slot with separator; (**b**) Pre-slot with triangular separator.

A material separator with low magnetic permeability (aluminum or similar) and a width of 1 mm, the same distance as the machine's air gap, is in the central part of the pre-slot. The purpose of the central separator is to prevent the above-mentioned flux leakage linked by the windings. As this is a non-magnetic separator, it prevents the pre-slot from closing the magnetic field lines and, therefore, preventing flux from circulating between two consecutive PMs.

In accordance with the developed FEMM model, inserting pre-slots with a separator manages to reduce the maximum cogging torque value by 37.9% compared with the original PMSG, as observed in the results shown in the graph in Figure 7, but it does not decrease induced voltage as using the separator reduces leakages.

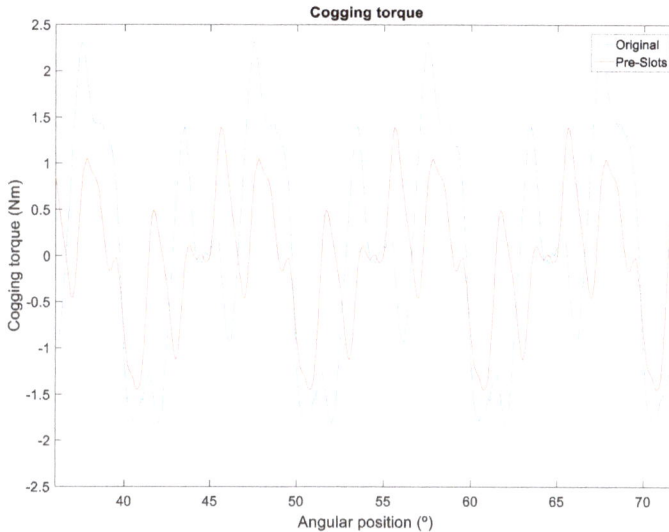

Figure 7. Comparison of cogging torque reduction with pre-slot method.

This cogging torque reduction can be improved by considering other alternative geometrical pre-slot configurations. A triangular separator, as shown in Figure 6b, can lessen the magnetic energy variation caused in the original teeth edges or in pre-slot separators, which decreases the machine's cogging torque.

Pre-slot geometry with a triangular separator prevents leakage flux through it, as occurs with the central separator model, thus preventing the undesired decrease in induced voltage and in linked flux through the machine windings. Similarly, it produces a higher reduction in maximum cogging torque, as shown in the graph in Figure 8, decreasing the cogging torque generated by over 47.8%.

An analysis of induced voltage harmonics was conducted to confirm that the installation of the proposed pre-slot system to reduce cogging torque does not affect the machine's technical characteristics. The PMSG is designed for small wind-power applications and, therefore, if an uncontrolled rectifier is used, the harmonics level is of no importance. However, if the connection is via a full converter, the opposite is the case. Figure 9 shows the frequency spectrum of the first 20 harmonics of each of the waves and the harmonic distortion rate (THD) for each model; therefore, the analysis was conducted from the fundamental frequency of 38.67 Hz to the twentieth harmonic of 773.4 Hz.

Table 2 shows how the THD values obtained remain low and similar across all cases and never exceed 2%. The pre-slot solution with a triangular separator presents the least harmonic distortion of the considered models; it is very similar to the closed-slot model and improves the original configuration.

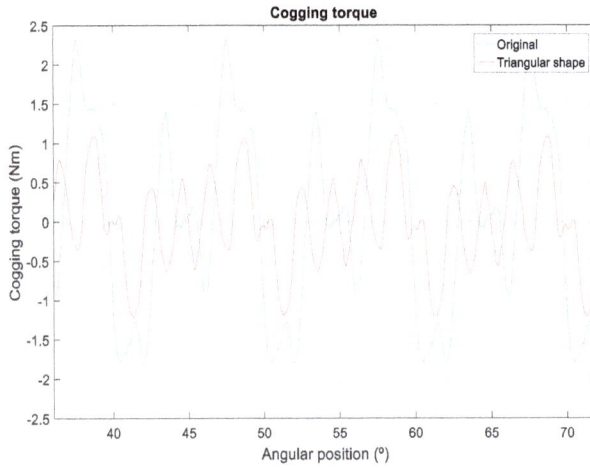

Figure 8. Comparison of cogging torque reduction with the triangular pre-slot method.

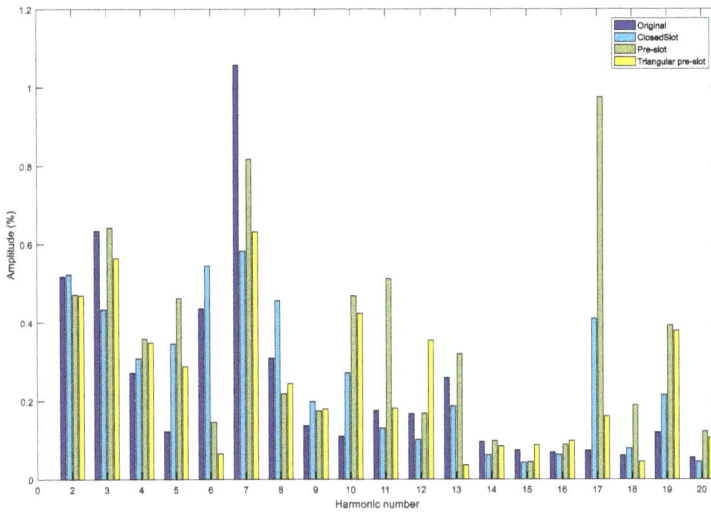

Figure 9. Harmonic spectrum for the proposed reduction method.

Table 2. Harmonic distortion rate (THD) of the different models.

Model	THD (%)
Original design	1.54
Design with closed slot	1.39
Design with pre-slot	1.88
Design with triangular pre-slot	1.33

5.2. Manufacturing Aspects to Reduce the Cogging Torque

Any change in the magnetic circuit alters its reluctance and, therefore, in accordance with Equation (4), it affects the cogging torque and must be considered to reduce it. Consequently, it was found that the holes for correctly aligning the rotor sheet metal with screws in the original design significantly influence the machine's cogging torque, depending on their position with respect to the PMs, Figure 1b.

The impact of these holes on the cogging torque was analyzed for their different positions with respect to the PMs. The conclusion is that the optimal position, which minimizes cogging torque, is a centered position with respect to the magnets. Figure 10b shows the case in which the rotor hole is centered with respect to the PMs. In this situation, the holes have virtually no influence on magnetic field lines linking one magnet with another. In contrast, when the hole is decentered, Figure 10a, the effect is a smaller effective area in the rotor through which the field lines circulate. Therefore, reluctance increases with respect to the case shown in Figure 10b (centered hole) and the magnetic energy in the rotor decreases as observed in Figure 11.

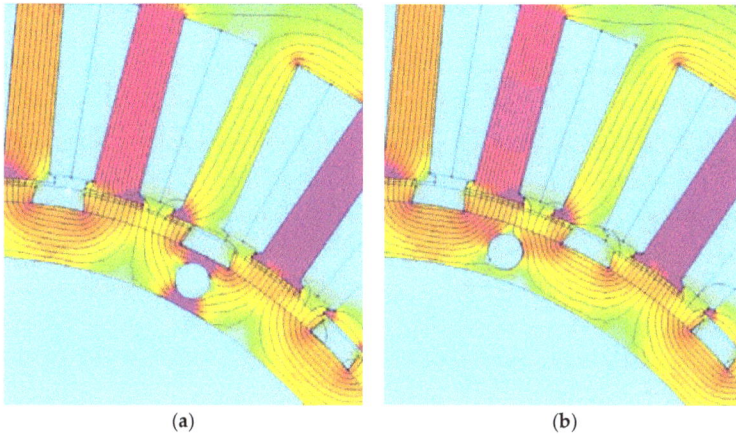

| (a) | (b) |

Figure 10. Magnetic field distribution ($\theta = 0°$): (**a**) Original model, without centered holes; (**b**) Model with centered holes.

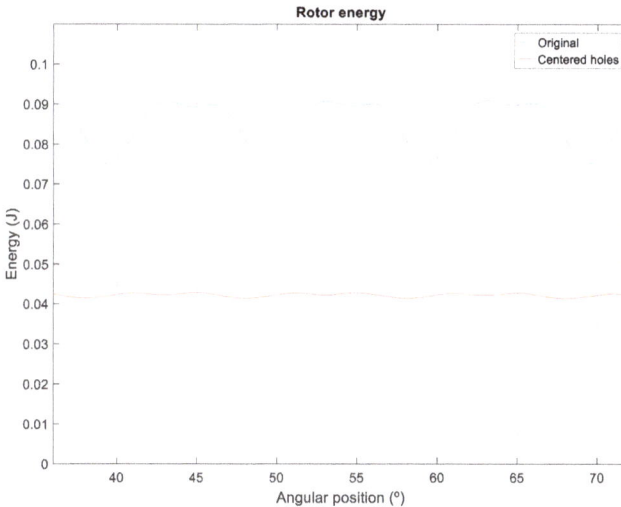

Figure 11. Magnetic energy in the rotor.

Figure 12 shows the magnetic energy stored in the machine with respect to the rotor during electrical 360° (mechanical 36°) for both PMSG models. If the hole is decentered, energy minimums occur when the hole is aligned with a stator tooth. In this position, the flux between two adjacent magnets would be the maximum if there was no hole. Consequently, as observed in Figure 12, and given that the teeth are distributed every mechanical 10° along the stator, the energy minimum occurs with this frequency.

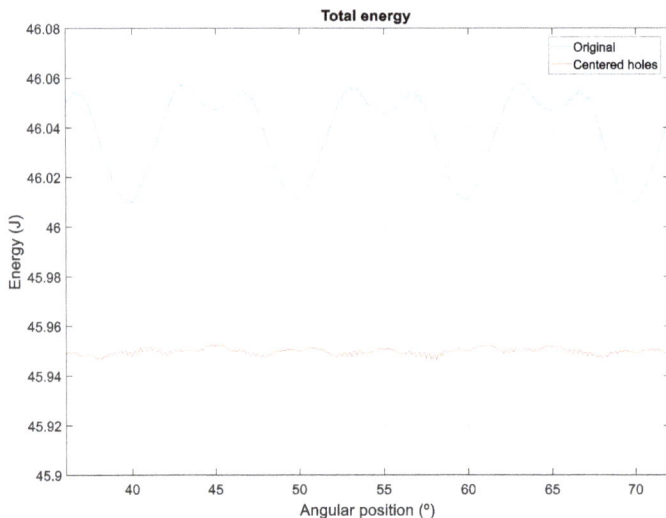

Figure 12. Total magnetic energy.

Figure 13 compares the cogging torque for both PMSGs, resulting in a higher value when the hole is decentered given that more magnetic energy variations occur, as presented in Figure 12. The cogging torque of the model with decentered holes is 2.32 Nm, while this value does not exceed 0.86 Nm when the holes are centered. Figure 13 also shows that the presence of decentered holes even produces a change in the cogging torque wave period.

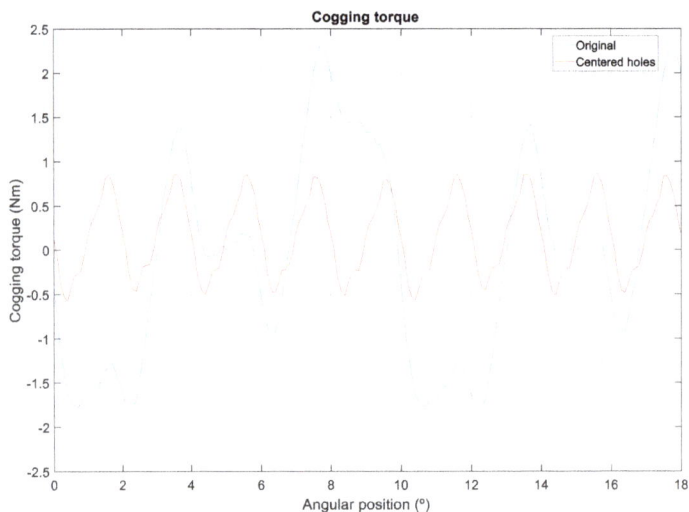

Figure 13. Cogging torque of the original PMSG and model with centered holes.

5.3. Comparative Results

Table 3 compares the maximum cogging torque values obtained with the original design with the two proposed pre-slot configurations and shows the reduction percentage with respect to the original design. The results of the different models are shown in the ideal case (with no manufacturing errors) and considering manufacturing errors in the magnets. Considering centered holes and the original design, the cogging torque is low (0.86 Nm), however in the real case when considering manufacturing errors this value is still too high in line with experimental measurements (3.31 Nm) with respect to the nominal torque (102 Nm) and needs to be reduced.

Table 3. Comparison of the maximum cogging torque values obtained for the prototype and the different models considered in the study.

Model	Without Manufacturing Errors		Considering Manufacturing Errors	
	Nm	Reduction (%)	Nm	Reduction (%)
	Original Design (without Centered Holes)			
Prototype	3.70	-	-	-
Original design	2.32	-	3.92	-
Pre-slot with separation	1.44	37.9	2.03	48.2
Triangular pre-slot	1.21	47.8	1.90	51.5
	Design with all Holes Centered			
Original design	0.86	-	3.31	-
Pre-slot with separation	0.61	29.1	1.80	45.6
Triangular pre-slot	0.59	31.4	1.76	46.8

Finally, the proposed pre-slot method was compared with the skewing technique and the combination of both is considered. The technique of fractional skewing in the rotor [24] comprises dividing the rotor and turning one division away from another for half the cogging torque period. Four divisions were considered in this analysis, as observed in Figure 14. Therefore, considering that the cogging torque period is mechanical 2°, the shift of one division with respect to another is half of this period, which equals 1°. As observed in Table 4, concerning the model with centered holes and in the ideal case of having no manufacturing errors, applying this combined technique manages to reduce the cogging torque to a peak value of 0.03 Nm or to 0.51 Nm if manufacturing errors are considered. In either of the two cases, the cogging torque reduction is very significant.

Table 4. Comparison of the maximum cogging torque values of PMSG with centered holes and the different models considering skewing.

Model	Without Manufacturing Errors		Considering Manufacturing Errors	
	Nm	Reduction (%)	Nm	Reduction (%)
Design with centered holes	0.86	-	3.31	-
Design with centered holes + Triangular pre-slot	0.59	31.4	1.76	46.8
Design with centered holes + Skewing	0.31	64.0	1.34	59.5
Design with centered holes + Triangular pre-slot + Skewing	0.03	96.5	0.51	84.6

Figure 15 shows the cogging torque waveform obtained for the PMSG with centered holes. Because of the reluctance periodicity, the cogging torque is a periodic waveform with a frequency given by:

$$f_{T_{cog}} = \frac{\omega \cdot LCM\left(N_{slots}, N_{poles}\right)}{360°} = 696 \text{ Hz} \qquad (6)$$

where ω is the mechanical speed (1392°/s), *LCM* the least common multiple of the number of slots ($N_{slots} = 36$) and the number of poles ($N_{poles} = 20$). The results in Figure 15 show the decrease in the cogging torque with the pre-slot triangular method and that this improvement is even better when combining this pre-slot installation technique with fractional skewing in the rotor, up to 84% less in the most realistic case of considering errors in manufacturing processes.

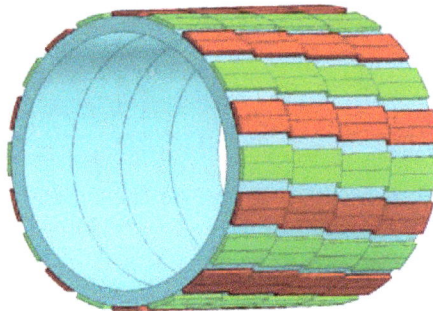

Figure 14. Fractional skewing in the rotor.

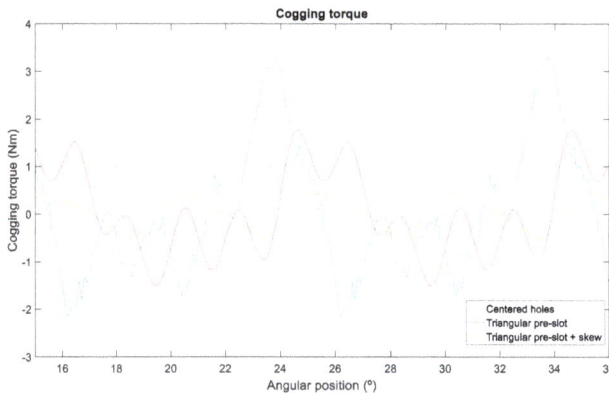

Figure 15. Cogging torque of the centered holes PMSG model, triangular pre-slot model and triangular pre-slot model with skewing (with manufacturing errors).

In addition, Figure 16 compares the air gap magnetic flux density in the proposed pre-slot method with the skewing technique, and the combination of both.

Finally, Table 5 shows how the cogging torque reduction affects the machine's induced voltage: the triangular pre-slot method proposed increases EMF due to the magnet flux canalization to the teeth, while skewing decreases it. When combining this pre-slot technique with fractional skewing in the rotor, again a decrease in EMF is observed but in a softer way. As a result, the skew of the rotor reduces the generator efficiency, but combined with the pre-slot technique this reduction will be lower. The cogging torque reduction techniques, as pre-slot technique or skewing, help reducing undesired ripple and allow achieving low cut-in speed in small wind systems; however they add complexity to the manufacturing process and thus may increase the product cost.

Table 5. Comparison of the electromotive force (EMF) and cogging torque values of PMSG with centered holes and the different models considering skewing.

Model	EMF (V)	(p.u.)	Cogging Torque (Nm)
Design with centered holes	241.52	1.000	3.31
Design with centered holes + Skewing	229.47	0.950	1.34
Design with centered holes + Triangular pre-slot	247.73	1.026	1.76
Design with centered holes + Triangular pre-slot + Skewing	233.04	0.965	0.51

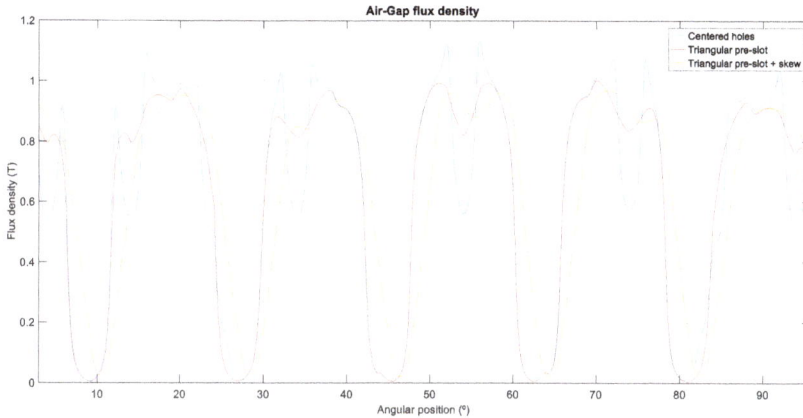

Figure 16. Air gap magnetic flux of the centered holes PMSG model, triangular pre-slot model, and triangular pre-slot model with skewing (with manufacturing errors).

6. Conclusions

This article presents a new cogging torque reduction technique that does not require changes to the machine's main geometry. It proposes placing pre-slots in the initial part of the stator slots. These pre-slots are made of the same ferromagnetic material as the stator, with a non-magnetic central separator (in two halves). The pre-slots are slid longitudinally between the slots after completing machine winding and, therefore, without altering the PMSG's fill factor.

Introducing a central part of non-magnetic material prevents leakage flux between the machine's teeth and also stops its induced voltage from reducing significantly with respect to the configuration without pre-slots.

The proposed method manages to reduce cogging torque in PMSGs with surface-mounted magnets by up to 47.8%. Additionally, the article analyzes how changing the magnetic circuits for construction reasons can affect the cogging torque, which can easily be optimized. The pre-slot technique is also compatible with other cogging torque reduction techniques, such as skewing. When the above-mentioned methods are combined, cogging torque is reduced by 84.6% considering manufacturing errors.

Author Contributions: M.G.-G. and Á.J.R. performed the simulations and wrote the paper. 4fores contributed with the prototype and valuable comments and corrections. All authors discussed the results and commented on the manuscript at all stages.

Funding: This research was funded by Spanish Economy, Industry and Competitiveness Ministry and by European Regional Development Fund (ERDF), grant number RTC-2016-5234-3, in the frame of the Project MHiRED "New technologies for wind-photovoltaic hybrid mini networks managed with storage in connection to the grid and with synchronous support in off-grid operation".

Acknowledgments: The authors wish to thank 4fores for granting their permission to publish some data presented in this article. Furthermore, the technical support from the Research Group on Renewable Energy Integration of the University of Zaragoza (funded by the Government of Aragon) is also gratefully acknowledged.

Conflicts of Interest: The authors declare no conflict of interest.

References

1. Fei, W.; Zhu, Z.Q. Comparison of Cogging Torque Reduction in Permanent Magnet Brushless Machines by Conventional and Herringbone Skewing Techniques. *IEEE Trans. Energy Convers.* **2013**, *28*, 664–674. [CrossRef]
2. Ose-zala, B.; Pugachov, V. Methods to Reduce Cogging Torque of Permanent Magnet Synchronous Generator Used in Wind Power Plants. *Elektron. Elektrotechnika* **2017**, *23*, 43–48. [CrossRef]

3. Leijon, J.; Sjölund, J.; Ekergård, B.; Boström, C.; Eriksson, S.; Temiz, I.; Leijon, M. Study of an Altered Magnetic Circuit of a Permanent Magnet Linear Generator for Wave Power. *Energies* **2018**, *11*, 84. [CrossRef]

4. Lejerskog, E.; Leijon, M. Detailed Study of Closed Stator Slots for a Direct-Driven Synchronous Permanent Magnet Linear Wave Energy Converter. *Machines* **2014**, *2*, 73–86. [CrossRef]

5. Raihan, M.A.H.; Smith, K.J.; Almoraya, A.A.; Khan, F. Interior Permanent Magnet Synchronous Machine (IPMSM) design for environment friendly Hybrid Electric Vehicle (HEV) Applications. In Proceedings of the Humanitarian Technology Conference IEEE Region 10, Dhaka, Bangladesh, 21–23 December 2017.

6. Öztürk, N.; Dalcali, A.; Çelik, E.; Sakar, S. Cogging Torque Reduction by Optimal Design of PM Synchronous Generator for Wind Turbines. *Int. J. Hydrog. Energy* **2017**, *42*, 17593–17600. [CrossRef]

7. Ozoglu, Y. New Stator Tooth for Reducing Torque Ripple in Outer Rotor Permanent Magnet Machine. *Adv. Electr. Comput. Eng.* **2016**, *16*, 49–56. [CrossRef]

8. Liu, C.; Lu, J.; Wang, Y.; Lei, G.; Zhu, J.; Guo, Y. Techniques for Reduction of the Cogging Torque in Claw Pole Machines with SMC Cores. *Energies* **2017**, *10*, 1541. [CrossRef]

9. Ito, T.; Akatsu, K. Electromagnetic Force Acquisition Distributed in Electric Motor to Reduce Vibration. *IEEE Trans. Ind. Appl.* **2017**, *53*, 1001–1008. [CrossRef]

10. Chen, Y.S.; Zhu, Z.Q.; Howe, D. Vibration of PM Brushless Machines Having a Fractional Number of Slots per Pole. *IEEE Trans. Magn.* **2006**, *42*, 3395–3397. [CrossRef]

11. Min, S.G.; Bramerdorfer, G.; Sarlioglu, B. Analytical Modeling and Optimization for Electromagnetic Performances of Fractional-Slot PM Brushless Machines. *IEEE Trans. Ind. Electron.* **2017**, *65*, 4017–4027. [CrossRef]

12. Tseng, W.; Chen, W. Design Parameters Optimization of a Permanent Magnet Synchronous Wind Generator. In Proceedings of the 2016 19th International Conference on Electrical Machines and Systems (ICEMS), Chiba, Japan, 13–16 November 2016.

13. Levin, N.; Orlova, S.; Pugachov, V.; Ose-Zala, B.; Jakobsons, E. Methods to Reduce the Cogging Torque in Permanent Magnet Synchronous Machines. *Electron. Electr. Eng.* **2013**, *19*, 23–26. [CrossRef]

14. Ma, G.; Li, G.; Zhou, R.; Guo, X.; Ju, L.; Xie, F. Effect of Stator and Rotor Notches on Cogging Torque of Permanent Magnet Synchronous Motor. In Proceedings of the 2017 20th International Conference on Electrical Machines and Systems (ICEMS), Sidney, Australia, 11–14 August 2017.

15. Liu, T.; Huang, S.; Gao, J.; Lu, K. Cogging Torque Reduction by Slot-Opening Shift for Permanent Magnet Machines. *IEEE Trans. Magn.* **2013**, *49*, 4028–4031. [CrossRef]

16. Dajaku, G.; Gerling, D. New Methods for Reducing the Cogging Torque and Torque Ripples of PMSM. In Proceedings of the 2014 4th International Electric Drives Production Conference (EDPC), Nuremberg, Germany, 30 September–1 October 2014.

17. Hsiao, C.; Yeh, S.; Hwang, J. A novel cogging torque simulation method for permanent-magnet synchronous machines. *Energies* **2011**, *4*, 2166–2179. [CrossRef]

18. Imamori, S.; Ohguchi, H.; Shuto, M.; Toba, A. Relation between Magnetic Properties of Stator Core and Cogging Torque in 8-Pole 12-Slot SPM Synchronous Motors. *IEEJ J. Ind. Appl.* **2015**, *4*, 696–702. [CrossRef]

19. Wu, D.; Zhu, Z.Q.; Chu, W.Q. Iron Loss in Surface-Mounted PM Machines Considering Tooth-Tip Local Magnetic Saturation. In Proceedings of the 2016 IEEE Vehicle Power and Propulsion Conference (VPPC), Hangzhou, China, 17–20 October 2011.

20. Wu, D.; Zhu, Z.Q. On-Load Voltage Distortion in Fractional Slot Surface-Mounted Permanent Magnet Machines Considering Local Magnetic Saturation. *IEEE Trans. Magn.* **2015**, *51*. [CrossRef]

21. Li, Y.X.; Li, G.J. Influence of Stator Topologies on Average Torque and Torque Ripple of Fractional-Slot SPM Machines with Fully Closed Slots. *IEEE Trans. Ind. Appl.* **2018**, *54*, 2151–2164. [CrossRef]

22. De Donato, G.; Capponi, F.G.; Caricchi, F. On the Use of Magnetic Wedges in Axial Flux Permanent Magnet Machines. *IEEE Trans. Ind. Electron.* **2013**, *11*, 4831–4840. [CrossRef]

23. Todorov, G.; Stoev, B.; Savov, G.; Kyuchukov, P. Effects of Cogging Torque Reduction Techniques Applied to Surface Mounted PMSMs with Distributed Windings. In Proceedings of the 2017 15th International Conference on Electrical Machines, Drives and Power Systems (ELMA), Sofia, Bulgaria, 1–3 June 2017.

24. Galfarsoro, U.; Parra, J.; McCloskey, A.; Zarate, S.; Hernández, X. Analysis of Vibration Induced by Cogging Torque in Permanent-Magnet Synchronous Motors. In Proceedings of the 2017 IEEE International Workshop of Electronics, Control, Measurement, Signals and Their Application to Mechatronics (ECMSM), Donostia-San Sebastian, Spain, 24–26 May 2017.

energies

MDPI

Article

Current Spikes Minimization Method for Three-Phase Permanent Magnet Brushless DC Motor with Real-Time Implementation

Mohamed Dahbi [1,2,*] , **Said Doubabi** [1] **and Ahmed Rachid** [2]

1   Applied Physics Department, Faculty of Sciences and Techniques Marrakech, Cadi Ayyad University, BP 549, Av Abdelkarim Elkhattabi, Gueliz, Marrakesh 4 000, Morocco; s.doubabi@uca.ac.ma
2   Laboratory of Innovative Technologies, Picardie Jules Verne University, 80025 Amiens, France; rachid.greenway@gmail.com
*   Correspondence: dahbi.mohamed2@gmail.com; Tel.: +212-6-40-27-68-64

Received: 10 October 2018; Accepted: 14 November 2018; Published: 19 November 2018

Abstract: Due to their high efficiency and low cost of maintenance, brushless DC motors (BLDCMs) with trapezoidal electromotive forces (back-EMFs), have become widely used in various applications such as aerospace, electric vehicles, industrial uses, and robotics. However, they suffer from large current ripples and current spikes. In this paper, a new method for minimizing current spikes appearing during BLDCM start-up or sudden set point changes is proposed. The method is based on controlling the MOSFET gates of the motor driver using R-C filters. These filters are placed between the PWM control signal generator and the MOSFET gates to smooth these control signals. The analysis of the proposed method showed that the R-C filter usage affects the BLDCM steady-state performances. To overcome this limitation, the R-C filter circuit was activated only during current spikes detection. The effectiveness of the proposed method was analytically analyzed and then validated through simulation and experimental tests. The obtained results allowed a reduction of 13% in current spikes amplitude.

Keywords: Brushless DC motors; current ripples; current spikes; modeling; back electromotive force; R-C filter

1. Introduction

Nowadays, brushless DC motors (BLDCMs) have become a preferable choice due to several advantages such as a high power to weight ratio, a high torque to current ratio, fast response, and above all high efficiency and less maintenance [1]. They are widely used and recommended in areas such as clean and explosive environments (where control induced sparks can cause undesirable damages), food and chemical industries, electric vehicles, and photovoltaic pumping systems [2]. Unlike DC motors, BLDCMs have no brushes, which provides for a long lifetime.

Generally, the BLDCM with trapezoidal electromotive forces (back-EMFs) is the most prevalent type [3], since it does not require complex control, expensive sensors, or high-resolution sensors when compared to brushless AC motors. The latter need a sinusoidal current waveform while BLDCMs require a square current waveform for proper operation [4]. The BLDCM drive control is based on a three-phase half-bridge structure that can be composed of six or four switches [4–6].

Torque ripples are considered one of the main drawbacks of the BLDCM. They are generated due to several reasons such as non-ideal form of the back-EMF, cogging, and reluctance torques [7], which have led researchers to investigate and propose several torque ripples reduction methods [3,4,7–10]. For instance, the analysis of torque ripples caused by the noncommutated phase is addressed in References [3,4], where the PWM_ON_PWM control strategy was implemented. An auxiliary DC voltage source, connected in series with the inverters' DC bus, was employed for torque ripple reduction in Reference [7]. A novel PWM method was established in Reference [8], where current spikes and current ripples, generated by the unipolar PWM control signals in the braking phases, were respectively source-illustrated and minimized. Optimal duty ratio calculation, to be applied to the incoming and outgoing phases during commutation intervals, was presented in References [9,10]. A current optimizing control method was investigated in Reference [11], where the three-phase current trajectories were set according to the torque reference.

Adding a DC-DC converter in with the three-phase inverter for torque ripple reduction was studied in References [12–20]. For instance, a buck converter was the implemented topology in References [12,13], with power factor correction based on PID fuzzy controller in Reference [14]. Other converter topologies were also used, such as a Z-source inverter in Reference [15], an integrated dual output DC-DC converter in Reference [16], and a CUK converter in Reference [17]. A SEPIC (Single Ended Primary Inductor Converter) DC-DC converter was implemented in Reference [18] with a three-level neutral-point-clamped (NPC) inverter. The output voltage of the SEPIC converter was regulated to be equal to four times the back-EMF voltage during the commutation period and was integrated using a switches selection circuit. This study [18] was improved in Reference [19] with a modified SEPIC converter to reduce the number of needed DC-DC converters with the usage of the same selection circuit. This selection circuit was eliminated in Reference [20] to minimize the implemented components. Other works were focused on studying the torque ripple induced from the motor itself, such as in References [21,22], where the torque ripples caused by non-ideal back-EMF were mitigated with pulses time calculation used for switches control.

Another issue regarding the BLDCM control is the current spikes. These spikes may lead to damage to the controller or the motor itself. The traditional all-turn-off current limit logic is the most used method for current limitation. However, the approach may damage the minimized DC link capacitor-based drive systems. In Reference [23], a detailed presentation of this logic with a novel current limitation strategy was presented to eliminate the oversize pumping-up voltage damage.

Based on these motivations, this paper adds further contributions by proposing a method to reduce the current spikes in the start-up and sudden set point changes. With this intention, an R-C filter was placed before the MOSFET gates, and its impact on transient and steady state regimes was analyzed. The proposed method is based on creating a delay in the MOSFET control signals before reaching the threshold voltage (progressive reaching to the threshold voltage). This induces a smooth current flow from the source to the MOSFET's drain compared to the all-turn-off current limitation technique, where the command of the gate MOSFETs does not change from discrete states. The proposed approach allows lower input power supply use without energy flow interruptions, since the high current demand is reduced. Moreover, complicated or bulky components and sophisticated platforms are not needed for practical implementation. The effectiveness of the proposed method was analytically analyzed and validated through simulation and experimental tests.

The rest of the paper is organized as follows. Section 2 gives details on the brushless DC motor operation, its model equations, and its torque ripple sources. In Section 3, the proposed method to minimize the current spikes is described along with its advantages and motors' behavior. Finally, simulation and experimental results are presented in Section 4 with and without implementing the proposed technique, followed by conclusions and perspectives in Section 5.

2. BLDC Motor Model Equation

The BLDCM is a synchronous motor with a three-phase winding stator and a permanent magnet rotor. Two motor phases are simultaneously conducting, which means that the gate drive signals are applied to two of the inverters' MOSFETs at the same time. Knowing which MOSFETs must be conducting is determined using Hall effect sensor signals denoted by H_a, H_b, and H_c. These signal combinations give an image of the rotor position.

Different simulation models of the BLDCM were proposed in References [24–27]. Some authors presented a global modeling using inverter mathematical equations [28], whereas others were based on motor equations [29,30].

The BLDCM phase can be modeled as a resistance in series with an inductor and a back-EMF, as shown in Figure 1. Figure 2 depicts the phase current waveforms, Hall effect sensor combinations (numbered as sectors), and the inverters' conducting MOSFETs.

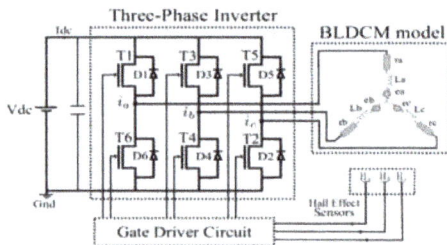

Figure 1. Brushless DC motor equivalent circuit with its inverter driver.

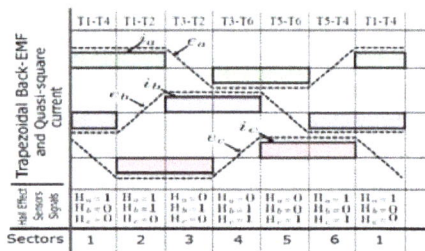

Figure 2. Trapezoidal electromotive forces (back-EMFs), phase currents, Hall effect sensor signals, and MOSFET states.

The model equations of the BLDC motor are presented as follows [24–30]:

$$
\begin{bmatrix} V_{an} \\ V_{bn} \\ V_{cn} \end{bmatrix} = \begin{bmatrix} r_a & 0 & 0 \\ 0 & r_b & 0 \\ 0 & 0 & r_c \end{bmatrix} \begin{bmatrix} i_a \\ i_b \\ i_c \end{bmatrix} + \begin{bmatrix} L_a & 0 & 0 \\ 0 & L_b & 0 \\ 0 & 0 & L_c \end{bmatrix} \frac{d}{dt} \begin{bmatrix} i_a \\ i_b \\ i_c \end{bmatrix} + \begin{bmatrix} e_a \\ e_b \\ e_c \end{bmatrix},
\tag{1}
$$

where V_{an}, V_{bn}, and V_{cn} are the stator winding phase voltages; r_a, r_b, and r_c are the stator winding resistances; L_a, L_b, and L_c are the phase inductances; i_a, i_b, and i_c are the phase currents; and e_a, e_b, and e_c are the back-EMFs.

The electromagnetic torque and the motor motion are given respectively by Equations (2) and (3):

$$
T_e = \frac{E_a i_a + E_b i_b + E_c i_c}{W},
\tag{2}
$$

$$
\frac{dW}{dt} = \frac{1}{J}(T_e - T_L - BW),
\tag{3}
$$

where W is the rotor angular velocity, B is the viscous friction coefficient, J is the moment of inertia, and T_L is the load torque.

In Figure 2, the transition from sector 2 to sector 3 induced an electromagnetic torque expressed by the following Equation (4) [11,14]:

$$
T_e = \frac{2 * E * i_c}{W},
\tag{4}
$$

where E is the flap-top value of the back-EMF.

Based on Equation (4), it could be understood that the electromagnetic torque was proportional to the noncommutated phase current during the commutation period. However, in practice, current shapes are not like those presented in Figure 2. There is a "dead time" between each commutation from one sector to another. This is due to the difference between the MOSFET rise and fall times. More details about such differences can be found in Reference [10].

3. Analysis of R-C Filter Implementation

Depending on the MOSFET features, the current flow from the drain to the source depends on the gate voltage V_{gs}.

For the BLDCM, the PWM mode is generally preferred for speed variation where the duty ratio is an image of an analogue signal. Using digital-to-analogue conversions, the duty ratio is calculated, and the appropriate PWM signal is delivered to the MOSFET gate to control the input voltage V_{dc}, which refers also to input current control.

The PWM signal is a succession of discrete states for well-chosen times. In the proposed control method, the discrete states transitions were avoided by adding an R-C filter between the PWM generator and the gate of the MOSFETs, which allowed smooth transients in the changing parts. This changing mode is explained in Figure 3a.

The encountered problem in this mode of control was the delay time added in the passage portions toward zero (t_2'). For the normal BLDCM control, this time equaled the fall MOSFET's time that can be achieved by adding a diode in parallel with the resistance, as shown in Figure 3b. The diode allowed the ability to discharge the capacitor quickly through the creation of a connection with the ground delivered by the PWM block.

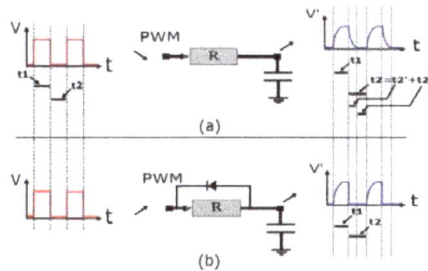

Figure 3. PWM signal with R-C filter: (**a**) R-C filter effect on the PWM signal, and (**b**) R-C filter with the added diode effect on the PWM signal.

In order to present the detailed impacts of the proposed control method on the motor performance, two current commutation cases must be distinguished, the upper-bridge current commutation and the lower-bridge current commutation. Detailed equations with the equivalent inverter schematic in both cases are presented as follows.

3.1. Lower-Bridge Current Commutation

At the lower-bridge current commutation, the analysis was done when switching off the MOSFET T_6 and switching on the MOSFET T_4. Figure 4 shows the MOSFET states before, during, and after the commutation interval for this case. A 100% duty ratio was chosen for simplification purposes.

The voltage equations can be written as:

$$\begin{bmatrix} 0 \\ 0 \\ V_{dc} \end{bmatrix} = \begin{bmatrix} r & 0 & 0 \\ 0 & r & 0 \\ 0 & 0 & r \end{bmatrix}\begin{bmatrix} i_a \\ i_b \\ i_c \end{bmatrix} + \begin{bmatrix} L & 0 & 0 \\ 0 & L & 0 \\ 0 & 0 & L \end{bmatrix}\frac{d}{dt}\begin{bmatrix} i_a \\ i_b \\ i_c \end{bmatrix} + \begin{bmatrix} E_a \\ E_b \\ E_c \end{bmatrix} + \begin{bmatrix} V_n \\ V_n \\ V_n \end{bmatrix}, \quad (5)$$

$$\begin{bmatrix} V_{dc} \\ 0 \\ V_{dc} \end{bmatrix} = \begin{bmatrix} r & 0 & 0 \\ 0 & r & 0 \\ 0 & 0 & r \end{bmatrix} \begin{bmatrix} i_a \\ i_b \\ i_c \end{bmatrix} + \begin{bmatrix} L & 0 & 0 \\ 0 & L & 0 \\ 0 & 0 & L \end{bmatrix} \frac{d}{dt} \begin{bmatrix} i_a \\ i_b \\ i_c \end{bmatrix} + \begin{bmatrix} E_a \\ E_b \\ E_c \end{bmatrix} + \begin{bmatrix} V_n \\ V_n \\ V_n \end{bmatrix}, \tag{6}$$

$$\begin{bmatrix} 0 \\ 0 \\ V_{dc} \end{bmatrix} = \begin{bmatrix} r & 0 & 0 \\ 0 & r & 0 \\ 0 & 0 & r \end{bmatrix} \begin{bmatrix} i_a \\ i_b \\ i_c \end{bmatrix} + \begin{bmatrix} L & 0 & 0 \\ 0 & L & 0 \\ 0 & 0 & L \end{bmatrix} \frac{d}{dt} \begin{bmatrix} i_a \\ i_b \\ i_c \end{bmatrix} + \begin{bmatrix} E_a \\ E_b \\ E_c \end{bmatrix} + \begin{bmatrix} V_n \\ V_n \\ V_n \end{bmatrix}. \tag{7}$$

In this case, $E_c = -E_b = K_eW$, and $E_a = \frac{6}{\pi}K_eW^2t - 12K_eW$ [30].
The motor phase currents were calculated and are presented as follows:

$$ri_a + L\frac{di_a}{dt} = -\frac{1}{3}[-V_{dc} - 24k_eW] - \frac{4}{\pi}K_eW^2t, \tag{8}$$

$$ri_b + L\frac{di_b}{dt} = \frac{1}{3}[-2V_{dc} - 9k_eW] + \frac{2}{\pi}K_eW^2t, \tag{9}$$

$$ri_c + L\frac{di_c}{dt} = \frac{1}{3}[V_{dc} - 15k_eW] + \frac{2}{\pi}K_eW^2t. \tag{10}$$

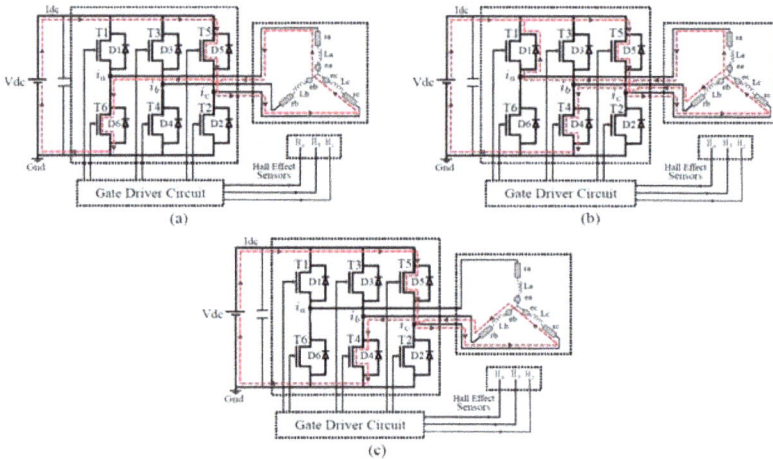

Figure 4. MOSFET states for lower-bridge current commutation: (a) Before the commutation interval, (b) during the commutation interval, and (c) after the commutation interval.

Using Equation (11), the inverse Laplace transform, and the initial value of the phase currents $i_a(0) = -i_c(0) = i_0$ and $i_b(0) = 0$, the resulting phase currents equations are expressed as Equations (12)–(14):

$$i_a + i_b + i_c = 0, \tag{11}$$

$$i_a(t) = \frac{-[-V_{dc} - 24K_eW]}{3r}\left(1 - e^{-(\frac{r}{L})t}\right) + \frac{Li_0}{r}e^{-(\frac{r}{L}t)} - \frac{4K_eW^2}{\pi r}\left(t - \frac{L}{r} + \frac{L}{r}e^{-(\frac{r}{L}t)}\right), \tag{12}$$

$$i_b(t) = \frac{-[2V_{dc} + 9K_eW]}{3r}\left(1 - e^{-(\frac{r}{L})t}\right) + \frac{2K_eW^2}{\pi r}\left(t - \frac{L}{r} + \frac{L}{r}e^{-(\frac{r}{L}t)}\right), \tag{13}$$

$$i_c(t) = \frac{[V_{dc} - 15K_eW]}{3r}\left(1 - e^{-(\frac{r}{L})t}\right) - \frac{Li_0}{r}e^{-(\frac{r}{L}t)} + \frac{2K_eW^2}{\pi r}\left(t - \frac{L}{r} + \frac{L}{r}e^{-(\frac{r}{L}t)}\right). \tag{14}$$

In this case, when using the R-C filter, the equations of the motor phase currents remained the same.

3.2. Upper-Bridge Current Commutation

During the upper-bridge current commutation, the analysis was done when switching off the MOSFET T_1 and switching on the MOSFET T_3. Figure 5 presents the MOSFET states before, during, and after the commutation interval for the upper-bridge current commutation.

The voltage equations can be written as:

$$
\begin{bmatrix} V_{dc} \\ 0 \\ 0 \end{bmatrix} = \begin{bmatrix} r & 0 & 0 \\ 0 & r & 0 \\ 0 & 0 & r \end{bmatrix} \begin{bmatrix} i_a \\ i_b \\ i_c \end{bmatrix} + \begin{bmatrix} L & 0 & 0 \\ 0 & L & 0 \\ 0 & 0 & L \end{bmatrix} \frac{d}{dt} \begin{bmatrix} i_a \\ i_b \\ i_c \end{bmatrix} + \begin{bmatrix} E_a \\ E_b \\ E_c \end{bmatrix} + \begin{bmatrix} V_n \\ V_n \\ V_n \end{bmatrix}, \tag{15}
$$

$$
\begin{bmatrix} 0 \\ V_{dc} \\ 0 \end{bmatrix} = \begin{bmatrix} r & 0 & 0 \\ 0 & r & 0 \\ 0 & 0 & r \end{bmatrix} \begin{bmatrix} i_a \\ i_b \\ i_c \end{bmatrix} + \begin{bmatrix} L & 0 & 0 \\ 0 & L & 0 \\ 0 & 0 & L \end{bmatrix} \frac{d}{dt} \begin{bmatrix} i_a \\ i_b \\ i_c \end{bmatrix} + \begin{bmatrix} E_a \\ E_b \\ E_c \end{bmatrix} + \begin{bmatrix} V_n \\ V_n \\ V_n \end{bmatrix}, \tag{16}
$$

$$
\begin{bmatrix} 0 \\ V_{dc} \\ 0 \end{bmatrix} = \begin{bmatrix} r & 0 & 0 \\ 0 & r & 0 \\ 0 & 0 & r \end{bmatrix} \begin{bmatrix} i_a \\ i_b \\ i_c \end{bmatrix} + \begin{bmatrix} L & 0 & 0 \\ 0 & L & 0 \\ 0 & 0 & L \end{bmatrix} \frac{d}{dt} \begin{bmatrix} i_a \\ i_b \\ i_c \end{bmatrix} + \begin{bmatrix} E_a \\ E_b \\ E_c \end{bmatrix} + \begin{bmatrix} V_n \\ V_n \\ V_n \end{bmatrix}. \tag{17}
$$

In this case, $E_b = -E_c = K_e W$ and $E_a = -\frac{6}{\pi} K_e W^2 t + 6 K_e W$ [30]. The motor phase currents are:

$$
r i_a + L \frac{d i_a}{dt} = -\frac{1}{3} [V_{dc} + 12 k_e W] + \frac{4}{\pi} K_e W^2 t, \tag{18}
$$

$$
r i_b + L \frac{d i_b}{dt} = \frac{1}{3} [2 V_{dc} + 3 k_e W] - \frac{2}{\pi} K_e W^2 t, \tag{19}
$$

$$
r i_c + L \frac{d i_c}{dt} = \frac{1}{3} [-V_{dc} + 9 k_e W] - \frac{2}{\pi} K_e W^2 t. \tag{20}
$$

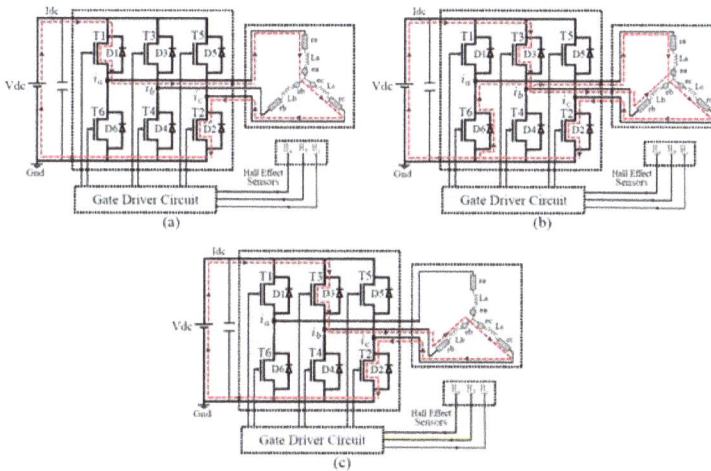

Figure 5. MOSFET states for upper-bridge current commutation: (**a**) Before the commutation interval, (**b**) during the commutation interval, and (**c**) after the commutation interval.

Using the inverse Laplace transform and the initial value of the phase currents, $i_c(0) = -i_a(0) = i_0$ and $i_b(0) = 0$, the calculated phase currents are expressed as Equations (21)–(23):

$$i_a(t) = \frac{-[V_{dc} + 12K_eW]}{3r}\left(1 - e^{-(\frac{r}{L})t}\right) + \frac{Li_0}{r}e^{-(\frac{r}{L}t)} + \frac{4K_eW^2}{\pi r}\left(t - \frac{L}{r} + \frac{L}{r}e^{-(\frac{r}{L}t)}\right), \quad (21)$$

$$i_b(t) = \frac{[2V_{dc} + 3K_eW]}{3r}\left(1 - e^{-(\frac{r}{L})t}\right) - \frac{2K_eW^2}{\pi r}\left(t - \frac{L}{r} + \frac{L}{r}e^{-(\frac{r}{L}t)}\right), \quad (22)$$

$$i_c(t) = \frac{[-V_{dc} + 9K_eW]}{3r}\left(1 - e^{-(\frac{r}{L})t}\right) - \frac{Li_0}{r}e^{-(\frac{r}{L}t)} - \frac{2K_eW^2}{\pi r}\left(t - \frac{L}{r} + \frac{L}{r}e^{-(\frac{r}{L}t)}\right). \quad (23)$$

The motor phase currents, when using the R-C filter, are:

$$i_a(t) = \frac{-[V_{dc} + 12K_eW]}{3r}\left(1 - e^{-(\frac{r}{L})t}\right) + \frac{Li_0}{r}e^{-(\frac{r}{L}t)} + \frac{4K_eW^2}{\pi r}\left(t - \frac{L}{r} + \frac{L}{r}e^{-(\frac{r}{L}t)}\right) + \left(\frac{V_{dc}\sigma_{RC}}{3r\sigma_{RC} - 3L}\left(e^{-\frac{t}{\sigma_{RC}}} - e^{-\frac{r}{\sigma_{RC}}}\right)\right), \quad (24)$$

$$i_b(t) = \frac{[2V_{dc} + 3K_eW]}{3r}\left(1 - e^{-(\frac{r}{L})t}\right) - \frac{2K_eW^2}{\pi r}\left(t - \frac{L}{r} + \frac{L}{r}e^{-(\frac{r}{L}t)}\right) - \left(\frac{2V_{dc}\sigma_{RC}}{3r\sigma_{RC} - 3L}\left(e^{-\frac{t}{\sigma_{RC}}} - e^{-\frac{r}{\sigma_{RC}}}\right)\right), \quad (25)$$

$$i_c(t) = \frac{[-V_{dc} + 9K_eW]}{3r}\left(1 - e^{-(\frac{r}{L})t}\right) - \frac{Li_0}{r}e^{-(\frac{r}{L}t)} - \frac{2K_eW^2}{\pi r}\left(t - \frac{L}{r} + \frac{L}{r}e^{-(\frac{r}{L}t)}\right) + \left(\frac{V_{dc}\sigma_{RC}}{3r\sigma_{RC} - 3L}\left(e^{-\frac{t}{\sigma_{RC}}} - e^{-\frac{r}{\sigma_{RC}}}\right)\right). \quad (26)$$

From the reported equations, the difference between upper-bridge and lower-bridge current commutation cases was observed at the incoming connected terminal voltage: When the lower-bridge commutation was activated, the terminal voltage of the incoming commutated phase (phase b) was equal to zero (connected to the ground). The R-C filter on the gate of the MOSFET T_4 did not present any difference because the incoming commutated phase voltage still equaled zero until the total conduction of the MOSFET T_4. In this case, the outgoing commutated phase was connected to the DC link voltage through the freewheeling diode D_1, as shown in Figure 5b, and was not controlled by the gate of the MOSFET T_1. This means that any difference from the normal controlling mode would not appear when the gate of the lower MOSFET T_4 was commutated using the R-C filter. Unlike the upper-bridge current commutation, the terminal voltage of the incoming commutated phase was the DC link voltage. Controlling the gate of the MOSFET T_3 using the R-C filter was the reason for the impact of this on the absorbed current. This was due to the delay added before the total conduction of the MOSFET T_3. This fact proved that the R-C filter impact was more important when it was implemented on the upper-bridge gate's MOSFETs.

It is worth mentioning that the phase currents equations, when adding the R-C filter, had higher values than without using an R-C filter. This can be seen in the additional "exponential" term added from controlling the gate of the MOSFET T_3 by the R-C filter (Equations (24)–(26)). It should be also emphasized that the R-C filter frequency should be well-chosen. It must be higher than the PWM control signal frequency to allow the motor to work without disturbances.

4. Simulation and Experiment Results

The simulation results are presented in Figures 6–13. Figure 6 shows the phases and absorbed currents during motor operation with a 100% duty ratio. The MOSFET control signals are presented in Figure 7. Theses curves corresponded to the case without using the R-C filter and where the R-C filter was activated for C = 0.6 µF and C = 1.5 µF.

Based on the obtained results, one can see that the current spike in the start-up phase without using the R-C filter (27 A) (Figure 6a) had a higher value than the one absorbed when using the R-C filter (19 A) (Figure 6c). This reduction in current spike value was obtained with an R-C filter frequency of 2.6 kHz. Changing the latter to 1 kHz, the current spike was reduced to 9.5 A (Figure 6e). Despite these decreases in current spikes, more current ripples appeared. The noncommutated phase induced a higher current ripple value during the commutation period in a steady state. Compared to the normal control strategy that induced 4.8 A, a 2.6 kHz R-C filter frequency induced a 9.5 A current ripple value, whereas the use of a 1 kHz R-C filter frequency decreased this current ripple to 8.5 A. The obtained results were in accordance with those of the absorbed current shown in Figure 6b,d,f. This confirms the obtained analytical results presented in the previous section, in terms of a higher current value during the commutation interval, when compared to the normal BLDCM control.

The corresponding MOSFET gate signals are illustrated in Figure 7 without using the R-C filter and when implementing the latter with a capacitor value of 0.6 μF and 1.5 μF. The difference is illustrated in the delay time added before the total MOSFET conduction. This R-C filter effect was applied only during front edges. During failing edges, the same signal waveform was applied for normal and R-C filter control methods.

In order to avoid the higher steady state current ripples caused by R-C filter usage, a combination of the two controlling modes, with and without R-C filter, was used. A relays-based selection circuit that allowed for activating or canceling the connection between the PWM block and the R-C filter was utilized. It must be taken into consideration that the proposed schematic must be able to discharge the R-C filter capacitor in nonconduction phases to allow reactivation of the R-C filter effect when renewing. The global schematic is presented in Figure 8.

Relays r1 and r3 were controlled by the same signal, which was the complement to that which ordered the relays r2 and r4. When the relays r1 and r3 were in conduction, the R-C filter effect was activated and connected to the gate of the MOSFET. During the R-C filter deactivation, relays r2 and r4 were in conduction. This combination connected the PWM signal, after the resistance, directly to the gate of the MOSFET and allowed the capacity to discharge by connecting it to the ground.

What followed was the activation of the R-C filter in the starting and sudden set point changes. This is explained by what this control method could cause in permanent regimes in terms of current ripple. Sudden changes in the set point were detected by the derivative of the latter and the comparison of it to a certain limit that must not be exceeded, after which the engine would see a significant current spike. Simulation results, using the programmed set point changes profile shown in Figure 9, are illustrated in Figure 10 without using an R-C filter and in Figure 11 with combined "with and without an R-C filter" controls. Figure 12 depicts the gate voltage signal of the R-C filter-controlled MOSFETs. The chosen set point variations, representing the PWM control signal duty ratio, were utilized to test the activation and deactivation of the R-C filter effect. The related simulation signal is presented in Figure 13.

From the reported results, without using an R-C filter control, the peak values of the phase current, and thus the input current, reached 26 A during the start-up phase and 13 A during sudden set point change, as illustrated in Figure 10a,b. However, when implementing the R-C filter with a 0.6 μF capacitor, the current values were reduced to reach 15 A and 11 A during the start-up and sudden set point change, respectively (Figure 11a,b). After changing the capacitor to 1.5 μF, the start-up and sudden set point changes currents were reduced to 11 A and 10 A, respectively (Figure 11c,d).

During gradual set point change, the current waveform did not show any spike, which did not necessitate R-C filter effect activation.

From these results, using relays on the six MOSFETs of the inverter seemed to be a satisfactory way to eliminate current ripples in the permanent regimes. The previous calculations, which explained that the R-C filter impact was important when the latter was implemented on the upper-bridge MOSFETs, may have reduced the number of relays and R-C filters from six to three.

The effectiveness of the proposed method was experimentally proven using the laboratory test rig illustrated in Figure 14. It was made up of an outrunner BLDCM structure with a neodymium magnet rotor, an inverter composed of IRFP260 MOSFETs, IR2110 gate drivers, a potentiometer, and an Arduino Mega2560 board used to generate BLDCM PWM control signals. The relays-based circuit contained three hfd2/003-m-l2-d relays, in addition to the basic components needed for their control (e.g., transistors, diodes). Based on the potentiometer value and Hall effect sensor signals, PWM control signals were provided to the IR2110 gate drivers through the proposed RC-filters and relays-based circuits for the high-bridge-controlled MOSFETs, and directly from the Arduino to the lower-bridge MOSFETs drivers. The potentiometer derivative was calculated and used for R-C filter activation or deactivation as illustrated previously in Figure 8. The IR2110 drivers were chosen due to their capability of isolating the high-power side from the low-power side, which ensured programming and interfacing cards security. The PWM frequency was set to 32 kHz, and a National Instrument card, NI USB-6259, was used to acquire input current waveforms with and without using an R-C filter through an LA 25-NP current sensor.

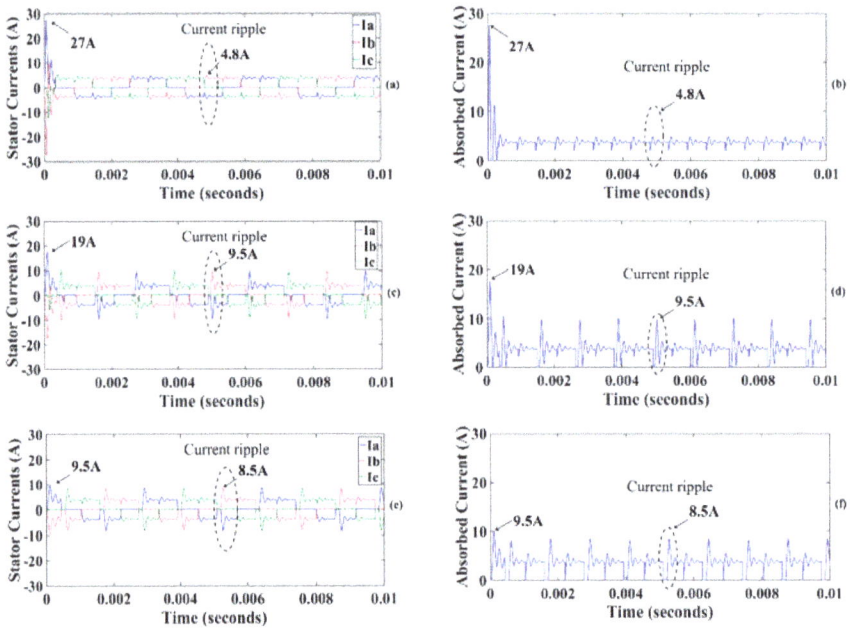

Figure 6. Current waveforms: (**a**) Phase currents without R-C filter, (**b**) absorbed current without R-C filter, (**c**) phase currents with R-C filter for C = 0.6 μF, (**d**) absorbed current with R-C filter for C = 0.6 μF, (**e**) phase currents with R-C filter for C = 1.5 μF, (**f**) absorbed current with R-C filter for C = 1.5 μF.

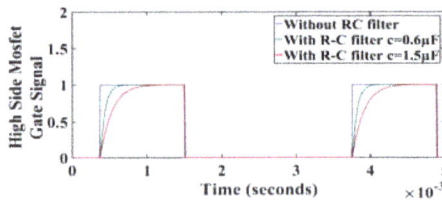

Figure 7. Gate control signal without R-C filter, with R-C filter for C = 0.6 μF, and with R-C filter for C = 1.5 μF.

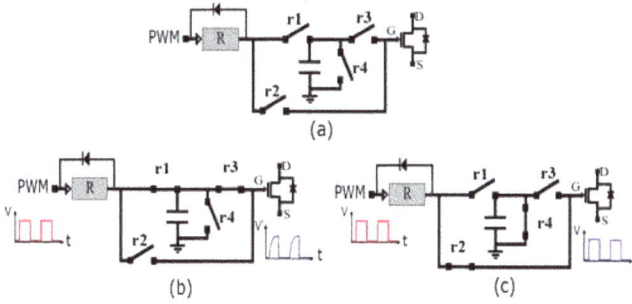

Figure 8. Global schematic of the proposed control method: (**a**) r1, r2, r3, and r4 relays are on; (**b**) r1 and r3 relays are on to activate the R-C filter effect; (**c**) r2 and r4 relays are on to deactivate the R-C filter effect.

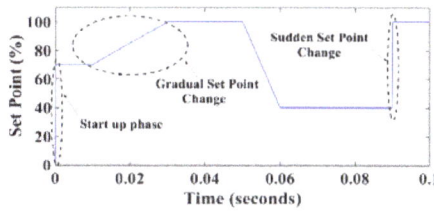

Figure 9. Set point changes.

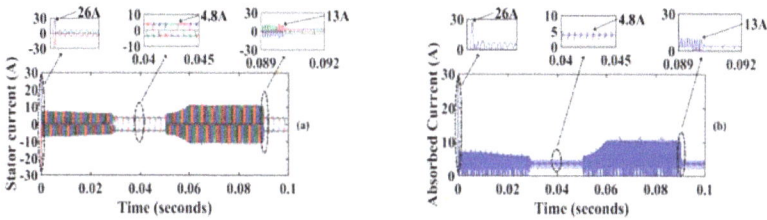

Figure 10. Current waveforms without R-C filter: (**a**) Phase currents, and (**b**) absorbed current.

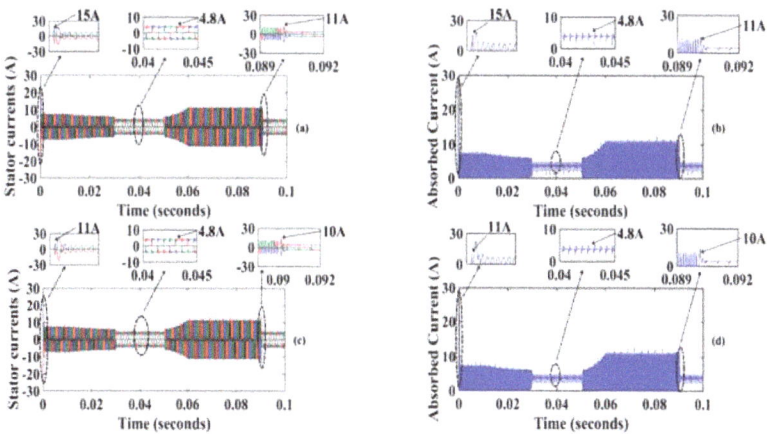

Figure 11. Phase and absorbed current waveforms for combined "with and without an R-C filter": (**a,b**) C = 0.6 µF, and (**c,d**) C = 1.5 µF.

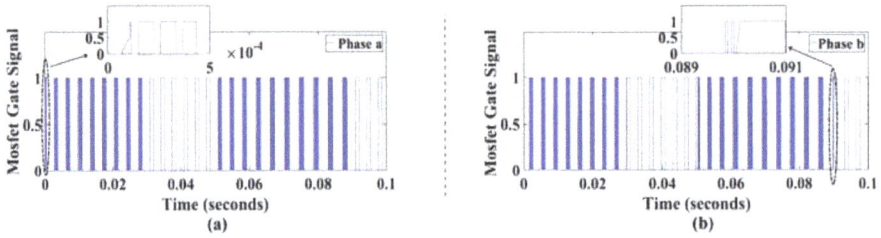

Figure 12. Gate control signal for combined "with and without an R-C filter" methods: (a) During start-up phases. (b) During set point change occurrence.

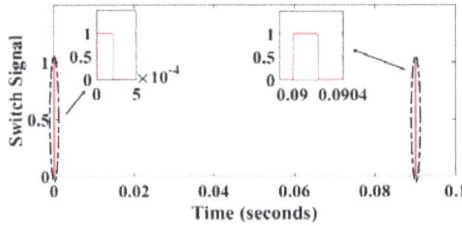

Figure 13. Activated switch signal.

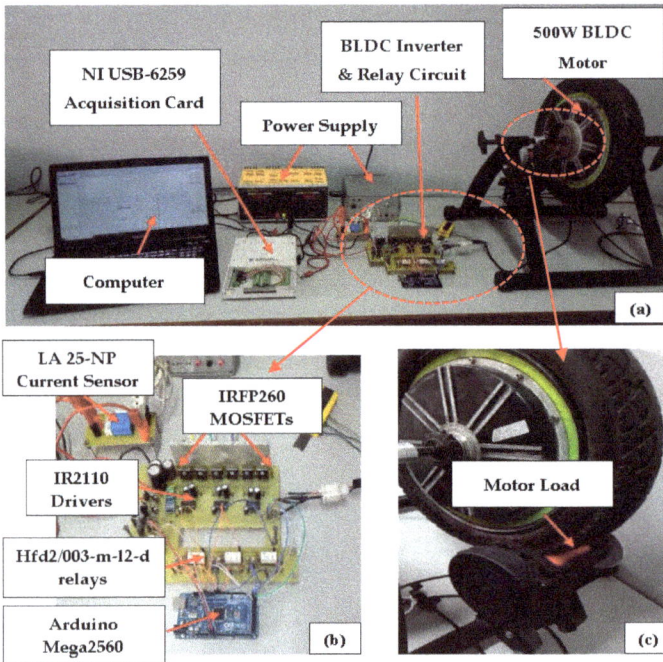

Figure 14. Experimental test rig for BLDC motor control: (**a**) Global test-rig, (**b**) BLDCM inverter, (**c**) motor load.

Figure 15 illustrates the experimental results with and without using an R-C filter during the start-up phase. It can be noticed that current spikes in normal BLDCM control (using the discrete PWM signal only) reached 16.5 A. In contrast, with the proposed method, the absorbed current spikes decreased to 14.3 A due to the induced delay in the MOSFET command before reaching the threshold voltage. These results confirm those illustrated through the first dashed line ellipsis in Figure 11a–d in terms of current spike reduction. It should be mentioned that the experimental results were obtained using a 15 kHz R-C filter frequency, which explains the current spike values difference from those in the simulation. In permanent regimes, the current waveforms are the same for both cases (with and without an R-C filter). The 15 kHz R-C filter frequency was selected to cope with some test rig limitations (such as program execution frequency and acquisition memory saturation), which were also the reason for presenting only start-up phase results. Improving the test rig performance should result in more of a decrease in the current spikes.

As a conclusion, one can see that these experimental results support the simulation and analytical analyses and also prove the effectiveness of the proposed current spikes minimization method.

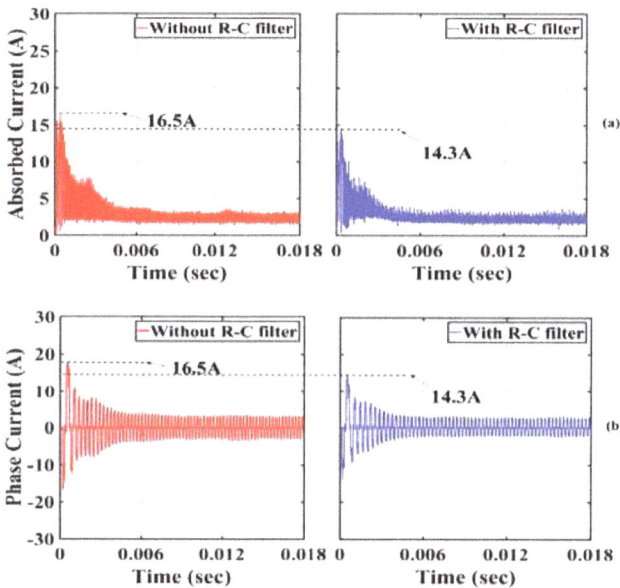

Figure 15. Current waveforms with and without an RC filter (R = 100 Ω, C = 0.1 µF): (**a**) Absorbed current, and (**b**) phase current.

5. Conclusions and Future Work

In this paper, BLDCM current spike limitations during start-up and set point sudden changes were presented by modifying the MOSFET gate control signals. An R-C filter was placed between the PWM signal generator and the MOSFET gate in order to smooth the discrete PWM control signal during transition phases. The proposed method induced more current ripples in the permanent regimes, which imposed a combination of the use of the R-C filter or not. This was achieved using relays-based circuits to activate and deactivate the R-C filter effect. Simulation results were carried out and showed that the proposed method attenuated the current spikes amplitude in both the start-up and sudden set point changes by 40% and 35%, respectively. An experimental test rig was designed to approve such results. In fact, the experimental results showed a decrease in current spikes amplitude during the start-up phase, with a percentage of 13%. This approach is appreciated since it did not require a high power supply and did not cause energy interruptions when using a limited current power supply.

It should be noted that the limitation of the proposed method, apart from higher current ripples during steady states, was the R-C filter frequency, which had to be well-chosen for adequate motor operation. Moreover, relays-based circuits had to be fast enough to respond when current spikes were detected. This fact can be addressed in further works by the use of high-speed switch-based circuits to control R-C filter activation and deactivation, such as MOSFETs. Additionally, basic and in-depth studies concerning the motor and converter efficiency, with and without an R-C filter, will be explored.

Author Contributions: Conceptualization, M.D., S.D., and A.R.; methodology, M.D. and A.R.; software, M.D.; validation, M.D., S.D., and A.R.; writing—original draft preparation, M.D.; writing—review and editing, M.D., A.R., and S.D.; supervision, S.D. and AR; funding acquisition, S.D. and A.R.

Funding: This research was funded by the Institute for Research in Solar Energy and New Energy (IRESEN)–VERES Project. Website: http://www.iresen.org/.

Acknowledgments: This work was supported by the Institute for Research in Solar Energy and New Energy (IRESEN)–VERES Project. The authors acknowledge their support and encouragement in carrying out this college work.

Conflicts of Interest: The authors declare no potential conflicts of interest with respect to the research, authorship, or publication of this article.

References

1. Chengde, T.; Mingqiao, W.; Baige, Z.; Zuosheng, Y.; Ping, Z. A Novel Sensorless Control Strategy for Brushless Direct Current Motor Based on the Estimation of Line Back Electro-Motive Force. *Energies* **2017**, *10*, 1384. [CrossRef]

2. Priyadarshi, N.; Padmanaban, S.; Mihet-Popa, L.; Blaabjerg, F.; Azam, F. Maximum Power Point Tracking for Brushless DC Motor-Driven Photovoltaic Pumping Systems Using a Hybrid ANFIS-FLOWER Pollination Optimization Algorithm. *Energies* **2018**, *11*, 1067. [CrossRef]

3. Wei, K.; Hu, C.S.; Zhang, Z.C. A Novel PWM Scheme to Eliminate the Diode Freewheeling in the Inactive Phase in BLDC Motor. *Front. Electr. Electron. Eng. China* **2006**, *1*, 194–198. [CrossRef]

4. Meilan, Z.; Zhi, L.; Quan, G.; Zeqing, X.; Huifeng, X. Influence of PWM modes on non-commutation torque ripple in brushless DC motor control system. In Proceedings of the 2013 2nd International Conference on Measurement, Information and Control, Harbin, China, 16–18 August 2013; pp. 1004–1008. [CrossRef]

5. Ben, R.A.; Masmoudi, A.; Elantably, A. On the analysis and control of a three-switch three-phase inverter-fed brushless DC motor drive. *COMPEL—Int. J. Comput. Math. Electr. Electron. Eng.* **2007**, *26*, 183–200. [CrossRef]

6. Ni, K.; Hu, Y.; Liu, Y.; Gan, C. Performance Analysis of a Four-Switch Three-Phase Grid-Side Converter with Modulation Simplification in a Doubly-Fed Induction Generator-Based Wind Turbine (DFIG-WT) with Different External Disturbances. *Energies* **2017**, *10*, 706. [CrossRef]

7. Baby, B.K.; George, S. Torque ripple reduction in BLDC motor with 120 degree conduction inverter. In Proceedings of the 2012 Annual IEEE India Conference (INDICON), Kochi, India, 7–9 December 2012; pp. 1116–1121. [CrossRef]

8. Kim, H.W.; Shin, H.K.; Mok, H.S.; Lee, Y.K.; Cho, K.Y. Novel PWM Method with Low Ripple Current for Position Control Applications of BLDC Motors. *J. Power Electron.* **2011**, *11*, 726–733. [CrossRef]

9. Kim, J.; Park, J.; Youn, M.; Moon, G. Torque ripple reduction technique with commutation time control for brushless DC motor. In Proceedings of the 8th International Conference on Power Electronics-ECCE Asia, Jeju, Korea, 30 May–3 June 2011; pp. 1386–1391. [CrossRef]

10. Wael, A.S.; Dahaman, I.; Basem, A.Z.; Amir, A.; Mohd, S.J.; Anees, A.S. Implementation of PWM Control Strategy for Torque Ripples Reduction in Brushless DC Motors. *Electr. Eng.* **2015**, *97*, 239–250. [CrossRef]

11. Tan, B.; Hua, Z.; Zhang, L. Chun, F. A New Approach of Minimizing Commutation Torque Ripple for BLDCM. *Energies* **2017**, *10*, 1735. [CrossRef]

12. Xiaofeng, Z.; Lu, Z. A New BLDC Motor Drives Method Based on BUCK Converter for Torque Ripple Reduction. In Proceedings of the 2006 CES/IEEE 5th International Power Electronics and Motion Control Conference, Shanghai, China, 14–16 August 2006; pp. 1–4. [CrossRef]

13. Kokawalage, H.R.S.; Madawala, U.K.; Liu, T. Buck converter based model for a brushless DC motor drive without a DC link capacitor. *IET Power Electron.* **2015**, *8*, 628–635. [CrossRef]

14. Jianli, J. A torque ripple suppression technique for brushless DC motor based on PFC buck converter. *IEICE Electron. Express* **2018**, *15*, 1–11.

15. Xinmin, L.; Changliang, X.; Yanfei, C.; Wei, C.; Tingna, S. Commutation Torque Ripple Reduction Strategy of Z-Source Inverter Fed Brushless DC Motor. *IEEE Trans. Power Electron.* **2016**, *31*, 7677–7690. [CrossRef]

16. Ramya, A.; Balaji, M. A new approach for minimizing torque ripple in a BLDC motor drive with a front end IDO dc-dc converter. *Turkish J Electr. Eng. Comput. Sci.* **2017**, *25*, 2910–2921. [CrossRef]

17. Wei, C.; Yapeng, L.; Xinmin, L.; Tingna, S.; Changliang, X. A Novel Method of Reducing Commutation Torque Ripple for Brushless DC Motor Based on Cuk Converter. *Trans. Power Electron.* **2017**, *32*, 5497–5508. [CrossRef]

18. Viswanathan, V.; Jeevananthan, S. Approach for Torque Ripple Reduction for Brushless DC Motor Based on Three-Level Neutral-Point-Clamped Inverter with DC–DC Converter. *IET Power Electron.* **2015**, *8*, 47–55. [CrossRef]

19. Viswanathan, V.; Jeevananthan, S. Commutation Torque Ripple Reduction in the BLDC Motor Using Modified SEPIC and Three-Level NPC Inverter. *IEEE Trans. Power Electron.* **2018**, *33*, 535–546. [CrossRef]

20. Viswanathan, V.; Jeevananthan, S. Hybrid Converter Topology for Reducing Torque Ripple of BLDC Motor. *IET Power Electron.* **2017**, *10*, 1572–1587. [CrossRef]

21. Haifeng, L.; Lei, Z.; Wenlong, Q. A New Torque Control Method for Torque Ripple Minimization of BLDC Motors with Un-Ideal Back EMF. *IEEE Trans. Power Electron.* **2008**, *23*, 950–958. [CrossRef]

22. Jiancheng, F.; Haitao, L.; Bangcheng, H. Torque Ripple Reduction in BLDC Torque Motor with Nonideal Back EMF. *IEEE Trans. Power Electron.* **2012**, *27*, 4630–4637. [CrossRef]

23. Wei, Y.; Xu, Y.; Zou, J.; Li, Y. Current Limit Strategy for BLDC Motor Drive with Minimized DC-Link Capacitor. *IEEE Trans. Ind. Appl.* **2015**, *5*, 3907–3913. [CrossRef]

24. Pillay, P.; Krishnan, R. Modeling, Simulation, and Analysis of Permanent-Magnet Motor Drives. Part II: The brushless DC motor drive. *IEEE Trans. Ind. Appl.* **1989**, *25*. [CrossRef]

25. Li, Y.; Rong, J. The study of new modeling method for permanent magnet brushless DC motor. In Proceedings of the 2011 International Conference on Electronic & Mechanical Engineering and Information Technology, Harbin, China, 12–14 August 2011; pp. 4713–4716. [CrossRef]

26. Ramasami, S.; Muhammad, K.Z.; Yadaiah, N. Modeling, simulation and analysis of controllers for brushless direct current motor drives. *J. Vib. Control* **2013**, *19*, 1250–1264. [CrossRef]

27. Varghese, L.; Kuncheria, J.T. Modelling and design of cost efficient novel digital controller for brushless DC motor drive. In Proceedings of the 2014 Annual International Conference on Emerging Research Areas: Magnetics, Machines and Drives (AICERA/iCMMD), Kottayam, India, 24–26 July 2014; pp. 1–5. [CrossRef]

28. Zhaojun, M.; Rui, C.; Changzhi, S.; Yuejun, A. The mathematical simulation model of brushless dc motor system. In Proceedings of the 2010 International Conference on Computer Application and System Modeling (ICCASM 2010), Taiyuan, China, 22–24 October 2010. [CrossRef]

29. BYOUNG-KUK, L.; MEHRDAD, E. Advanced Simulation Model for Brushless DC Motor Drives. *Electr. Power Compon. Syst.* **2010**, *31*, 841–868. [CrossRef]

30. Chuang, H.S.; Ke, Y.; Chuang, Y.C. Analysis of commutation torque ripple using different PWM modes in BLDC motors. In Proceedings of the Conference Record 2009 IEEE Industrial & Commercial Power Systems Technical Conference, Calgary, AB, Canada, 3–7 May 2009; pp. 1–6. [CrossRef]

![energies logo] *energies*

MDPI

Article

On Field Weakening Performance of a Brushless Direct Current Motor with Higher Winding Inductance: Why Does Design Matter?

Ozgur Ustun [1,†] ![ORCID], Omer Cihan Kivanc [2,*,†,‡] ![ORCID], Seray Senol [3,†] and Bekir Fincan [1,†]

1   Electrical Engineering Department, Istanbul Technical University, Istanbul 34467, Turkey; oustun@itu.edu.tr (O.U.); fincan@itu.edu.tr (B.F.)
2   Electrical and Electronics Engineering Department, Istanbul Okan University, Istanbul 34959, Turkey
3   ABB UK Engineering Centre, Leicestershire LE67 4JP, UK; seray.senol@gb.abb.com
*   Correspondence: cihan.kivanc@okan.edu.tr; Tel.: +90-216-677-1630
†   These authors contributed equally to this work.
‡   Current address: Akfirat, Istanbul 34959, Turkey.

Received: 14 October 2018; Accepted: 8 November 2018; Published: 12 November 2018

Abstract: This paper comprises the design, analysis, experimental verification and field weakening performance study of a brushless direct current (BLDC) motor for a light electric vehicle. The main objective is to design a BLDC motor having a higher value d-axis inductance, which implies an improved performance of field weakening and a higher constant power speed ratio (CPSR) operation. Field weakening operation of surface-mounted permanent magnet (SMPM) BLDC motors requires a large d-axis inductance, which is characteristically low for those motors due to large air gap and PM features. The design phases of the sub-fractional slot-concentrated winding structure with unequal tooth widths include the motivation and the computer aided study which is based on Finite Element Analysis using ANSYS Maxwell. A 24/20 slot–pole SMPM BLDC motor is chosen for prototyping. The designed motor is manufactured and performed at different phase-advanced currents in the field weakening region controlled by a TMS320F28335 digital signal processor. As a result of the experimental work, the feasibility and effectiveness of field weakening for BLDC motors are discussed thoroughly and the contribution of higher winding inductance is verified.

Keywords: brushless dc motor; phase-advanced method; winding inductance; sub-fractional slot-concentrated winding; field weakening; periodic timer interrupt

---

1. Introduction

BLDC motor drive systems have been greatly used in a wide range of applications, ranging from servo applications to electric vehicle (EV) propulsion systems due to their higher torque capability, minimum maintenance requirement, better controllability, and higher efficiency [1]. Implementing BLDC motors in electric power trains requires an exploitation of their superior features by means of various control strategies. In EV drive systems, higher torque at starting and higher speed at cruising are essential targets to satisfy the requirements of a road vehicle. This feature is named the constant power speed ratio (CPSR) and is the ratio of the maximum speed to the ultimate speed value of the constant torque region, i.e., base speed [2,3]. For most of the passenger cars, propulsion electric motor drive systems have 5:1 or higher CPSR values to provide both extreme operation modes in smaller electric motor structures, i.e., high torque at starting and high speed at flat road cruising [4]. Even for in-wheel applications, higher cruising speed is required to fulfill the driver requirements [5]. However, it is not possible to obtain higher torques at higher speeds for a certain power value without a dedicated control method [2,3]. In addition, terminal voltage is a fundamental constraint for electric motor drives in EV applications [6]. To obtain a design of a desirable electric motor for EV applications with a given output power, flux weakening performance is essential, i.e., a higher flux for a higher torque, and a lower flux for a higher speed. Due to the operational principles of an electric motor, the flux weakening control is substantially different [2,7]. However, the field weakening operation is highly applicable in interior permanent magnet synchronous motors (IPMSMs) and induction motors [8]. The field weakening performance of surface mounted permanent magnet (SMPM) motors is not satisfactory because of non-varying *dq*-axis inductance, whilst IPM synchronous motor field weakening performance relies mainly on the difference between the inductance values of *dq*-axes [9]. For BLDC motors, the fundamental principle is simple; the flux weakening is accomplished by the help of high inductance. However, in BLDC motors, winding inductance is generally low due to large air gaps and the permeability of PM materials. Thus, the winding inductance value, which seems to be a relatively unimportant design parameter formerly, becomes an objective for motor design studies [10].

Studies generally focus on control methods and motor design in order to increase the field weakening capability of BLDC motors in the literature. Previous studies on SMPM motors propose that using fractional-slot-concentrated windings with specific slot–pole combinations enhances the value of winding inductance [11,12]. The designs show the practical upper limit of the winding inductance value. It is not possible to design a winding inductance exceeding a certain value due to the physical constraints of motor dimensions and materials. Flux weakening of conventional SMPM machines with traditional winding distributions is mostly inefficient for achieving higher CPSR values [13,14]. Additionally, the main drawback of the conventional surface PM machines is the well-known low winding inductance leading to a high characteristic current [15]. The prior intention has been to reduce the characteristic current to rated value by increasing *d*-axis inductance for an optimum field weakening operation [3,16]. Studies in the literature indicate several approaches based on improved winding configurations and stator pole geometries in order to enhance the *d*-axis inductance of surface PM machines, to improve the field weakening capability [9]. Additionally, some alternative approaches have been developed to increase the BLDC motor field weakening performance in the literature. In the study described in [17], a current profiling scheme using the instantaneous power method is implemented with respect to an SMPM motor. In addition, the phase current value will increase with a higher speed; in this case, there will be inevitable stresses on the power electronic circuit and some limitations due to supply. The low frequency current components cause a reduction in the system efficiency due to higher power losses, which also provoke overheating and performance instability. Moreover, as shown in [16], including the addition of external inductance in series with the stator winding enhances the field weakening capability of the machine. However, an external inductance is not a reliable option for electric vehicles because of its bulky structure. In [18], an angular displacement estimation is presented using hall-effect sensors in order to operate at an above rated speed for a BLDC motor. However, measurement sensitivity and estimation error depending on motor speed are serious problems. In the study described in [19], a double-bridge winding switching technique is used to widen the speed range of an SMPM synchronous motor. Performing flux weakening operation is also feasible by applying a fractional-slot-concentrated winding structure to SMPM machines. Increasing *d*-axis inductance also lowers copper losses due to an alternate teeth-wound-concentrated winding structure, which enables more space for winding. Moreover, using unusual winding layouts and unequal tooth widths maximizes the machine performance substantially [20].

In this study, a field weakening controller was designed for a specially designed BLDC motor. The design phase includes the development of a motor with sub-fractional slot-concentrated winding and unequal tooth geometry for a direct drive EV application. In fact, the proposed technique based on stator structure and winding design enable high torque density, decreased copper losses, a reduction in cogging torque, and an overall enhancement of machine performance. The remainder of this paper is organized as follows: The principles of the proposed sub-fractional slot-concentrated windings technique and design approach are presented in Section 2 with simulation results. The design phase includes the development of a motor with sub-fractional slot-concentrated winding and unequal tooth geometry for an outer rotor BLDC motor. In Section 3, the control approach and experimental results of the proposed method are presented. Additionally, the findings and comparative results are provided in Section 4. Finally, the conclusion is presented in Section 5. The results of the experimental study on field weakening operations provide some important clues for the driving range of the designed motor.

2. Materials and Methods

2.1. Principle of the Field Weakening Operation of SMPM Motors

Because they have no rotor saliency, SMPM machines are subjected to some limitations arising from terminal voltage constraint. Hence, within voltage-limited operation, an SMPM motor provides a higher value electromotive force (emf) [13]. Due to the lack of a flux control parameter, SMPM motors can achieve higher speeds only by providing a negative inductance voltage that is contributing to the back-emf, and this slope depends on the d-axis inductance. The basic voltage equation of a BLDC motor regarding its equivalent circuit is given in Equation (1) [5].

$$V_a = R_s i_a + L_s \frac{di_a}{dt} + E_a \tag{1}$$

where R_s and L_s are equivalent stator resistance and inductance, respectively; i_a is the current of phase-a, and E_a is back-emf. It is evident that BLDC motors with low winding inductance values can only demand fast decaying advanced currents which hamper higher CPSR values [2]. Thus, the field weakening capability of SMPM depends on its characteristic current, which is inversely proportional to d-axis inductance in Equation (2) [21].

$$I_{ch} = \psi_m / L_d \tag{2}$$

where ψ_m, L_d, and I_{ch} are the magnet flux linkage, the d-axis inductance, and the motor characteristic current, respectively. The characteristic current and its role in field weakening are shown in Figure 1. The intersection area of the voltage limit circle and current limit circle represents useful output power. The only way to provide a wider operation region for higher speeds is to make the characteristic current smaller, i.e., to make the d-axis inductance value higher [3,13]. However, the main bottleneck of the traditional SMPM motor design is the well-recognized low and non-changing winding inductance which leads to a higher characteristic current [5,16].

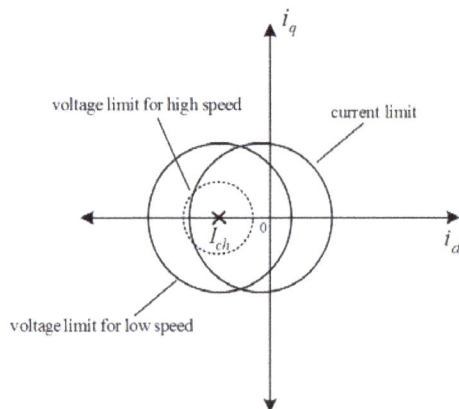

Figure 1. Current limit circle and characteristic current.

2.2. The Proposed Sub-Fractional Slot-Concentrated Windings Structure

In order to overcome the lower winding inductance problem, the sub-fractional design approach shown in Figure A1 is introduced. This design brings the above-mentioned advantages for a certain range of q value, i.e., the number of stator slots per pole per phase ratio, which is less than or equal to 1/2 for the increased number of poles [5]. In addition, this specific range of q gives the highest fundamental component of the winding factor, K_{w1}, which results in a higher torque production capability. Because a suitable design for field weakening operation has been obtained, the design study is focused on machines with $q = 2/5$ and an alternate teeth-wound stator [22]. Figure 2 is the demonstration of a sub-fractional slot-concentrated winding with all teeth-wound and alternate teeth-wound stator structure versions, where FF is the fill factor, and A is the slot area.

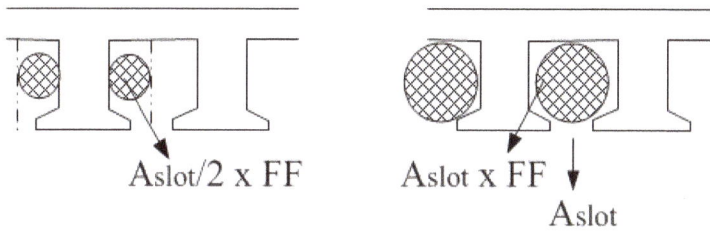

Figure 2. The demonstration of stator slots with alternate teeth winding and all teeth winding configuration.

"All teeth-wound" means each stator tooth has a winding around it, and "alternate teeth-wound" means that only one tooth in a two-teeth sequence has a winding around it. The winding fills all stator slot area and does not leave space for another winding. A larger slot space enables an increased number of turns, as demonstrated in Figure 2, and allows a wider range of field weakening operation by leading to a higher value of winding inductance. All teeth-wound motors are promoted as enabling slimmer motor designs because of their smaller winding overhangs. However, besides having larger winding overhangs, all teeth-wound motors outperform them due to their superior performances in field weakening and regenerative braking operations, which mainly rely on higher winding inductances. Using unequal teeth design, i.e., making thinner teeth alternatively, provides winding space, which contributes to higher inductance. The flux linkage per phase due to the magnet poles is given in Equations (3) [23].

$$\hat{\psi}_{ph} = c(NK_{w1})\hat{B}_1\left(\frac{1}{p}\right)dl_a \tag{3}$$

where N, c, K_{w1}, $\hat{B}_1$, p, d, and l_a are the number of turns per coil, the number of coils per phase connected in series, the maximum fundamental winding factor, the peak fundamental flux density, the pole pair, the air gap diameter, and the machine active length. Various winding configurations are compared to each other by assuming that they have the same magnet flux-linkage limitation per phase for the same rotor design [20]. A 24/20 slot–pole machine with alternate teeth-wound, non-overlapping, concentrated windings having $c_1 = 4$ coils/phase and N_1 number of turns, a 24/20 slot–pole motor with an all teeth wound non-overlapping concentrated winding with $c_2 = 8$ coils/phase and N_2 number of turns, and a 60/20 slot–pole motor having $q = 1$ overlapping integer-slot-concentrated winding with $c_3 = 10$ coils/phase and N_3 number of turns can be compared due to their winding inductances. Assuming all three designs have the same flux linkage value, $\hat{\psi}_{ph}$ in Equation (4). The winding number of turns of these designs can be found via Equation (5).

$$\hat{\psi}_1 = \hat{\psi}_2 = \hat{\psi}_3 = \hat{\psi}_m. \tag{4}$$

$$N_1 = 1.932N_2 = 2.588N_3. \tag{5}$$

The winding inductance is proportional to the square of the number of turns by assuming no saturation in the stator core where the reluctance is not a function of excitation level. The theoretical inductance values of alternate teeth and all teeth-wound machines, i.e., L_1 and L_2, respectively, are given in Equation (6).

$$L_1 = 1.866L_2. \tag{6}$$

From the general theory of PM machines, it is obvious that the highest torque production can be accomplished by using designs which have an entirely overlapped coil pitch and pole pitch [22]. However, this type of design cannot produce a rotating motion. An optimized solution for higher torque production capability includes a maximum possible value of overlapping between those measures. This combination plays a critical role in balanced three-phase winding designs shown in Figure 3 and in machine electromagnetic performance [12,24]. In addition, the rectangular shape stator teeth enable easy insertion of coils to the slots. A comparative finite element method study on back-emf waveforms of the same rotor flux linkage limitation is presented in Figure 4.

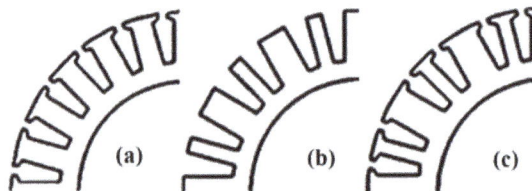

Figure 3. (a) Equal teeth, (b) rectangular shaped unequal teeth, (c) unequal teeth windings with pole shoe.

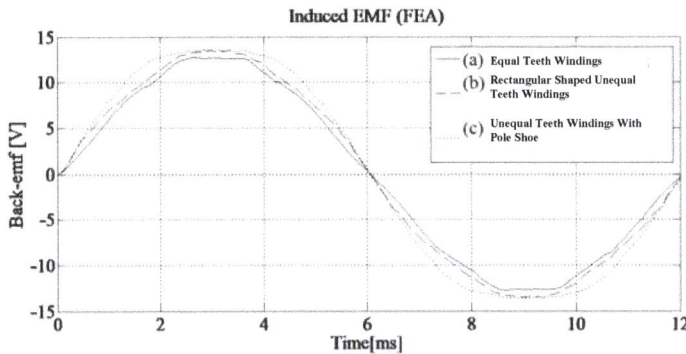

Figure 4. Finite element analysis back-emf waveforms for Figure 3 designs.

At this point, there is a trade-off between the motor electromagnetic performance and manufacturability. While the rectangular shaped tooth ensures a simple structure, the flat portion of the induced back-emf is relatively reduced. In light of above approach, for stator design, a 24/20 slot–pole combination is chosen in order to achieve a higher winding factor and simplified winding layers. A 2D magnetic analysis using Ansys Maxwell (ANSYS, Inc., Canonsburg, PA, USA) is performed with an adaptive mesh option, which is presented in Figure 5. The considerable advantage of the adaptive mesh option enables the determination of optimal mesh numbers and types for the machine structure. A 3D magnetic analysis using Ansys Maxwell is implemented by means of a 3D model of the designed motor, which is demonstrated in Figure 6. Moreover, the design specifications and machine materials are given in Tables A1 and A2, respectively.

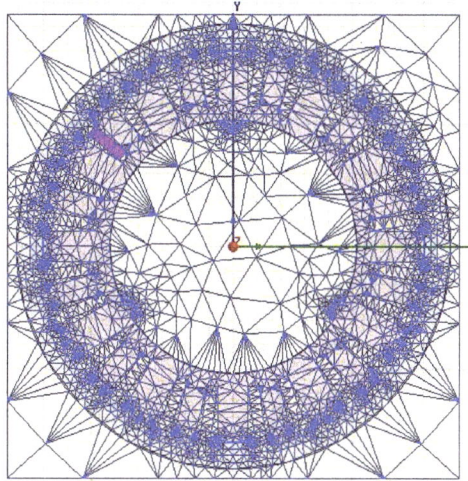

Figure 5. Adaptive mesh plot for magnetostatic analysis.

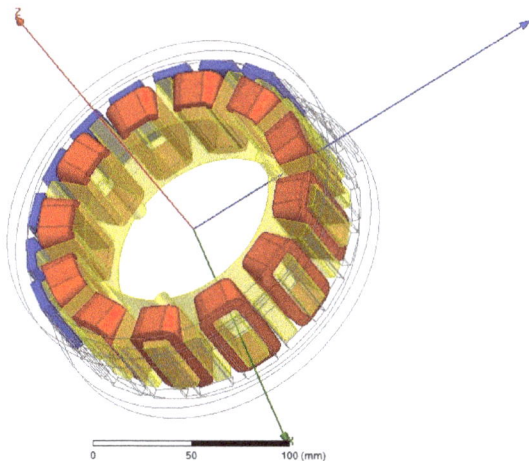

Figure 6. 3D model of the 1 kW 24/20 slot–pole machine.

To investigate the key parameter, i.e., winding inductance, numerous calculations, analyses, and experiments are conducted. As a result of the 3D analysis, winding inductance values of the unequal-teeth winding with pole shoes and rectangular unequal-teeth winding are calculated as 157.5 and 133.4 μH, respectively. Moreover, as a result of the 2D analysis, equal-teeth winding and unequal-teeth winding with pole shoes are calculated as 90.8 and 132.4 μH, respectively. Measured rectangular unequal-teeth winding is obtained as 130.3 μH. The magnetic field distribution analysis of the motor is presented in Figure 7. Moreover, the stator pole design is based on the rectangular shaped unequal teeth pole geometry for simplicity in manufacturing. Some manufacturing steps of the designed motor are presented in Figure 8.

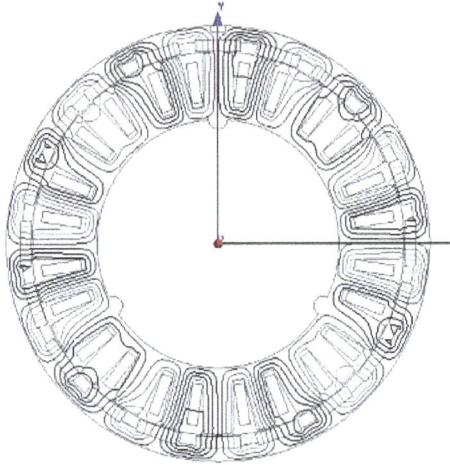

Figure 7. Magnetic flux lines of unequal teeth stator with rectangular shaped stator poles.

Figure 8. Manufacture of a 1 kW 24/20 slot–pole surface mounted permanent magnet (SMPM) motor.

3. Results

The experimental setup consists of a motor prototype, a PM synchronous generator, a torque-speed measurement system, an eZdsp board with TMS320F28335 DSP, the signal conditioning boards, and an inverter. The proposed subfractional slot-concentrated winding structure motor is verified using the virtual hall signal generation method, which is coupled with the overall system, as shown in Figures 9 and 10. The BLDC motor is loaded (to 20 N·m) by a generator, providing a variable braking torque. The overall control strategy is applied using TMS320F28335 DSP. All required voltage and current waveforms are measured using voltage and current sensors, and these values are transmitted to the controller by the signal conditioning circuits. The proposed method is illustrated in Figure 11. The PI controller is operated solely up to the base speed to satisfy the controlling requirements. If the error indicates that the reference speed is higher than the base speed, the system changes the commutation signals due to the field weakening demand.

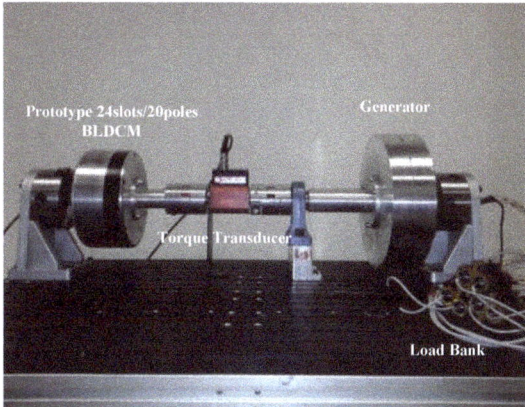

Figure 9. Experimental test-bed: prototype 24/20 slot–pole BLDC motor coupled to generator.

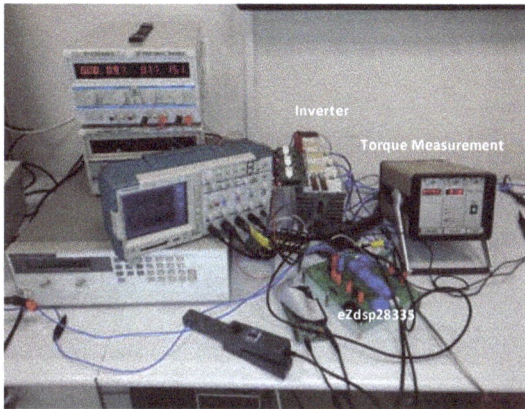

Figure 10. Experimental test-bed: inverter, DSP control unit and torque transducer.

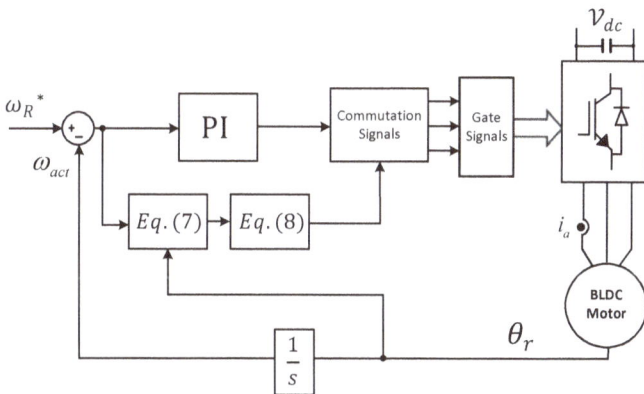

Figure 11. Overall block diagram of the proposed BLDC motor drive in the field weakening operation.

The field weakening control is implemented by considering two different operation regions: first, a conventional control for below base-speed and, second, a phase-advanced control for constant power region [25,26]. The operation region is defined by using an $f(\theta)$ function which reflects the speed information according to reference speed, ω_{ref}, and rated speed, ω_{rated}, values. Up to the base speed, the motor speed is controlled by adjusting PWM according to the controller output. In this situation, the phase current is in phase with an allegedly flat portion of back-emf, which is assumed an interval duration of $2\pi/3$. The captured waveform of the phase current is presented in Figure 12 for the constant torque operation region. As can be seen in Figure 12, the phase current flows through a high inductance winding. The measured efficiency and torque-speed characteristics of the motor in the constant torque region are illustrated in Figures 13 and 14, respectively. The analyses and tested results are in agreement.

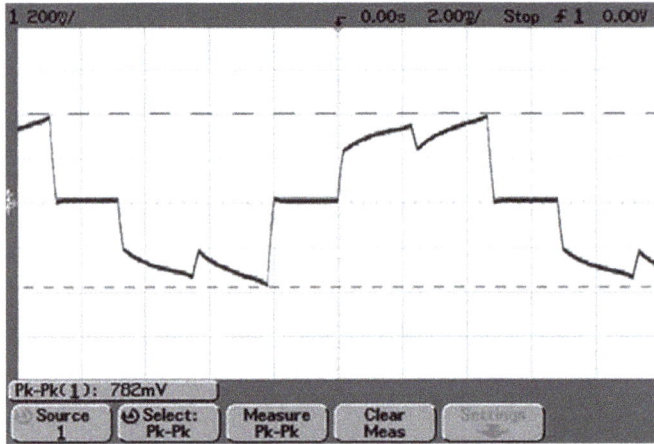

Figure 12. Constant torque region experimental results: motor phase current waveform (10 A/div, 2 ms/div).

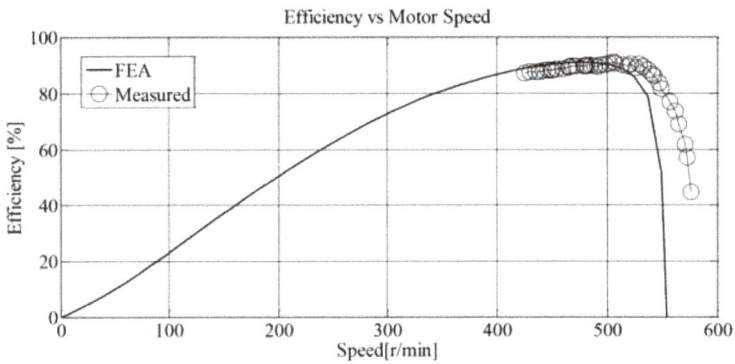

Figure 13. Efficiency versus motor speed.

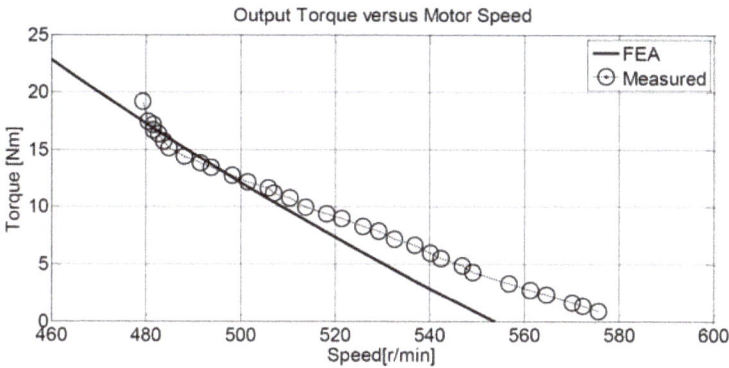

Figure 14. Output torque versus motor speed.

When the electric machine reaches its constant power region, the advance in phase current causes a current leap because of the lack of back-emf. This situation is the essence of field weakening operation: The current will decay slowly due to the higher valued phase inductance, and the back-emf will exceed the terminal voltage. If the ω_{ref} value is chosen in the field weakening operation region, the shifting angle, i.e., time shift, is obtained by using the algorithm given in Figure A2. The real hall signals and the actual hall sensor data representing the aimed speed are used as inputs of the algorithm. The output of the algorithm is a set of virtual hall sensor signals which provide the advanced commutation. This task is accomplished by using a look-up table with the relation of speed and angle. After this operation, the hall signals are shifted, as $(360 - \theta_{fw})$ the new virtual hall signals are transmitted to the switching logic circuit [14]. A subroutine to update the process is given in Equation (7).

$$\phi = n\frac{360 - \theta_{fw}}{360} \tag{7}$$

where n is a factor which defines the ratio between the electrical period and the loop period of the sub-routine. ϕ is a conversion ratio carrying the information of new virtual hall signals. Additionally, θ_{fw} and θ_{act} are the shifted hall signal angle information and the actual hall signal information, respectively. The real hall signals are stored in a vector presented in Equation (8) in order to create the virtual hall signals. The vector must be at least in an n-dimension. The real hall signals, H_s, are assigned to the shifted indices, Z_i, as illustrated in the vector according to the variation of $L[Z_i]$ and ϕ. Moreover, as mentioned before, over-currents are highly possible due to the lack of a sufficient back-emf value at advanced angle switching. Thus, the current must be checked continuously to overcome any problem due to the over-current operation.

$$L[Z_i] = \frac{\omega_{rated}\theta_{fw}n}{360\omega_{act}}. \tag{8}$$

In Figure 15, the hall signal, phase current, and back-emf waveforms are given for the speed of 1012 min^{-1} and the shaft torque of 5 N·m, whose waveform is shown in Figure 16. The current starts with a leap because of the low back-emf, and it decays slowly because of the high inductance. This slow decay causes an increased back-emf due to the negative current change rate. Thus, the fundamental idea underlying the field weakening operation of the BLDC motor can be comprehended via this figure. Similar results are given in Figure 17. The hall signal and phase current waveforms are given for the speed of 960 min^{-1} and the shaft torque of 7 N·m, whose waveform is shown in Figure 18, while the phase-advanced angle is 25°. As the advanced angle increases the speed of motor, the shaft torque is decreased to satisfy the constant power value. A nonlinear relation is observed between the speed and the angle as is expected. After 20° of phase-advanced angle, the speed is increased dramatically. At the value of 30°, the motor has reached a 1047 min⁻¹ speed under the loading of 9 N·m. The maximum speed obtained is 1624 min⁻¹ at the loading of 4 N·m and with 60°.

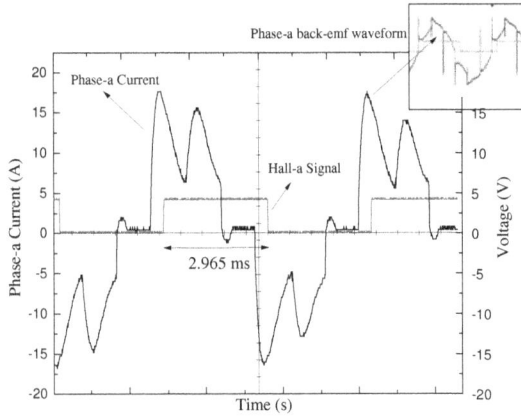

Figure 15. Phase–*a* current and hall–*a* signal under 5 N·m load with 25° reference delay angle.

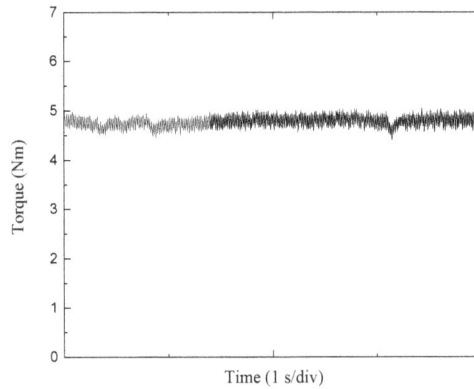

Figure 16. Torque response under 5 N·m load with 25°.

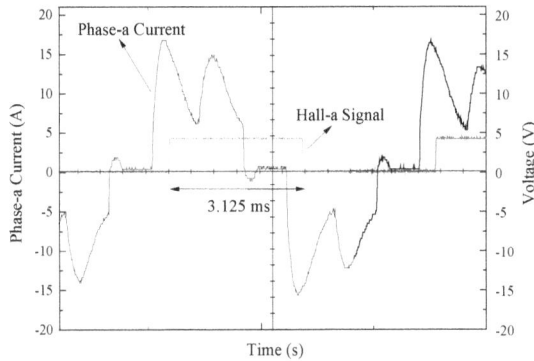

Figure 17. Phase–*a* current and hall–*a* signal under 7 N·m load with 25° reference delay angle.

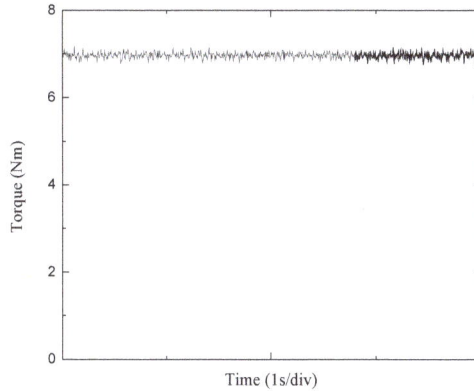

Figure 18. Torque response under 7 N·m load with 25°.

4. Discussion

The results show that there are some essential limitations in the field weakening control of SMPM BLDC motors. As can be seen, the difference of current waveforms given in Figures 15 and 17 shows the effect of the phase-advanced current drive, which causes a high start-up current. These current pulses are somewhat limited by phase inductance. Another limitation is that a substantially high value of phase inductance is not applicable by considering the limitations of the electric motor structure and materials. Another important diminishing factor is the circulating current which occurs as a bumping current waveform. These current lumps deteriorate the field weakening operation and cause very high RMS currents, which are increased by ascending values of phase angles. As can be seen in Figure 19, the CPSR value, a critical criterion for EV propulsion applications, is smaller than those of IPMSM and induction motors.

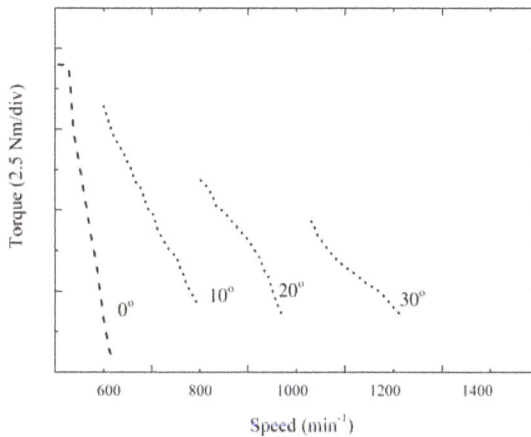

Figure 19. Output torque versus motor speed for different advanced angles.

The findings also uncover the features of unequal tooth structure: a low contribution to winding inductance improvement and, on the other hand, an enhanced torque production resulting from an inherent high winding factor . The experimental results show an adequate motor design that allows for higher winding inductance. However, it is also revealed that the field weakening capability of SMPM BLDC motors has certain limits. In Figure 20, the field weakening performance of an industrial BLDC motor with relatively lower winding inductance is shown. As can be seen, the current is decayed quickly after the advanced commutation due to a lower inductance. If the advanced angle is increased, it causes an abrupt current jump which is dangerous for semiconductor devices and winding heating. Moreover, the similar advanced angle method is implemented in [18]. In that study, the authors achieved a CPSR value of 2.94. The CPSR value is limited by the low winding inductance and the current is almost decayed to zero at the end of the 60° commutation interval. This performance shows the importance of higher winding inductance that can be achieved by a sub-fractional concentrated winding design.

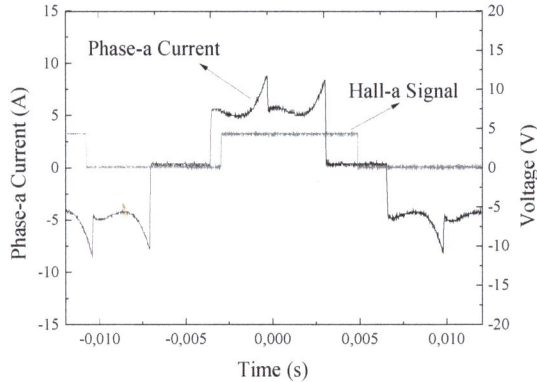

Figure 20. Phase–*a* current and hall–*a* signal under 5 N·m load with 10° reference delay angle.

For the studied motor, the obtained CPSR value after the performance tests was found to be 3.24, which shows slight incompetence for this type of motor. For electrical vehicle applications, these motors can function in limited speed-light electric vehicles such as electric scooters and e-tricycles rather than conventional electric passenger cars. Some studies which show optimistic results rely on theoretical bases. However, these results show limited performance.

5. Conclusions

The presented study includes the design, analysis, and experimental work of a sub-fractional slot-concentrated winding BLDC motor with higher winding inductance. The designed SMPM motor possesses unequal rectangular tooth widths, which infer easy manufacturing and enhanced torque production. The object of the study is to develop a BLDC motor design that contributes to drive performance, i.e., field weakening control. This contribution is yielded by the implementation of inductance value of higher stator windings. For a given motor topology, the higher winding inductance can be accomplished by using an alternate teeth-wound sub-fractional slot-concentrated winding structure. The value of winding inductance was calculated by FEA, and an experimental measurement was performed. The test results verify the intended design study. The implemented field weakening control strategy is presented in detail. The results of the experimental study on field weakening operations provide some important clues for the driving range of the designed motor. Considering all of the increased prospects from drive systems, the controller design effort is tightly correlated to the electric motor design.

Author Contributions: O.U. conceived of the idea of the research, participated to all study phases, and provided guidance and supervision. O.C.K. developed software and wrote the original draft preparation; S.S. performed FEM analysis; B.F. developed software. All authors have contributed significantly to this work.

Funding: This research received no external funding.

Conflicts of Interest: The authors declare no conflict of interest.

Appendix A

Definition of Problem
(1) determination of the field weakening constraints
(2) development of solution technique

⇩

Motor Parameters
(1) development of design criteria
(2) determination of dimension and material
(3) winding design approach

⇩

Finite Element Analysis
(1) ANSYS® RMxprt design: stator pole design etc.
(2) ANSYS® MAXWELL 2D and 3D design: magnetostatics, magnetic field distribution analysis etc.
(3) configuration of parameters (rotor, stator, windings etc.)
(4) evaluation of steady-state performance of the BLDC motor
(5) evaluation of transient performance of the BLDC motor

⇩

Motor Design Optimization
(1) optimization of parameters
(2) optimization of winding inductance value

⇩

Motor Manufacturing
(1) rotor yoke and magnets mechanical drawing
(2) laminated stator mechanical drawing
(3) stator and shaft drawing
(4) manufacturing of the designed motor
(5) winding and bearing

⇩

Field Weakening Controller Algorithm Design and Implementation
(1) controller design for constant torque region: six-step BLDC controller
(2) controller design for constant power region: phase advance method etc.
(3) controller parameters determination
(4) evaluation of steady-state performance: speed and torque response
(5) evaluation of transient performance: operation under different advance angles with various torque conditions

⇩

Controller Optimization
(1) determination of constraints under field weakening operation
(2) optimization of control parameters

⇩

System Overall Performance Verification
(1) evaluation of performance of drive system and proposed technique

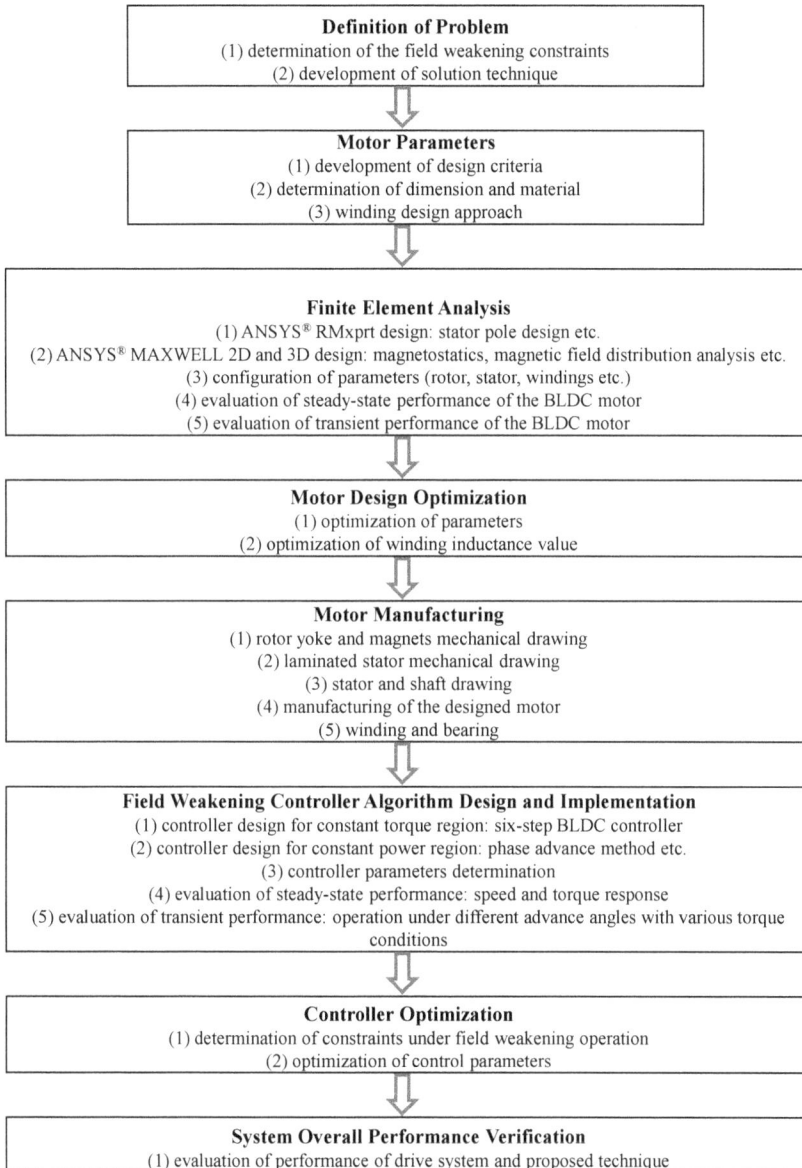

Figure A1. Design methodology flowchart.

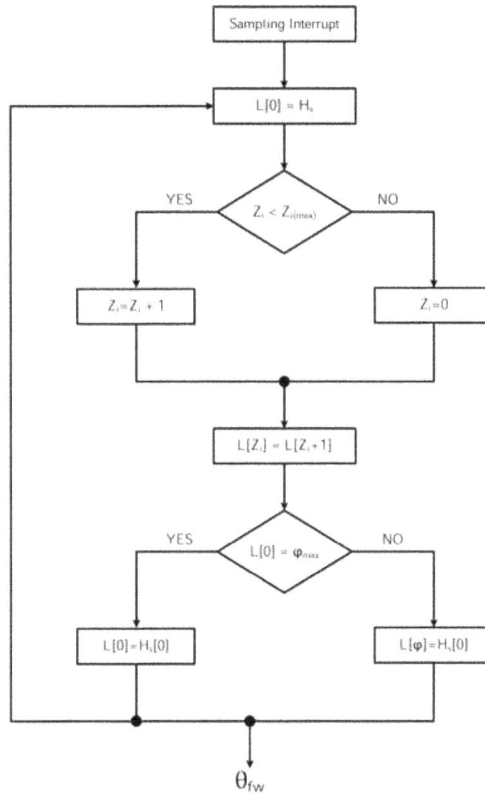

Figure A2. Proposed field weakening algorithm.

Table A1. Specifications and parameters of the manufactured BLDC motor.

Parameters	Value
Rated Power	1 (kW)
Rotor Length	40 (mm)
Rated Voltage	24 (V)
Outer Diameter	180.5 (mm)
Rated Speed	500 (min^{-1})
Winding Factor of the Stator	0.965
Fill Factor of the Stator	65.94 (%)
d-axis Inductance	98.41 $(L_1 + L_{ad})$ (μH)

Table A2. Material Description.

Part of BLDC Motor	Material
Rotor Yoke	Stell 1010
Magnets	NdFe35
Stator	M36-29G
Coils	Copper
Inner and Outer Regions	Vacuum

References

1. He, C.; Wu, T. Permanent magnet brushless DC motor and mechanical structure design for the electric impact wrench system. *Energies* **2018**, *11*, 1360. [CrossRef]
2. Jahns, T.M. Flux-weakening regime operation of an interior permanent-magnet synchronous motor drive. *IEEE Trans. Ind. App.* **1987**, *IA-23*, 681–689. [CrossRef]
3. Schiferl, R.; Lipo, T.A. Power Capability of Salient Pole Permanent Magnet Synchronous Motors in Variable Speed Drive Applications. In Proceedings of the Industry Applications Society Annual Meeting, Pittsburgh, PA, USA, 2–7 October 1988.
4. Patil, N.A.; Lawler, J.S.; McKeever, J.W. Determining Constant Power Speed Ratio of the Induction Motor from Equivalent Circuit Parameters. In Proceedings of the IEEE SoutheastCon, Huntsville, AL, USA, 3–6 April 2008.
5. Senol, S.; Ustun, O. Study on a BLDC Motor Having Higher Winding Inductance: A Key to Field Weakening. In Proceedings of the XXth International Conference on Electrical Machines (ICEM), Marseille, France, 2–5 September 2012.
6. Chaithongsuk, S. Optimal design of permanent magnet motors to improve field-weakening performances in variable speed drives. *IEEE Trans. Ind. Electron.* **2012**, *59*, 2484–2494. [CrossRef]
7. Kong, H.; Cui, G.; Zheng, A. Field-Weakening Speed Extension of BLDCM Based on Instantaneous Theory. In Proceedings of the International Conference on Electrical and Control Engineering, Wuhan, China, 25–27 June 2010.
8. Jung, S.; Mi, C.C.; Nam, K. Torque control of IPMSM in the field-weakening region with improved DC-link voltage utilization. *IEEE Trans. Ind. Electron.* **2015**, *62*, 3380–3387. [CrossRef]
9. Junlong, L.; Yongxiang, X.; Jibin, Z.; Baochao, W.; Qian, W.; Weiyan, L. Analysis and design of SPM machines with fractional slot concentrated windings for a given constant power region. *IEEE Trans. Magn.* **2015**, *51*, 1118–1125. [CrossRef]
10. Hanselman, D.C. *Brushless Permanent Magnet Motor Design*, 2nd ed.; The Writers' Collective: Columbus, OH, USA, 2003; ISBN 978-1932133639.
11. El-Refaie, A.M.; Jahns, T.M.; Novotny, D. Analysis of surface permanent magnet machines with fractional-slot concentrated windings. *IEEE Trans. Energy Convers.* **2006**, *21*, 34–43. [CrossRef]
12. Cros, J.; Viarouge, P. Synthesis of high performance PM motors with concentrated windings. *IEEE Trans. Energy Convers.* **2002**, *17*, 248–253. [CrossRef]
13. Soong, W.L.; Miller, T. Field-weakening performance of brushless synchronous AC motor drives. *IEE Proc. Electr. Power Appl.* **1994**, *141*, 331–340. [CrossRef]
14. Dorrell, D.G. A Review of the design issues and techniques for radial-flux brushless surface and internal rare-earth permanent-magnet motors. *IEEE Trans. Ind. Electron.* **2011**, *58*, 3741–3757. [CrossRef]
15. Chan, C.C.; Jiang, J.Z.; Xia, W.; Chan, K.T. Novel wide range speed control of permanent magnet brushless motor drives. *IEEE Trans. Pow. Electron.* **1995**, *10*, 539–546. [CrossRef]
16. Bianchi, N.; Bolognani, S.; Chalmers, B.J. Salient-rotor PM synchronous motors for an extended flux-weakening operation range. *IEEE Trans. Ind. Electron.* **2000**, *36*, 1118–1125. [CrossRef]
17. Miti, G.K.; Renfrew, A.C.; Chalmers, B.J. Field-weakening regime for brushless DC motors based on instantaneous power theory. *IEE Proc. Electr. Power Appl.* **2001**, *148*, 265–271. [CrossRef]
18. Yang, Y.; Ting, Y. Improved angular displacement estimation based on hall-effect sensors for driving a brushless permanent-magnet motor. *IEEE Trans. Ind.* **2014**, *61*, 504–511. [CrossRef]
19. Hemmati, S.; Lipo, T.A. Field Weakening of a Surface-Mounted Permanent Magnet Motor by Winding Switching. In Proceedings of the International Symposium on Power Electronics Power Electronics, Electrical Drives, Automation and Motion, Sorrento, Italy, 20–22 June 2012.
20. El-Refaie, A.M.; Zhu, Z.Q.; Jahns, T.M.; Howe, D. Winding Inductances of Fractional Slot Surface-Mounted Permanent Magnet Brushless Machines. In Proceedings of the Industry Applications Society Annual Meeting IAS'08, Edmonton, AB, Canada, 5–9 October 2008.
21. Soong, W.; Ertugrul, N. Field-weakening performance of interior permanent-magnet motors. *IEEE Trans. Ind. Appl.* **2002**, *38*, 1251–1258. [CrossRef]
22. Li, G.J.; Zhu, Z.Q. Analytical modeling of modular and unequal tooth width surface-mounted permanent magnet machines. *IEEE Trans. Magn.* **2015**, *51*, 8107709. [CrossRef]

23. Ishak, D.; Zhu, Z.; Howe, D. Permanent Magnet Brushless Machines With Unequal Tooth Widths and Similar Slot and Pole Numbers. In Proceedings of the 39th IAS Annual Meeting Industry Applications Conference, Seattle, WA, USA, 3–7 October 2004.

24. Ishak, D.; Zhu, Z.; Howe, D. Comparison of PM brushless motors, having either all teeth or alternate teeth wound. *IEEE Trans. Energy Convers.* **2006**, *21*, 95–103. [CrossRef]

25. Lawler, J.; Bailey, J.M.; McKeever, J.W.; Pinto, J. Extending the constant power speed range of the brushless DC motor through dual-mode inverter control. *IEEE Trans. Power Electron.* **2004**, *19*, 783–793. [CrossRef]

26. Rong, L.; Weiguo, L. A Novel PM BLDC Motors Inverter Topology for Extending Constant Power Region. In Proceedings of the 33rd Annual Conference of the IEEE Industrial Electronics Society, Taipei, Taiwan, 5–8 November 2007.

energies

MDPI

Article

Node Mapping Criterion for Highly Saturated Interior PMSMs Using Magnetic Reluctance Network

Damian Caballero [1,2,*] [ID]**, Borja Prieto [1,2], Gurutz Artetxe [1,2]** [ID]**, Ibon Elosegui [1,2]**
and Miguel Martinez-Iturralde [1,2]

[1] Ceit, Manuel Lardizabal 15, 20018 Donostia/San Sebastian, Spain; bprieto@ceit.es (B.P.);
    gartetxe@ceit.es (G.A.); ielosegui@ceit.es (I.E.); mmiturralde@ceit.es (M.M.-I.)
[2] Universidad de Navarra, Tecnun, Manuel Lardizabal 13, 20018 Donostia/San Sebastian, Spain
* Correspondence: dcaballero@ceit.es

Received: 19 July 2018; Accepted: 27 August 2018; Published: 31 August 2018

Abstract: Interior Permanent Magnet Synchronous Machine (IPMSM) are high torque density machines that usually work under heavy load conditions, becoming magnetically saturated. To obtain properly their performance, this paper presents a node mapping criterion that ensure accurate results when calculating the performance of a highly saturated IPMSM via a novel magnetic reluctance network approach. For this purpose, a Magnetic Circuit Model (MCM) with variable discretization levels for the different geometrical domains is developed. The proposed MCM caters to V-shaped IPMSMs with variable magnet depth and angle between magnets. Its structure allows static and dynamic time stepping simulations to be performed by taking into account complex phenomena such as magnetic saturation, cross-coupling saturation effect and stator slotting effect. The results of the proposed model are compared to those obtained by Finite Element Method (FEM) for a number of IPMSMs obtaining excellent results. Finally, its accuracy is validated comparing the calculated performance with experimental results on a real prototype.

Keywords: interior permanent magnet synchronous machines; magnetic reluctance network

1. Introduction

The demand for Permanent Magnet Synchronous Machines (PMSM) is rapidly increasing in high-performance applications, such as electric vehicles [1–4], due to their high power density. In particular, the configuration of buried magnets inside the rotor is becoming very popular, because of the additional torque made available to saliency, their wide constant power speed region, and their high demagnetization withstand capability, among others [5,6].

The design process of a PMSM frequently involves the aid of software based on Finite Elements Methods (FEM) [7–13]. Although accuracy of the results is very high, it requires a high computational cost, together with an elevated amount of time to define the problem. This makes FEM more suitable for validation purposes rather than for preliminary machine design by iterative process. Consequently, the use of analytical design tools that rely on magnetic circuits instead of FEM is becoming widespread.

Different authors have presented simple magnetic circuits for different PMSM topologies [14–16]. The results are acceptable as a first estimation of the machine performance, but they are not comprehensive enough if a transient analysis or a deeper study is required. For this purpose, complex magnetic circuit models based on magnetic reluctance network, known as Magnetic Circuit Model (MCM), have been proposed [17]. The methodology is very similar to simple magnetic equivalent circuits; the main difference lies in the larger number of elements that the machine's geometry is discretized.

In the literature, different MCMs have been proposed [18–23]. However, a clear meshing criterion that guarantees accurate results regardless of the size or geometry of the machine, i.e., a node mapping criterion for different machine regions, is not provided. In general, the reluctance element's distribution is set according to the main flux paths [20,24]. However, in machines that work at heavy load conditions, the flux paths are in most cases unpredictable, especially in IPMSMs, where the complex geometry of the rotor makes difficult their analytical modelling [8,25,26]. In addition, the fact that the magnetic flux can only pass through an element in a unique defined direction makes it advisable to establish a general node mapping criteria for IPMSMs.

In this paper, a general node mapping criterion for IPMSMs of any geometry and size is presented. To that end, MCM with variable discretization levels for the different geometric domains is developed. The proposed MCM models V-shaped IPMSMs, with variable magnet depth and angle between magnets. A suitable MCM structure composed of generic cells, named nodal elements, as modelling element for any part of the machine is proposed. The nodal elements contain geometric and electromagnetic information of the modelled physical domain. The model contemplates rotor motion, allowing dynamic analysis, i.e., time stepping simulations. The MCM takes into account the effects of stator slotting, the airgap magnetomotive force (MMF) waveform due to armature current, and cross-magnetizing saturation effects due to an improved magnetic flux path definition. This, in turn, leads to more accurate results than other simpler magnetic circuit models shown in the bibliography.

This paper is organized as follows. First, the MCM developed for IPMSMs is described. Next, the magnetic phase flux linkage at heavy load conditions for different V-shaped IPMSMs is calculated. Then, an analysis of the accuracy of the results from using different node mappings in the MCM is conducted, and the results are compared to those obtained by the FEM software. This analysis provides a node mapping criterion to ensure sufficient accuracy when calculating the performance of an IPMSM with a magnetic reluctance network. Finally, once the node mapping criterion is described, the presented methodology is validated by comparing the calculated performance to tests on a real traction IPMSM working under heavy load conditions.

2. Definition of the Magnetic Circuit Model

In this section, the MCM developed for IPMSM is described. The MCM allows the number of nodes the machine is discretized into to be selected. The proposed MCM is divided into three regions: stator, rotor and airgap. A generic nodal element, shown in Figure 1, is proposed.

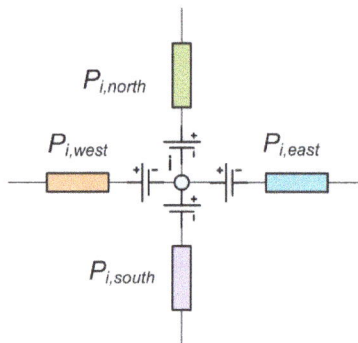

Figure 1. Nodal element structure.

The node is set in the center of the nodal element. Each nodal element has its own coordinates (i,j), so it may be stored in a nodal elements matrix. Appropriate row and column are assigned to each nodal element according to its location on the associated discretized geometrical domain being modelled. Information regarding the associated geometry (identifier, row, column, region, position, etc.) is stored as well. Each nodal element has four sub-elements that capture both the radial and orthoradial magnetic fluxes and, therefore, possible cross-coupling effects [27,28]. The sub-element geometrical domain can be rectangular, trapezoidal or even triangular. Sub-element information that is held is fundamental for the solving process: orientation length L, transversal section A_i, permeance, permeability μ, MMF source, etc. Magnetic permeance, $P_{i,orientation}$, is employed for convenience in the solving process; it is the inverse of magnetic reluctance, $\Re_{i,orientation}$, which is calculated by Equation (1) [29].

$$\Re_{i,orientation} = \int_0^L \frac{dx}{\mu(x) \cdot A_{i,orientation}(x)} \tag{1}$$

To reduce the computational effort, if periodicities exist within the machine, just part of it is modelled. The number of periodicities is obtained according the number of pole pairs and slots (Equation (2)).

$$N_{sim} = GCD(Q_s, p) \tag{2}$$

where Q_s is the number of stator slots, p the number of pole pairs and GCD stands for Greatest Common Divisor.

The geometric positioning of the nodal elements that model different regions of the machine, is established according to the existing main magnetic flux paths. In addition, it facilitates the connection between nodal elements, and therefore, the different regions of the machine can be accurately meshed.

2.1. Stator

The stator yoke, slots, slot-openings, teeth, and tooth-tip are modelled separately. The MCM parameters that define the stator node mapping are presented in Table 1. As an illustrative example, the values given in the table result in the node mapping shown in Figure 2.

Table 1. Stator reluctance network parameters.

Definition	Parameter	Value (See Figure 2)
Yoke rows	$N_{row,y,s}$	1
Tooth rows	$N_{row,t,s}$	4
Tooth-tip rows	$N_{row,s0}$	1
Tooth columns	$N_{col,t,s}$	3
Tooth-tip columns (One side)	$N_{col,t0}$	2
Slot-opening columns	$N_{col,s0}$	2

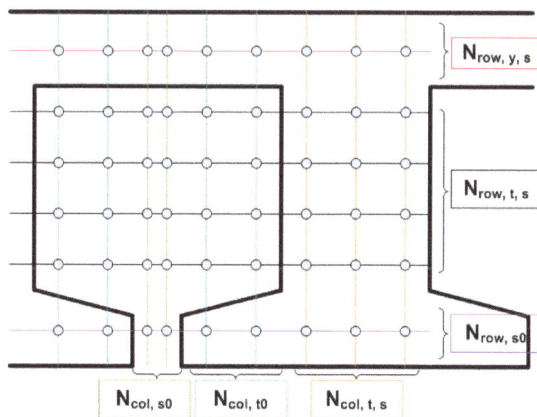

Figure 2. Stator nodal model.

Presented parameters in Table 1 define the number of node columns and rows in stator regions, as can be observed in Figure 2. For open slot machines, $N_{col,t0}$ is null. The armature current MMF sources are located at the north and south sub-elements of each element belonging to the teeth and are calculated by Equation (3).

$$MMF_w = \frac{Z_{slot,ph,s}}{N_{w,paral,s} \cdot N_{row,t,s} \cdot 2} \cdot [Ws] \cdot [I_{UVW}] \tag{3}$$

where $[Ws]$ is the winding sequence matrix that relates the teeth wound by each coil to the corresponding phase winding [30]. $[I_{UVW}]$ is the phase current vector. $Z_{slot,ph,s}$ is the number of turns per slot and layer. Finally, $N_{w,paral,s}$ is the number of parallel connected winding groups.

2.2. Rotor

The parameters that define the reluctance network for IPMSM rotors are presented in Table 2. Due to the existing symmetry, only the mapping for half a pole needs to be defined. To ease the comprehension, the values given in the table result in the node mapping shown in Figure 3.

Table 2. IPMSM rotor reluctance network parameters.

Definition	Parameter	Value (See Figure3)
PM rows	$N_{row,PM}$	1
Bridge rows	$N_{row,bridge}$	1
Bridge columns	$N_{col,bridge}$	1
Bridge inter pole columns	$N_{col,bridge,q}$	1
PM columns	$N_{col,PM}$	3

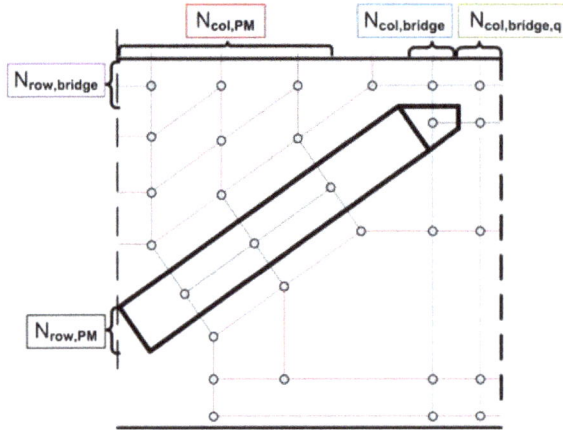

Figure 3. Half pole rotor nodal model for V-shaped PMSMs.

In Figure 4, the whole pole node mapping scheme for V-shaped IPMSMs is shown.

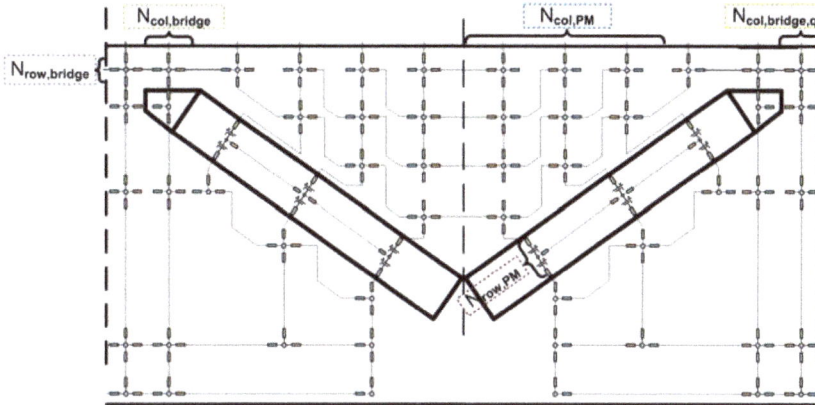

Figure 4. Whole pole rotor nodal model for V-shaped PMSMs.

Regarding the rotor nodal mapping, some nuances must be taken into consideration:

- $N_{row,PM}$ also define the number of node rows in the non-magnetic material and in the q-axis magnetic bridge.
- $N_{col,PM}$:
    - It defines rotor yoke's node mapping: over and below PM side. These node distributions are "triangular", with $N_{col,PM}$ columns and as we move away from d-axis, the number of rows decreases from $N_{col,PM}$ to one.
    - It defines with other parameters as $N_{col,bridge}$ and $N_{col,bridge_q}$ and one necessary extra node (for node connection purposes), the number of columns in the $N_{row,bridge}$ upper rotor nodal rows, belonging to an arc whose thickness is the bridge height.

It is remarkable that the same nodal mapping defined by Table 2 may be established for extreme cases of IPMSM. As an example, V-shaped in Figure 5 and embedded in Figure 6 are shown.

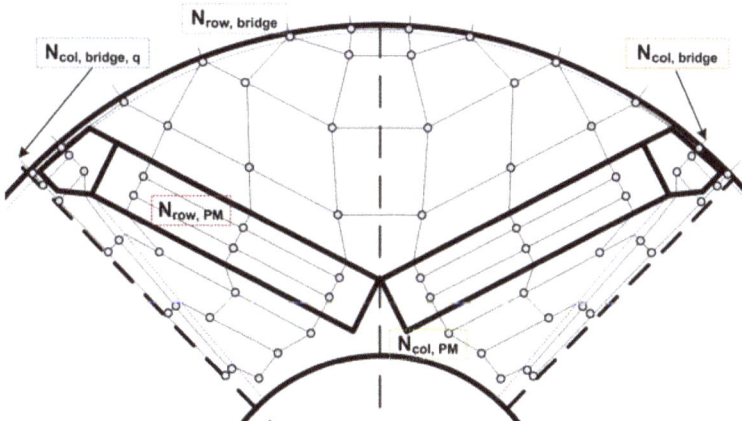

Figure 5. Rotor nodal model for V-shaped PMSMs.

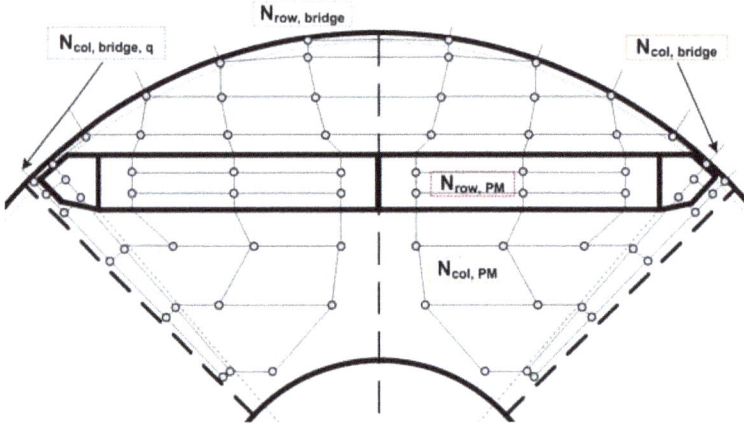

Figure 6. Rotor nodal model for embedded PMSMs.

The magnets, the rotor yoke, the magnetic bridge and the rotor slot non-magnetic material, which is responsible for preventing a large magnetic flux leakage, are modelled by a selectable number of elements. The MMF contribution is computed by Equation (4) and assigned to the north and south sub-elements corresponding to permanent magnets.

$$MMF_{PM} = h_{PM} \cdot H_c \tag{4}$$

h_{PM} is the magnet height, and H_c the coercive field strength.

2.3. Airgap

This region is the most significant part of the MCM and the overall precision of the model depends on its modelling. Moreover, it is the link between the stator and rotor models, thus a correct modelling of the airgap is necessary. Therefore, the airgap model is built by taking into account the relative position between the stator and the rotor. Since machine rotation is considered, airgap node mapping is varied for each time step. An airgap element is placed at the same angular position of each stator and rotor nodal element in contact with the airgap, as shown in Figure 7. It is important to note that more finely discretized stator and/or rotor models entail a more detailed airgap region node mapping.

The number of airgap row elements is determined by the parameter $N_{row,agap}$. In Figure 7, the proposed airgap node mapping is visually displayed for a $N_{row,agap}$ equal to 2.

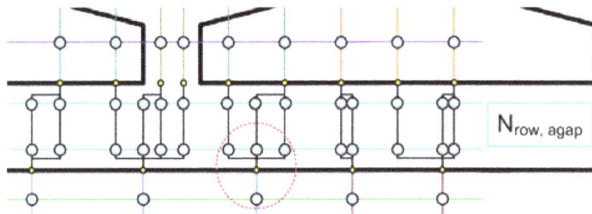

Figure 7. Airgap nodal model.

To link the airgap with stator and rotor entities, auxiliary nodes are placed in the airgap region boundaries, as shown in Figure 7 with small yellow circles. They only have north and south sub-elements, and an infinite permeance. Therefore, it is a mathematical element that is introduced in the circuit matrix system. Its necessity is due to the inter-entities borders, where the airgap sub-elements are connected to the nearest auxiliary node (Figure 8). If they are connected simultaneously to various airgap nodes, the existence of these auxiliary nodes guarantees that each nodal element belonging to any machine entity is connected to other nodes by no more than four sub-elements. Thus, it is possible to generate the flux and permeanace matrix, according to the defined nodal element structure (Figure 1).

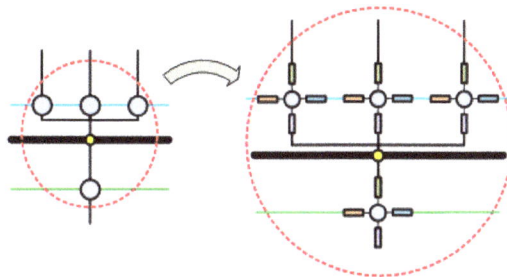

Figure 8. Airgap multinode linking.

2.4. Solving Process

Once all the nodes are linked and the branches defined, the MCM is completely set. Owing to the non-linear magnetic behavior of the stator and rotor materials, whose B-H curve are imported from materials database and used as a lookup table, the MCM needs to be solved iteratively. Firstly, permeance and magnetic source flux matrices, $[P]$ and $[\phi]$, are calculated for each sub-element. The matrix circuit is solved in terms of Kirchoff's Voltage Law (KVL), and the scalar magnetic potential at each node is obtained by:

$$[V_m] = [\phi] \cdot [P]^{-1} \tag{5}$$

The magnetic flux that crosses each corresponding branch from node i to node j is calculated by:

$$\phi_m(i,j) = (V_m(i) - V_m(j) + MMF(i,j)) \cdot P(i,j) \tag{6}$$

where $MMF(i,j)$ is the addition of different MMF sources at each branch. A weighted average of the updated and previous iteration magnetic permeability is used in the following iteration step by the corresponding permeance sub-element, rewriting data in $[P]$ and recalculating Equations (5) and (6) [20].

The iteration process is finished when the following convergence criterion is met:

$$\mu_{error,k} - \mu_{error,k-1} \le \epsilon \tag{7}$$

where $\mu_{error,k}$ is the committed mean error in sub-element permeability at iteration k.

2.5. Data Post-Processing

A key parameter for determining machine behavior is the phase magnetic flux linkage, Ψ_{ph} [31]. The magnetic flux linked by each phase winding can be computed by Equation (8).

$$\Psi_{ph} = \frac{N_{sim} \cdot Z_{slot,ph,s}}{N_{w,paral}} \cdot [\Phi_{t,s}] \cdot [Ws] \tag{8}$$

where N_{sim} is the number of symmetries which the problem has been divided due to the existing geometry periodicity, $Z_{slot,ph,s}$ is the number of turns per slot and layer, and $N_{w,paral}$ is the number of pole groups in parallel. The matrix $[\Phi_{t,s}]$ is the magnetic flux that crosses each pair of slot-teeth. This is easily obtained once the MCM is solved.

Depending on whether the MCM is solved for a load operating point or at no-load condition, the phase magnetic flux linkage is denoted as $\Psi_{load,ph}$ and $\Psi_{PM,ph}$, respectively. Using Park's transformation matrix, it is possible to work at d-q rotational reference system, whose main advantage is the fact that the different variables are time invariant. The electromagnetic torque is computed by Equation (9).

$$T_{em} = \frac{3}{2} \cdot p \left(\Psi_d \cdot I_q - \Psi_q \cdot I_d \right) \tag{9}$$

The phase back EMF (Equation (10)) and the phase voltage (Equation (11)) are obtained by the time derivatives of the no-load and load fluxes:

$$E_{ph}(t) = \frac{d\Psi_{PM,ph}(t)}{dt} \tag{10}$$

$$U_{ph}(t) = I_{ph}(t) \cdot R_{ph} + \frac{d\Psi_{load,ph}(t)}{dt} \tag{11}$$

where I_{ph} is the phase current and R_{ph} is the stator phase winding resistance.

3. Results Using MCM with Different Node Mapping

The described MCM is implemented in MATLAB. To obtain a MCM node mapping criterion that guarantee an acceptable balance between results accuracy and computational costs, an analysis of the influence of the discretization level obtained by the model was conducted.

Furthermore, to assess the validity of the proposed MCM, three different size IPMSMs, whose characteristics are presented in Table 3, were evaluated.

Table 3. V-shaped IPMSM characteristics.

Parameter	Motor A	Motor B	Motor C
Number of poles	4	6	8
Number of slots	48	54	24
Airgap length [mm]	2	2	1
Stator outer diameter [mm]	820	370	150
Rotor outer diameter [mm]	570	276	108
Stack length [mm]	820	404	300
V-shaped PM depth [mm]	100	24	10
V-shaped PM height [mm]	9	5	3.5
Stator yoke height [mm]	64	28	10
Stator slot width [mm]	15	6.5	6
Magnet remanence [T]	1.304	1.304	1.13
Magnet relative permeability	1.06	1.06	1.04

The analysed machines have very different geometries to validate the proposed MCM and establish a node mapping criterion, regardless of the size of the machine.

The load conditions for each machine are presented in Table 4.

Table 4. Load operation point.

Parameter	Motor A	Motor B	Motor C
Output power [kW]	147	180	50
Speed [r/min]	700	1250	2500
Phase current RMS [A]	160	400	130
q-component current [A]	148	392	125
d-component current [A]	−60	−79	−34

As Table 5 and Figure 9 illustrate, the three machines operate under heavy load conditions for the considered operating points. In the case of Motor C (Figure 9c), although the stator is not highly saturated, it can be observed that the armature current MMF is comparable to permanent magnet MMF, thereby establishing the magnetic axis almost in the q-axis. Besides, the machine comprises wide slots that behaves as barriers to the magnetic flux and, therefore, increasing the magnetic saturation at the rotor bridges. Altogether, this makes not only the magnetic flux paths more unpredictable but also the cross-coupling effect appreciable.

Table 5. FEM measured B_{max}.

Region	Motor A	Motor B	Motor C
Stator Tooth [T]	1.8	2.0	1.5
Stator Core [T]	1.9	1.9	1.6
Rotor Bridge [T]	2.3	2.5	2.6
Rotor Core [T]	1.6	1.5	1.4

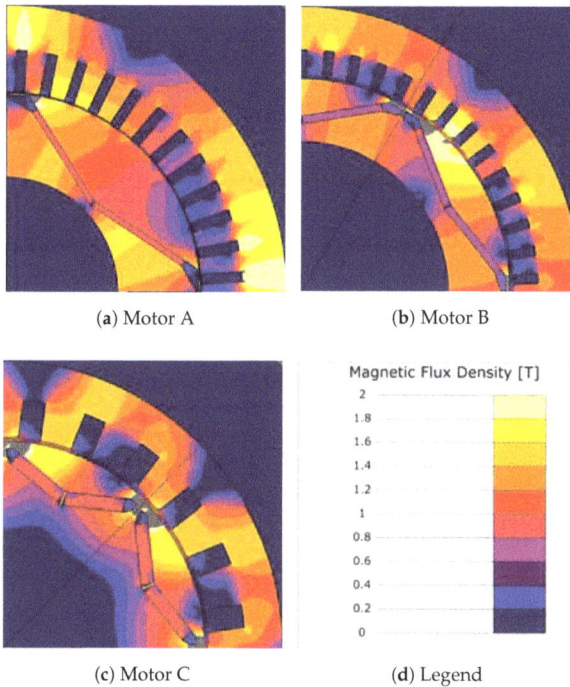

(a) Motor A

(b) Motor B

(c) Motor C

(d) Legend

Figure 9. Magnetic flux density distribution for the three motors.

To establish a general node mapping criterion, the presented motor geometries have been evaluated with the aid of the developed MCM and the discretization level presented in Table 6. Each column defines a whole machine node mapping and is defined so that a specific machine's modelled part is studied, i.e., in the case of *Stator X*, the node mapping parameters that define the stator are modified while the rest of parameters are kept at their lowest possible value. The parametric analysis carried out allows to separately analyse the influence of each model. The Xs in Table 6 refer to values that have been evaluated in each analysis, ranging from 1 to 6. For the parameter controlling the rotor discretization (Rotor X), the Xs values are evaluated in threes due to the complexity of the IPMSM rotor geometry and the difficulties to predict the magnetic flux paths at heavy load conditions.

Table 6. MCM node mapping employed in the analysis.

Parameter	StatorX	AirgapX	BridgeX	RotorX
$N_{row,s0}$	X	1	1	1
$N_{col,s0}$	X	1	1	1
$N_{row,ts}$	X	1	1	1
$N_{col,ts}$	X	1	1	1
$N_{row,ys}$	X	1	1	1
$N_{row,bridge}$	1	1	X	1
$N_{row,PM}$	2	2	2	2
$N_{col,PM}$	3	3	3	$3 \cdot X$
$N_{col,bridge}$	1	1	X	1
$N_{col,bridge_q}$	1	1	X	1
$N_{row,agap}$	2	X	2	2

As explained in Section 2.5, the key parameter to determine the machine performance is the magnetic flux linkage, $\Psi_{load,ph}$. Thus, it is carried out a further analysis on this parameter obtaining. In Table 7, $\Psi_{load,ph}$ results obtained by FEM are presented.

Table 7. $\Psi_{load,ph}$ FEM results.

	1st Harmonic	3rd Harmonic
Motor A	2988 mWb	250 mWb
Motor B	702 mWb	29.4 mWb
Motor C	246 mWb	14 mWb

In Table 8, the relative errors between the results of the MCM and FEM are presented for the first and third time harmonics of the $\Psi_{load,ph}$ waveforms. Each row corresponds to results obtained when a MCM node mapping defined in Table 6 is employed with a determined X value. For example, Stator 1 means one value for all the stator parameters, and the other parameters are set to the default value.

Table 8. MCM node mapping $\Psi_{load,ph}$ relative error values.

	1st Harmonic [%]			3rd Harmonic [%]		
Motor	A	B	C	A	B	C
Stator1	5.3	9.3	21.7	9.1	12.2	15.1
Stator2	2.5	3.0	13.5	7.7	12.9	11.8
Stator3	1.0	0.8	4.2	2.6	15.3	7.4
Stator4	1.1	0.4	3.2	0.8	13.7	8.4
Stator5	0.5	0.5	1.0	2.1	15.2	6.3
Stator6	0.6	0.4	0.4	0.4	13.8	7.2
Rotor3	5.4	9.3	21.8	9.1	12.6	15.0
Rotor6	5.2	7.0	18.8	7.2	25.5	14.2
Rotor9	5.3	8.6	14.5	1.3	15.8	9.5
Rotor12	5.2	9.0	12.3	10.0	22.9	8.1
Rotor15	4.9	8.8	11.0	12.3	27.2	12.6
Rotor18	4.6	9.4	10.1	13.9	28.1	6.1
Bridge1	5.4	9.3	21.8	9.2	12.6	15.0
Bridge2	4.8	6.2	22.5	16.7	13.5	15.0
Bridge3	4.6	8.0	23.0	12.3	7.2	15.1
Airgap1	5.4	9.4	21.8	9.1	12.0	14.7
Airgap2	5.3	9.3	21.8	9.2	12.2	15.0
Airgap3	5.3	9.3	21.7	9.3	12.2	15.0

In Figure 10, the error tendency for the first harmonic is graphically plotted.

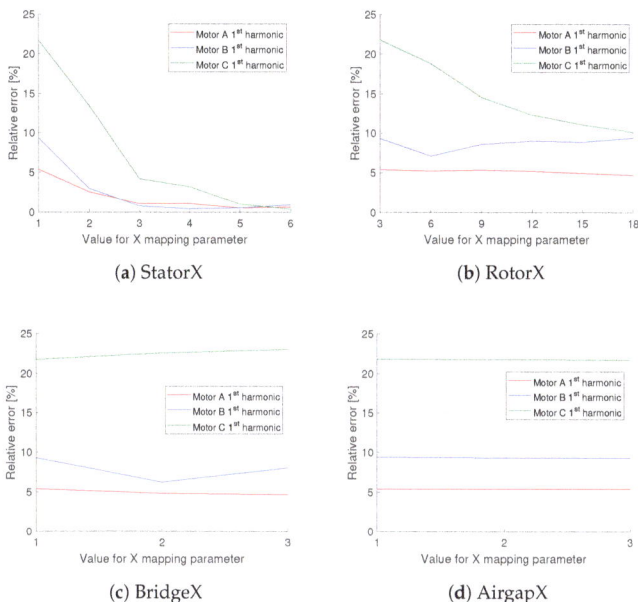

(a) StatorX

(b) RotorX

(c) BridgeX

(d) AirgapX

Figure 10. MCM node mapping $\Psi_{load,ph}$ relative error.

The discrepancies between the results obtained by the MCM and the FEM are fairly low, especially taking into account the deep saturation of the iron parts for the considered operating points.

To show graphically the great accuracy obtained by the developed MCM, some important magnetic calculated characteristics are presented and compared with obtained by FEM simulation.

In Figure 11, the $\Psi_{load,ph}$ corresponding to Motor A is presented. Its time distribution at an electric cycle for a given stator phase is shown.

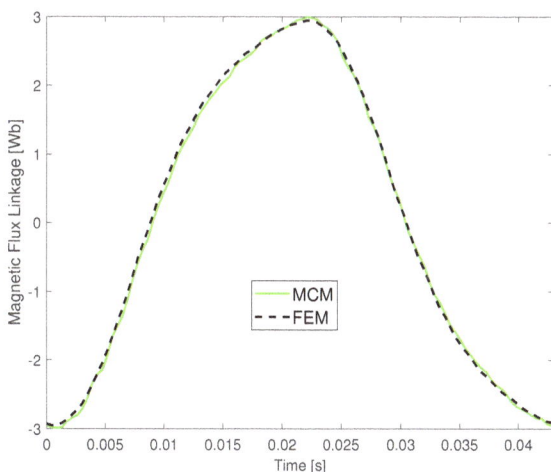

Figure 11. Phase magnetic flux linkage (Motor A).

In Figure 12, the spatial distribution of airgap magnetic flux density along a pole for Motor A at no load operation point is shown. It reflects the accuracy obtained in terms of taking into account the slotting effect, noticing the well calculated magnetic flux density ripple, and also, in terms of the well calculated beginning and ending of the mentioned waveform, which means that the bridge's model works correctly.

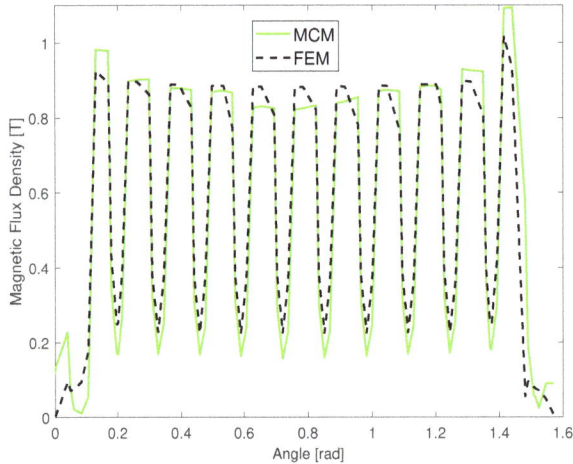

Figure 12. Airgap magnetic flux density at no load condition (Motor A).

The MCM node mapping used is the one termed "*Stator6*", due to the excellent results obtained.

Employing MCM magnetic information, the main electromagnetic characteristics at load conditions have been calculated as explained in Section 2 for the three machines. These are presented in Table 9.

Table 9. Load performance.

		Value			Relative Error [%]		
Motor		A	B	C	A	B	C
Torque [N· m]		2005	1370	173	0.2	−0.5	−4.8
U_{ph} [V]	1st harm	434.0	285.8	285.2	−0.7	3.6	6.3
	3rd harm	103.1	38.9	42.8	−7.1	5.0	−2.7

4. Results Discussion and Node Mapping Criterion

When a machine is modelled by a magnetic reluctance network, it is usually not clear which node mapping configuration offers the optimal balance between accuracy and simulation time. With the aim of establishing a node mapping criterion, the results collected in Table 8 and plotted in Figure 10 are examined in greater detail in the following paragraphs:

Stator (Figure 10 a)

As can be observed, the higher is the number of stator nodal elements, the better is the accuracy of the results. The main reason is that, when the number of stator elements increases, so does the number of airgap elements.

Rotor (Figure 10 b)

Although the rotor has a considerable number of nodal elements, the stator is poorly meshed. This means that the linking between airgap and stator is not good enough, especially in terms of taking the slotting effect into account.

For Motor A and Motor B, the number of airgap nodes does not increase as much when $N_{col,PM}$ is increased, unlike the case where the stator is more finely discretized. Nevertheless, for Motor C, the number of airgap nodes is significantly increased, because the number of slots per pole is low, and therefore the effect of stator discretization in the airgap meshing is lower.

Bridge (Figure 10 c)

Because of the high magnetic saturation in the bridge, the number of nodes does not affect the results.

Airgap (Figure 10 d)

The airgap meshing is controlled by the number of existing nodal elements at the stator–airgap border, and the rotor–airgap border. A different number of $N_{row,agap}$, does not involve a thorough whole machine physical domain modelling. Therefore, the obtained results are invariant.

Additionally, the computational time was measured for the different mapping configurations. In Figure 13, the normalized simulation time is displayed for Motor B using different node densities.

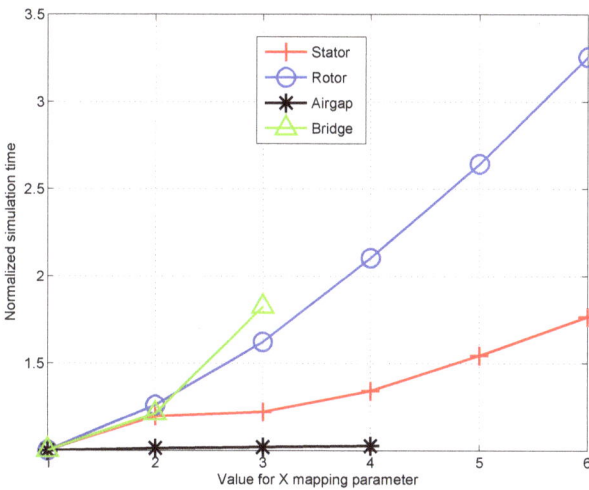

Figure 13. Normalized time simulation vs MCM node map.

For the analysis called BridgeX, the defined MCM has a slightly higher computational cost due to the increasing number of deeply saturated permeance sub-elements. In addition, it should be noted that the high computational cost for the RotorX analysis is due to the fact that the number of rotor elements increases three times faster than parameter X.

The reflected data in Table 10 show the required computation time using FEM and MCM corresponding to Stator1 node mapping. The time data relate to one time step solving process.

Table 10. Computational time MCM vs FEM.

	Number of Elements		Computation Time [s]	
	MCM	**FEM**	**MCM**	**FEM**
Motor A	424	53154	1.3	9.1
Motor A	364	21149	1.2	3.6
Motor A	208	25067	1.4	5.1

It should be highlighted that the required computation time for solving MCM shown in Table 10 is up to seven times less than the required time for solving FEM. This together with the fact that, in contrast to FEM, MCM is instantaneously generated, makes our proposed MCM a suitable and fast designing tool.

Based on the accuracy of the results and the required solving time, it can be stated that a good node mapping choice is in the range of configurations *Stator*3 to *Stator*6.

In Table 11, the airgap node mapping densities are presented for all three machine configurations under study. The information corresponds to a single airgap row. Various airgap node row details are given: number of nodes per pole pair, average distance between two consecutive nodes, and number of nodes per slot pitch.

Table 11. Airgap region nodal information.

Node Map	Nodes per Pole Pair			Average Nodes Distance [mm]			Nodes per Slot Pitch		
	A	**B**	**C**	**A**	**B**	**C**	**A**	**B**	**C**
Stator1	72	60	36	12.4	4.8	2.4	3	3.3	6
Stator2	120	96	48	7.5	3.0	1.8	5	5.3	8
Stator3	168	132	60	5.3	2.2	1.4	7	7.3	10
Stator4	216	168	72	4.2	1.7	1.2	9	9.3	12
Stator5	264	204	84	3.4	1.4	1.0	11	11.3	14
Stator6	312	240	96	2.9	1.2	0.9	13	13.3	16
Rotor3	72	60	36	12.4	4.8	2.4	3	3.3	6
Rotor6	84	72	48	10.6	4.0	1.8	3.5	4	8
Rotor9	96	84	60	9.3	3.5	1.4	4	4.6	10
Rotor12	108	96	72	8.3	3.0	1.2	4.5	5.3	12
Rotor15	120	108	84	7.5	2.7	1.0	5	6	14
Rotor18	132	120	96	6.8	2.4	0.9	5.5	6.6	16

Next, an analysis to obtain a general node mapping criterion was carried out.

- The number of nodes per pole pair is increased because of the higher value of X. Additionally, the difference between different geometries is due to the fact that the number of slots per pole are different, and the stator mapping is modelled for each slot-tooth pair. In this regard, setting a number of nodes per pole is not sufficient to establish a general node mapping criterion.
- The average distance between consecutive nodes depends on the number of airgap nodes and especially on the airgap diameter, i.e., motor size. As can be seen in Motor C, despite having very low values, it does not guarantee excellent results.
- The number of nodes per slot pitch for Motor A and Motor B with at least a value of 5 is sufficient for obtaining acceptable results. Nevertheless, when analysing rotor mappings, although this ratio is reached, the stator magnetic flux paths are not very well defined. In the case of Motor C, due to the lower number of nodes per pole pair, a value of at least 10 (Stator3) is needed. This value is also reached at rotor analysis but, as explained previously, the stator node mapping is not very well defined.

The analysis presented here leads to the next node mapping criterion for modelling IPMSM by MCM.

1. *The key point is that the number of nodes per slot in the airgap must be at least seven.*
    As stated in Table 8 and the airgap region nodal information in Table 11, specifically those displayed in the third column (defined with the heading Nodes per slot and pitch), the best results are obtained for MCM node mappings with a ratio value of at least seven nodes per slot in the airgap region.
2. *There must be at least three equidistant nodes in the stator slot and the tooth span.*
    To consider the slotting effect and hence obtain accurate results, according to Table 8, the nodal configuration named Stator 3, that guaranties having at least three equidistant nodes in the stator slot and the tooth span, is enough.
3. *At least three rows of nodal elements show be employed at both rotor yokes.*
    In consonance with Figure 10, to increase the stator nodal density (Figure 10a), it is fundamental to obtain accurate results, even with the minimum number of rotor nodal rows ($N_{col,PM}$), which in this case has been established as three.

By applying these criteria, an adequate connection between the stator and rotor models can be generated, leading to accurate results.

5. Experimental Validation

With the aim of validating the proposed methodology, a real traction motor with embedded magnets was tested. Its main characteristics are shown in Table 12.

Table 12. Tested IPMSM characteristics.

Parameter	Value
Output power [kW]	225
Voltage [V]	318
Stator outer diameter [mm]	420
Rotor outer diameter [mm]	317
Stack length [mm]	380
Magnet quality	N40UH

The prototype machine has been previously simulated via FEM at rated load. In Figure 14, it can be appreciated that the motor is magnetically saturated, reaching more than 2 T at some points.

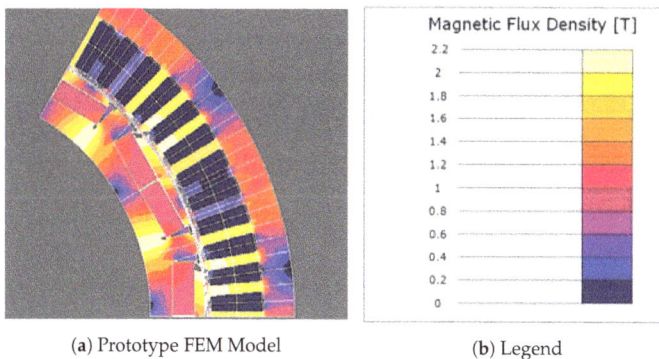

(a) Prototype FEM Model   (b) Legend

Figure 14. Prototype's Magnetic flux density distribution.

After applying the node mapping criteria, the main characteristics were obtained and compared with FEM simulations, as shown in Table 13. The operation point is set by defining the same current values for both numerical models.

Table 13. Prototype's performance calculated via proposed MCM and FEM.

Parameter	MCM	FEM	Relative Error [%]
Back EMF [V]	329	315	4.4
Electromagnetic Power [kW]	230	224	2.7
Phase current [A]	434	434	0.0
Power factor	0.99	0.99	0.9
Line Voltage [V]	316	305	3.3
D-axis load Magnetic flux linkage [mWb]	316	300	5.4
Q-axis load Magnetic flux linkage [mWb]	271	273	0.8
Stator Tooth load Magnetic flux density [T]	2.2	2.1	6.3
Stator Core load Magnetic flux density [T]	1.4	1.4	2.9
Rotor Bridge load Magnetic flux density [T]	2.5	2.8	10.7
Rotor Core load Magnetic flux density [T]	2.0	2.1	3.4

The prototype performance was obtained by standard IEC testing in the test bench shown in Figure 15.

Figure 15. IPMSM Prototype at test bench.

In Table 14, the experimental results at rated load conditions are presented. To allow for a fair comparison, the MCM was evaluated at an operating point corresponding to the rated power.

Table 14. Prototype measured performance at test bench.

Parameter	MCM	Test	Relative Error [%]
Back EMF [V]	329	302	8.9
Mechanical Power [kW]	225	223	0.9
Phase current [A]	434	425	2.1
Power factor	0.99	0.99	0.8
Line Voltage [V]	318	318	0.0

Finally, to visually compare the carried out measurements, in Figure 16, the obtained Back electromotive force waveform at tests and MCM is shown.

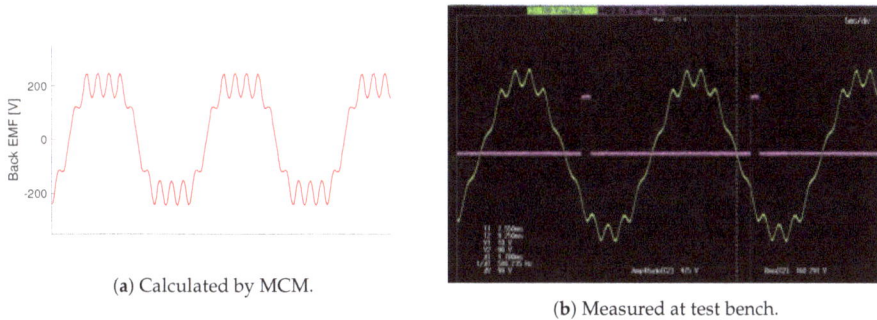

(**a**) Calculated by MCM.

(**b**) Measured at test bench.

Figure 16. Protype's Back EMF at 50% rated speed.

As it can be observed in Table 13, and especially in Table 14 and Figure 16, both experimental and MCM results match largely, showing that the presented approach is suitable for designing IPMSM and predicting its performance even under heavy saturation.

6. Conclusions

Nowadays, the use of PMSMs is exponentially increasing, with a specific interest in its use at very demanding conditions. Consequently, taking into account the different electromagnetic phenomena that take place inside the machine is crucial to predict in a very precise manner machine performance.

To this aim, in this paper, a general node mapping criterion for modelling highly saturated IPMSM using a magnetic reluctance network has been proposed. A MCM model, based on a magnetic reluctance network, is developed to model V-shaped interior mounted magnet rotors. The proposed MCM model allows selecting the discretization levels for the different machine parts. Furthermore, it not only allows simulating rotor motion but also considers the magnetic cross-coupling effect, the slotting effect and the iron magnetic saturation.

To validate the MCM model, Several V-shaped machines of different sizes and geometries were used together with FEM simulations. The results from these validations were remarkably accurate and efficient, requiring less time to complete the process. Finally, aiming to validate the MCM model with the suggested node mapping criterion, a real IPMSM prototype was tested. The comparison of the obtained results shows a great correspondence, proving the validity of the proposed method to determine highly saturated Interior PMSMs performance.

Overall, the present study expands the field of magnetic circuit models for highly saturated machines. Specifically, it provides the means to evolve within this area towards using the node mapping criterion to apply it to other highly demanded PMSM rotor solutions such as the Spoke type or the multilayer IPMSM. In addition, it would be interesting to analyse the way to optimize the implemented software code, as well as to study the use of other numerical analysis methods for solving nonlinear systems since it could lead to the development of an even faster machine designing tool.

Author Contributions: The presented work was carried out through the cooperation of all authors. D.C., B.P. and G.A. conducted the research and wrote the paper. I.E. and M.M.I. edited the manuscript and supervised the study.

Acknowledgments: This work was financially supported by the Basque Country Government Economic Development and Infraestructure Department, by means of grants program "ELKARTEK" (project KK20170095 *Electromagnetic, mechanical and thermal study of light motors-MOTLIG*).

Conflicts of Interest: The authors declare no conflict of interest.

Abbreviations

The following abbreviations are used in this manuscript:

IPMSM	Interior mounted Permanent Magnet Synchronous Machine
MCM	Magnetic Circuit Model
FEM	Finite Elements Methods
EMF	Electromotive Force

References

1. Burress, T.; Campbell, S. Benchmarking EV and HEV power electronics and electric machines. In Proceedings of the 2013 Transportation Electrification Conference and Expo (ITEC), Detroit, MI, USA, 16–19 June 2013.
2. Sun, L.; Cheng, M.; Jia, H. Analysis of a Novel Magnetic-Geared Dual-Rotor Motor With Complementary Structure. *IEEE Trans. Ind. Electron.* **2015**, *62*, 6737–6747. [CrossRef]
3. Zheng, P.; Wang, W.; Wang, M.; Liu, Y.; Fu, Z. Investigation of the Magnetic Circuit and Performance of Less-Rare-Earth Interior Permanent-Magnet Synchronous Machines Used for Electric Vehicles. *Energies* **2017**, *10*, 2173. [CrossRef]
4. Gu, W.; Zhu, X.; Quan, L.; Du, Y. Design and optimization of permanent magnet brushless machines for electric vehicle applications. *Energies* **2015**, *8*, 13996–14008. [CrossRef]
5. Wu, W.; Zhu, X.; Quan, L.; Du, Y.; Xiang, Z.; Zhu, X. Design and Analysis of a Hybrid Permanent Magnet Assisted Synchronous Reluctance Motor Considering Magnetic Saliency and PM Usage. *IEEE Trans. Appl. Supercond.* **2018**, *28*, 1–6. [CrossRef]
6. Yue, L.; Yulong, P.; Yanjun, Y.; Yanwen, S.; Feng, C. Increasing the saliency ratio of fractional slot concentrated winding interior permanent magnet synchronous motors. *Electr. Power Appl. IET* **2015**, *9*, 439–448. [CrossRef]
7. Bai, J.; Zheng, P.; Tong, C.; Song, Z.; Zhao, Q. Characteristic Analysis and Verification of the Magnetic-Field-Modulated Brushless Double-Rotor Machine. *IEEE Trans. Ind. Electron.* **2015**, *62*, 4023–4033. [CrossRef]
8. Sizov, G.; Ionel, D.; Demerdash, N. Modeling and Parametric Design of Permanent-Magnet AC Machines Using Computationally Efficient Finite-Element Analysis. *IEEE Trans. Ind. Electron.* **2012**, *59*, 2403–2413. [CrossRef]
9. Parasiliti, F.; Villani, M.; Lucidi, S.; Rinaldi, F. Finite-Element-Based Multiobjective Design Optimization Procedure of Interior Permanent Magnet Synchronous Motors for Wide Constant-Power Region Operation. *IEEE Trans. Ind. Electron.* **2012**, *59*, 2503–2514. [CrossRef]
10. Ruuskanen, V.; Nerg, J.; Pyrhonen, J.; Ruotsalainen, S.; Kennel, R. Drive Cycle Analysis of a Permanent-Magnet Traction Motor Based on Magnetostatic Finite-Element Analysis. *Veh. Technol. IEEE Trans.* **2015**, *64*, 1249–1254. [CrossRef]
11. Zheng, P.; Zhao, J.; Liu, R.; Tong, C.; Wu, Q. Magnetic characteristics investigation of an axial-axial flux compound-structure PMSM used for HEVs. *Magn. IEEE Trans.* **2010**, *46*, 2191–2194. [CrossRef]
12. Chong, L.; Rahman, M. Saliency ratio derivation and optimisation for an interior permanent magnet machine with concentrated windings using finite-element analysis. *IET Electr. Power Appl.* **2010**, *4*, 249–258. [CrossRef]
13. Cavagnino, A.; Bramerdorfer, G.; Tapia, J.A. Optimization of Electric Machine Designs—Part I. *IEEE Trans. Ind. Electron.* **2017**, *64*, 9716–9720. [CrossRef]
14. Hwang, C.C.; Cho, Y. Effects of leakage flux on magnetic fields of interior permanent magnet synchronous motors. *IEEE Trans. Magn.* **2001**, *37*, 3021–3024. [CrossRef]

15. Zhu, L.; Jiang, S.; Zhu, Z.; Chan, C. Analytical Modeling of Open-Circuit Air-Gap Field Distributions in Multisegment and Multilayer Interior Permanent-Magnet Machines. *IEEE Trans. Magn.* **2009**, *45*, 3121–3130. [CrossRef]

16. Chen, Q.; Liu, G.; Zhao, W.; Shao, M. Nonlinear adaptive lumped parameter magnetic circuit analysis for spoke-type fault-tolerant permanent-magnet motors. *Magn. IEEE Trans.* **2013**, *49*, 5150–5157. [CrossRef]

17. Perho, J. *Reluctance Network for Analysing Induction Machines*; Helsinki University of Technology: Otaniemi, Finland, 2002.

18. Vincent, R.; Emmanuel, V.; Lauric, G.; Laurent, G. Optimal sizing of an electrical machine using a magnetic circuit model: application to a hybrid electrical vehicle. *IET Electr. Syst. Transp.* **2015**, *6*, 27–33. [CrossRef]

19. Raminosoa, T.; Rasoanarivo, I.; Meibody-Tabar, F.; Sargos, F.M. Time-Stepping Simulation of Synchronous Reluctance Motors Using a Nonlinear Reluctance Network Method. *IEEE Trans. Magn.* **2008**, *44*, 4618–4625. [CrossRef]

20. Tangudu, J.K.; Jahns, T.M.; El-Refaie, A.; Zhu, Z. Lumped parameter magnetic circuit model for fractional-slot concentrated-winding interior permanent magnet machines. In Proceedings of the Energy Conversion Congress and Exposition, San Jose, CA, USA, 20–24 September 2009; pp. 2423–2430.

21. Farooq, J.; Srairi, S.; Djerdir, A.; Miraoui, A. Use of permeance network method in the demagnetization phenomenon modeling in a permanent magnet motor. *Magn. IEEE Trans.* **2006**, *42*, 1295–1298. [CrossRef]

22. Kuttler, S.; Benkara, K.; Friedrich, G.; Vangraefschepe, F.; Abdelli, A. Analytical model taking into account the cross saturation for the optimal sizing of IPMSM. In Proceedings of the 2012 XXth International Conference on Electrical Machines (ICEM), Marseille, France, 2–5 September 2012; pp. 2779–2785.

23. Aden, A.; Amara, Y.; Barakat, G.; Hlioui, S.; De La Barriere, O.; Gabsi, M. Modeling of a radial flux PM rotating machine using a new hybrid analytical model. In Proceedings of the 2014 International Conference on Electrical Sciences and Technologies in Maghreb (CISTEM), Tunis, Tunisia, 3–6 November 2014; pp. 1–5.

24. Amrhein, M.; Krein, P. Induction Machine Modeling Approach Based on 3-D Magnetic Equivalent Circuit Framework. *IEEE Trans. Energy Convers.* **2010**, *25*, 339–347. [CrossRef]

25. Rasmussen, C.; Ritchie, E. A magnetic equivalent circuit approach for predicting PM motor performance. In Proceedings of the Conference Record of the 1997 IEEE Industry Applications Conference Thirty-Second IAS Annual Meeting, New Orleans, LA, USA, 5–9 October 1997; pp. 10–17.

26. Lovelace, E.; Jahns, T.; Lang, J.H. A saturating lumped-parameter model for an interior PM synchronous machine. *IEEE Trans. Ind. Appl.* **2002**, *38*, 645–650. [CrossRef]

27. Yamazaki, K.; Kumagai, M. Torque Analysis of Interior Permanent-Magnet Synchronous Motors by Considering Cross-Magnetization: Variation in Torque Components With Permanent-Magnet Configurations. *IEEE Trans. Ind. Electron.* **2014**, *61*, 3192–3201. [CrossRef]

28. Lee, S.; Jeong, Y.S.; Kim, Y.J.; Jung, S.Y. Novel Analysis and Design Methodology of Interior Permanent-Magnet Synchronous Motor Using Newly Adopted Synthetic Flux Linkage. *IEEE Trans. Ind. Electron.* **2011**, *58*, 3806–3814.

29. Ostovic, V. *Dynamics of Saturated Electric Machines*; Springer: New York, NY, USA, 1989.

30. Han, S.H.; Jahns, T.; Soong, W. A Magnetic Circuit Model for an IPM Synchronous Machine Incorporating Moving Airgap and Cross-Coupled Saturation Effects. In Proceedings of the Electric Machines & Drives 2007 International Conference, Antalya, Turkey, 3–5 May 2007; pp. 21–26.

31. Ong, C.M. *Dynamic Simulation of Electric Machinery: Using MATLAB/SIMULINK*; Upper Saddle River: Bergen, NJ, USA, 1998.

Article

Quantitative Comparisons of Six-Phase Outer-Rotor Permanent-Magnet Brushless Machines for Electric Vehicles

Yuqing Yao [1], Chunhua Liu [1,*] and Christopher H.T. Lee [2]

[1]   School of Energy and Environment, City University of Hong Kong, 83 Tat Chee Avenue, Kowloon Tong, Hong Kong, China; Yuqing.Yao@my.cityu.edu.hk

[2]   Research Laboratory of Electronics, Massachusetts Institute of Technology, Cambridge, MA 02139, USA; chtlee@mit.edu

*   Correspondence: chunliu@cityu.edu.hk; Tel.: +852-3442-2885

Received: 20 July 2018; Accepted: 11 August 2018; Published: 16 August 2018

Abstract: Multiphase machines have some distinct merits, including the high power density, high torque density, high efficiency and low torque ripple, etc. which can be beneficial for many industrial applications. This paper presents four different types of six-phase outer-rotor permanent-magnet (PM) brushless machines for electric vehicles (EVs), which include the inserted PM (IPM) type, surface PM (SPM) type, PM flux-switching (PMFS) type, and PM vernier (PMV) type. First, the design criteria and operation principle are compared and discussed. Then, their key characteristics are addressed and analyzed by using the finite element method (FEM). The results show that the PMV type is quite suitable for the direct-drive application for EVs with its high torque density and efficiency. Also, the IPM type is suitable for the indirect-drive application for EVs with its high power density and efficiency.

Keywords: permanent-magnet machine; brushless machine; Vernier machine; flux switching machine; multiphase machine; outer rotor; electric vehicle

1. Introduction

In recent years, interest in electrical vehicles (EVs) is growing rapidly driven by the concerns about environment issues [1,2]. EVs are environmentally friendly and enjoy many attractive merits compare to their tradition fossil-fuel counterparts [3]. In particular, EVs offer definite advantages for fossil-fuel use reduction, which causes less harmful gases, lower cost of driving, cost-effective maintenance, less noise and vibration. Also, EVs provide more comfort at low speed than their conventional fossil-fuel counterparts.

The electric machines for EVs are regarded as one of the key components of the whole EV system [4]. In order to ensure the driving quality of EVs, there are many strict requirements for the EV machines [5,6]. First of all, high reliability is essential for the challenging environment and frequent starting and stopping during driving. Secondly, a high power density is desirable for weight reduction. Moreover, high efficiency is particularly important for a better economic outcome. Nowadays, there are several types of electric machines available for EV applications. In the early years of EV development, DC motors were considered as the mainstream because of their easy control and excellent speed regulation [7]. However, they suffer from fragile mechanical structures, poor thermal dissipation and low efficiency, which are not desirable features for EV application. Then, AC asynchronous machines attracted much attention because of their high efficiency and simple structure [8,9]. However, they suffer from problems such as complex control algorithms and low torque density. Permanent-magnet synchronous (PMS) brushless machines which enjoy the high efficiency, high power density and high reliability, are considered an attractive option for EV applications [10,11]. In particular, the permanent-magnet flux-switching (PMFS) type and permanent-magnet vernier (PMV) type, as the representatives of PMS brushless machines, are extraordinary for EV use. The PMFS type is a combination of a synchronous machine and a switched reluctant machine [12], which enjoys high torque density and a robust rotor structure [13], so it can enhance the high performance for EVs. The PMV type integrates a coaxial magnetic gear into a PM brushless machine. In this way, it is capable of achieving a high electric frequency at low rotation speed [14]. Meanwhile, it always adopts a concentric winding layout in order to have short end windings, which is more suitable for the limited space of EVs., The general advantages and disadvantages of the four different types of machines are summarized in Table 1 [15–17].

Table 1. Comparison of Four Proposed Types of Machines.

Machine Type	Advantages	Disadvantages
IPM	• Simple and mature	• High speed operation
	• Easy manufacture	• Relatively large torque ripple
SPM	• Mature control strategy	• PM fixing problem
PMFSM	• Robost rotor structure	• Difficult manufacturing
	• Good heat dissipation	• Special design
	• Easy winding	• Relatively large torque ripple
PMVM	• Low speed operation	• Complex structure
	• High torque density	• Special design

Moreover, multiphase machines with a higher degree of freedom have fault-tolerant capability, so they are suitable for EVs, and can also be considered a great option [18,19]. In addition, the outer-rotor structure of these machines, which are designed with a rotating outer structure and a fixed inner structure, is quite attractive for EV applications [20]. In this way, the structure has better performance for EV propulsion.

This paper is focused on four different types of six-phase outer-rotor PM machines, which are suitable for EV applications. These four types of machines combine the advantages of PM machines and multiphase machines, thus achieving high power density, high torque density, better fault tolerance capability, and lower torque ripple. The key is to quantitatively compare the operation performance of the four different types of machines, namely the inserted PM (IPM) type, surface PM (SPM) type, PM flux-switching (PMFS) type, and PM vernier (PMV) type. First, the design criteria and operating principles are elaborated in Section 2. Then, a quantitative comparison of the machines is described in Section 3, which includes the key characteristics, normal operation performance, and the fault-tolerant operation. Finally, the conclusions of the quantitative comparison are summarized in Section 4.

2. Proposed Six-Phase Machine Topologies and Operation Principles

The proposed topologies are presented with both cross-section and exploded views in Figure 1. These four machine types adopt the six-phase arrangement and outer-rotor structure. Figure 1a shows the IPM topology with PMs inserted inside the rotor, while Figure 1b shows the SPM type with PMs mounted on the surface of the rotor. The main difference between the IPM machine and the SPM machine is their arrangements of PMs. Figure 1c shows the topology of PMFS machine, while Figure 1d shows the PMV machine topology. The PMFS machine has all PMs inserted inside the stator teeth, while the PMV machine has PMs mounted on both stator side and rotor side.

The operating speed n of the machine is governed by the pole pairs and the electric frequency of the proposed machines:

$$n = \frac{60f}{p} \tag{1}$$

where f is the electric frequency and p is the pole-pairs of the machines.

The turns of armature coil N_a can be determined by the following relationship:

$$N_a = \sqrt{\frac{\alpha_a R_a S_a}{4\rho L_a}} \tag{2}$$

where α_a is the factor of slots filling, S_a is the slot area of the coil, and L_a is the average coil length. In this paper, the turns of armature coil of each phase and slot filling factor for the proposed machines are set to be 200 and 40%, respectively.

To achieve the fair comparison, the peripheral dimension, PM volume, air gap length and current density, are set as equal.

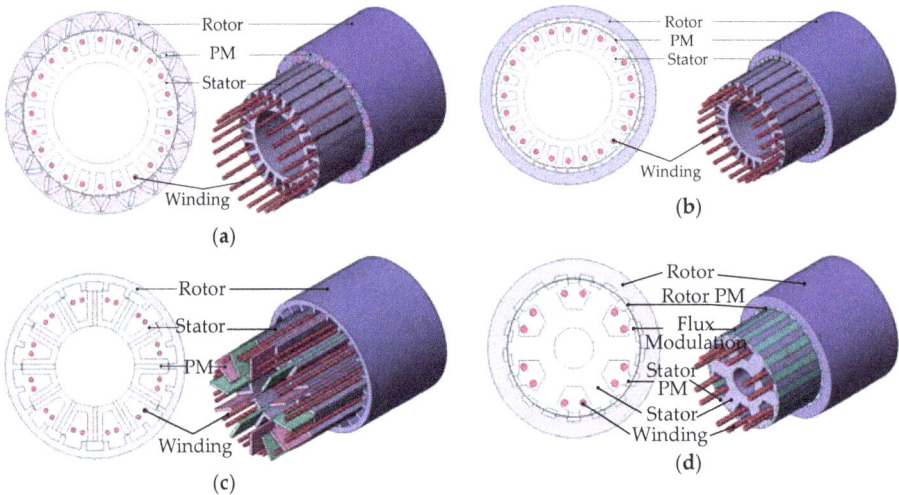

Figure 1. Proposed six-phase outer-rotor machine topologies: (a) IPM type; (b) SPM type; (c) PMFS type; (d) PMV type.

2.1. Proposed IPM Type and SPM Type

The IPM machine adopts the V-shape PM poles inserted within its rotor. As investigated, it has been suggested that the V-shape or W-shape PM poles would have the better operation performance than the traditional radial PM arrangement [21]. Moreover, in order to prevent the short-circuiting PM fluxes, the magnetic insulation material is purposely implemented. The IPM type adopts the single-layer fractional slot concentrated winding, which can reduce the end winding length and decrease the copper loss [22]. In fact, the IPM machine consists of 24 stator slots and 10 pole pairs, while the mechanical and electrical degrees between two slots can be calculated as [23]:

$$\alpha_m = \frac{360°}{z} = \frac{360°}{24} = 15° \tag{3}$$

$$\alpha_e = p\frac{360°}{z} = 10 \times \frac{360°}{24} = 150° \tag{4}$$

where z is the number of stator slots and p is the pole pairs. According to Equations (3) and (4), the mechanical angle is 15° and the electrical angle is 150° for each adjacent stator slot. In this case, the stator is able to produce a rotating magnet field with six-phase current. Referring to the principle of minimum magnetoresistance, namely the flux tends to flow along the path of minimum magnetoresistance and the rotor will rotate synchronously with the rotating magnetic field produced by the stator.

The SPM machine shares the same stator structure with IPM type, while the PM arrangement of the two machines are different. Unlike the IPM type with installed PMs within its stator, the SPM accommodates the PMs on the rotor surface. The SPM machine adopts the single layer distributed winding layout and consists of 24 stator slots and 20 pole pairs. Consequently, its mechanical degree and electrical degree between two slots can be found to be 15° and 300°, respectively.

2.2. Proposed PMFS Type

The proposed six-phase outer-rotor PMFS machine consists of 12 stator slots and 22 salient rotor poles. It combines the special features of a switched reluctance machine and a synchronous machine, hence enjoying the advantages of robust rotor structure, high power density and high torque density. The PMs are inserted in each stator tooth with the opposite direction in each two adjacent ones. The proposed machine adopts the double-layer concentrated winding method, where each coil is around a stator tooth with PM inserted in the middle of it.

The machine follows the flux switching principle [24], which is shown in Figure 2. Four typical positions are shown. First, as shown in Figure 2a, the right part of the stator tooth aligns directly with one of the rotor teeth, so it provides the largest flux linkage. Second, as shown in Figure 2b, the stator tooth is completely misaligned with the rotor tooth. Consequently, no magnetic flux flows in-between them. Also, the flux linkage of the coil is reduced to zero. Third, as shown in Figure 2c, the left part of the stator tooth aligns directly again to another rotor tooth, so it provides the largest flux linkage. Finally, as shown in Figure 2d, the flux linkage of the coil decreases to zero again.

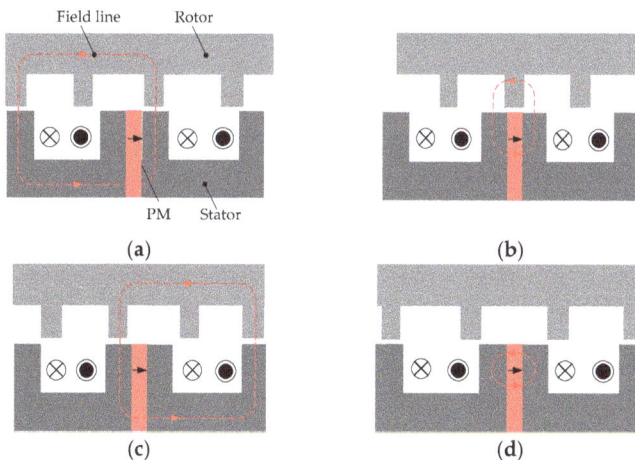

Figure 2. Working principle of PMFS type at four typical positions: (**a**) Positon 1; (**b**) Position 2; (**c**) Position 3; (**d**) Position 4.

Actually, the number of stator slots and rotor pole-pairs are critical for the PMFS machine. The flux linkage and steady torque can be influenced substantially [25]. Since the magnetization direction of each two adjacent PMs are opposite, the number of PMs should be an even number. Thus, the number of stator slots N_s should be an even number as well. As a result, the number of stators of a six-phase machine should be a multiple of six, i.e., $N_s = 2\,km$, $k = 1, 2, 3, \dots$ etc. With this setting, N_s can be determined as twelve. The relationship between stator pole number N_s and rotor pole number N_r is governed by [25]:

$$N_r = \frac{(12 \pm n)N_s}{6} \tag{5}$$

where n is a positive integer and should not be the multiple times of three. For the proposed six-phase machine, it adopts the combination of 12 stator slots and 22 rotor teeth.

2.3. PMV Type

The proposed six-phase outer-rotor PMV machine consists of six stator teeth each with three flux modulation poles. The machine adopts the double-layer concentrated winding with each phase coil fitting around one stator tooth. It installs PMs on both its stator side and rotor side. Meanwhile, the consequent pole structure is employed to improve the utilization of PM materials.

The PMV machine utilizes the concept of magnetic-gearing effect, hence achieving the high-torque low-speed capability. The flux modulation poles (FMPs) are introduced into the stator teeth of the proposed machine. The lower rotating speed PM field is modulated to couple with the higher rotating speed stator air gap magnetic field. Consequently, a self-governing speed effect results. Thus, when the outer rotor PM moves a little angle at a low speed, there will be a great change of flux interacting with the magnet field generated by the rotating armature winding. According to the magnetic gearing effect, the number of pole pairs of flux density distribution in the space harmonic is produced by the high speed stator winding rotating magnetic field and low speed rotor pole magnetic field as [26]:

$$
\begin{aligned}
p_{m,k} &= mp_r + kQ_s \\
m &= 1, 3, 5, \dots, \infty \\
k &= 0, \quad \pm 1, \pm 2, \pm 3, \dots \pm \infty
\end{aligned}
\tag{6}
$$

where Q_s is the number of FMP and p_r is the number of rotor PM pole-pairs. Furthermore, the operating speed of the flux density in space harmonic Ω_r can be described as:

$$\Omega_{m,k} = \frac{mp_r}{mp_r + kQ_s}\Omega_r \tag{7}$$

when $m = 1$ and $n = 1$, the harmonic magnetic field of the air gap results in the largest modulated magnitude. Thus, the number of rotor magnet poles becomes:

$$p_r = Q_s - p_s \tag{8}$$

where p_s is the number of winding pole-pairs. For the proposed six-phase outer-rotor PMV machine, the number of rotor pole pairs is selected as 17, the number of winding pole pairs is 1, and the number of flux modulation poles as 18. Consequently, the gear ratio is given as:

$$G_r = \frac{p_r - Q_s}{p_r} = \frac{17 - 18}{17} = -\frac{1}{17} \tag{9}$$

3. Performance Comparison

In this section, the key operation characteristics, such as the magnetic flux distribution line, air gap flux density, flux linkage, no-load EMF, cogging torque, steady torque and power loss of the proposed machines, are analyzed and compared by using the finite element method (FEM) with the JMAG Designer tool.

Moreover, the fault-tolerant operation performance of each machine is demonstrated and compared with the normal operation. Table 2 shows the key parameters of the four proposed six-phase outer-rotor PM machines, which follow the abovementioned comparison conditions.

Table 2. Key Parameters of Proposed Six-phase Outer-rotor PM Machines.

Items	IPM	SPM	PMFSM	PMVM
Outer rotor diameter	220 mm	220 mm	220 mm	220 mm
Stator diameter	179 mm	179 mm	197 mm	169 mm
Air gap	0.5 mm	0.5 mm	0.5 mm	0.5 mm
Stack length	100 mm	100 mm	100 mm	100 mm
PM volume	406 cm^3	406 cm^3	406 cm^3	406 cm^3
PM thickness	5 mm	5 mm	5 mm	5 mm
Stator slots	24	24	12	18
Rotor pole-pairs	10	20	22	17

3.1. Basic Characteristics

Firstly, the magnetic flux distribution and winding distribution of these four proposed machines at generating mode are presented in Figure 3. The IPM type and SPM type both adopt the single-layer winding arrangement, while the PMFS type and PMV type adopt the double-layer winding arrangement. The air gap flux density of the four proposed machines are shown in Figure 4, the maximum of the air gap flux-density are 0.8 T, 1.4 T, 2.5 T and 2.0 T, respectively. In general, the higher flux density will provide a higher torque density [27], so among the four machines the proposed PMV type and the PMFS type have the potential to produce the better performance. In addition, the flux-linkage of the four proposed machines are shown in Figure 5, which indicates that all the proposed machines can offer the balance flux-linkage among the six-phase patterns.

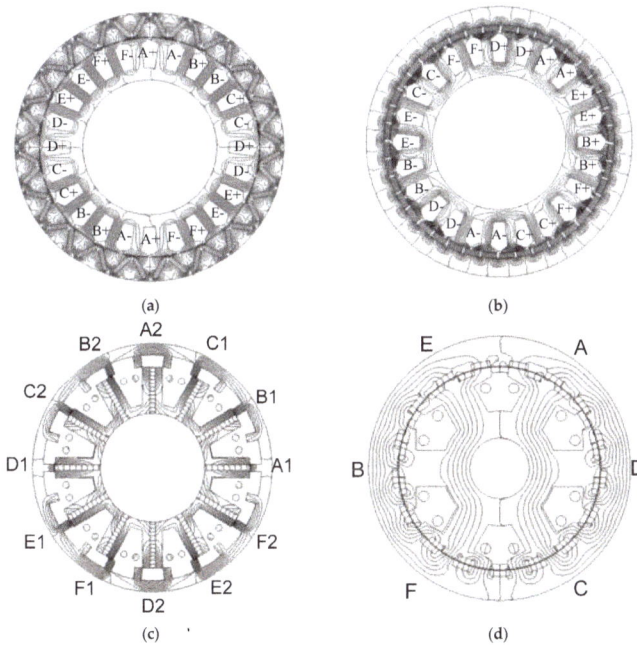

Figure 3. Magnetic flux distribution of proposed six-phase outer-rotor PM machines: (**a**) IPM type; (**b**) SPM type; (**c**) PMFS type; (**d**) PMV type.

Second, the no-load EMF is governed by [28]:

$$E = \frac{d\psi}{dt} \tag{10}$$

where Ψ is no-load flux-linkage shown in Figure 5. Hence, the no-load EMF should be high, since the no-load flux-linkage sinusoidal wave changes rapidly with the time and has a high amplitude. The RMS values of no-load EFM at different rotation speeds are presented in Figure 6. It can be seen that the output voltage has the positive correlation with the rotation speed. The PMV type has the highest no-load voltage over the four machines due to the magnetic gearing effect. Hence, the rated rotation speed of the four machines are determined as 2000, 1000, 1000 and 600 r/min. The waveforms of the no-load EMFs are presented in Figure 7. The amplitudes of the no-load EMF of IPM type, SPM type, PMFS type and PMV type are 582, 615, 524 and 576 V, respectively.

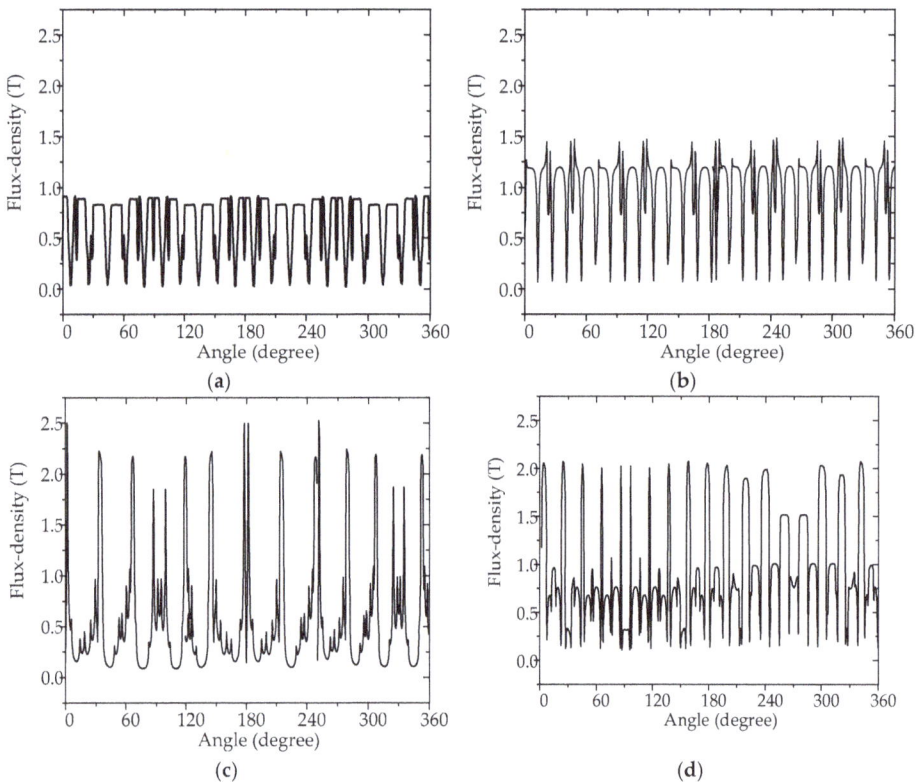

Figure 4. Airgap flux-density of proposed six-phase outer-rotor PM machines: (**a**) IPM type; (**b**) SPM type; (**c**) PMFS type; (**d**) PMV type.

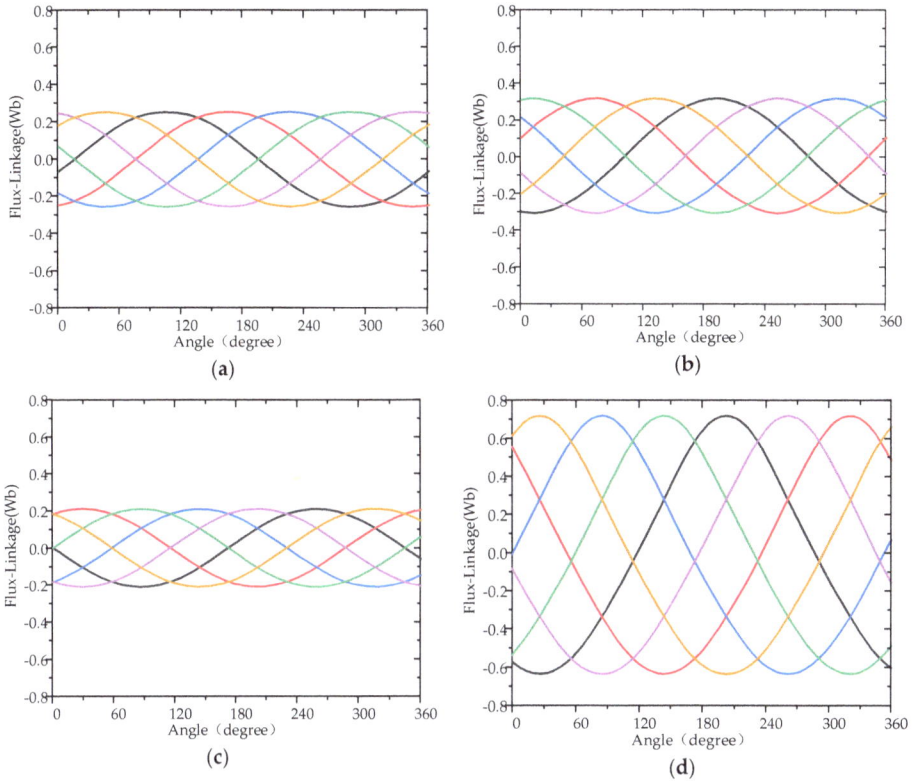

Figure 5. No-load flux-linkage of proposed six-phase outer-rotor PM machines: (**a**) IPM type; (**b**) SPM type; (**c**) PMFS type; (**d**) PMV type.

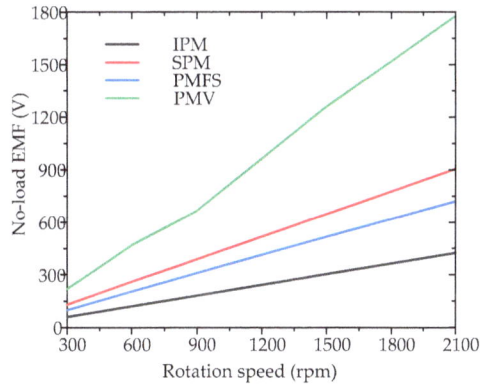

Figure 6. RMS of no-load EMFs of proposed six-phase outer-rotor PM machines under different speeds.

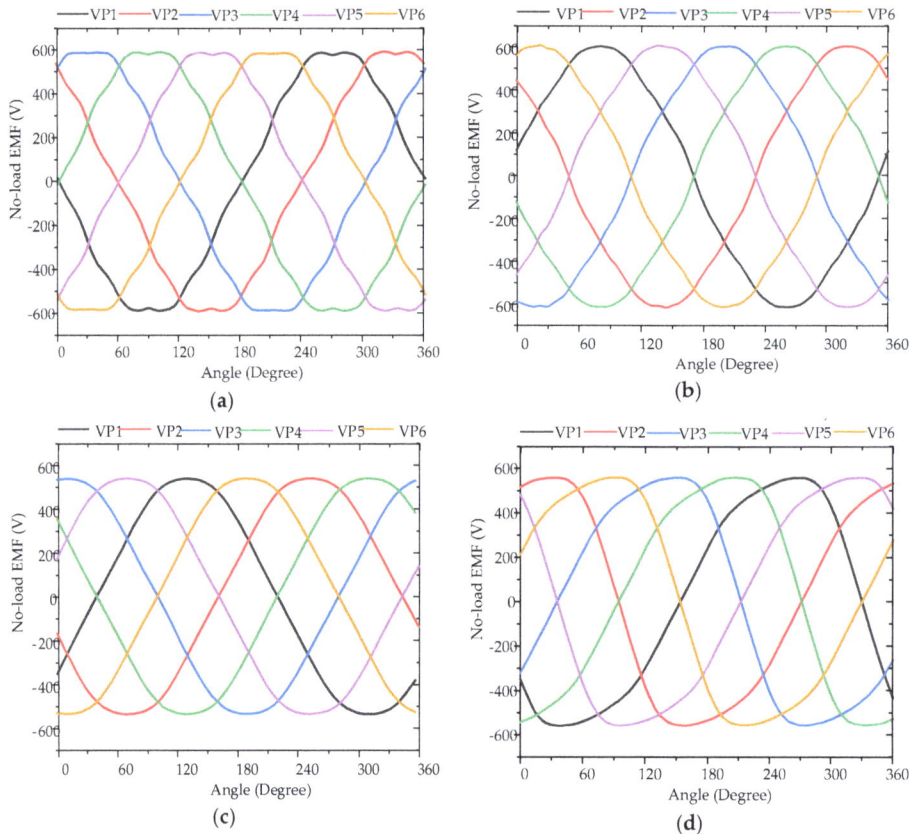

Figure 7. No-load EMF of proposed six-phase outer-rotor PM machines: (**a**) IPM type; (**b**) SPM type; (**c**) PMFS type; (**d**) PMV type.

3.2. Normal Operation

The normal operations of the proposed machines are analyzed and discussed. These machines operate at 2000, 1000, 1000 and 600 r/min, respectively, which runs with the current density of 6 A/mm². The steady torque of the machines are given in Figure 8, which are 93.93, 91.14, 113.38 and 199.91 N·m, respectively. It can be found that the steady torque of PMV type is the largest, which is nearly 50% larger than the three counterparts. Moreover, the steady torque of IPM type and SPM type are very similar. However, the torque ripple of IPM type is much smaller than SPM type. In addition, the PMFS type has the smallest torque ripple among the four machines. The torque-speed characteristic of the four proposed machines are shown in Figure 9.

Furthermore, the torque density of each machine can be calculated as:

$$S_T = \frac{T_{max}}{V} \tag{11}$$

where V is the volume of electric machine. In this way, the torque density of the four proposed machines are 24.71, 23.98, 29.83 and 52.58 kN·m/m³, respectively, so the PMV type has the best torque density among the four types.

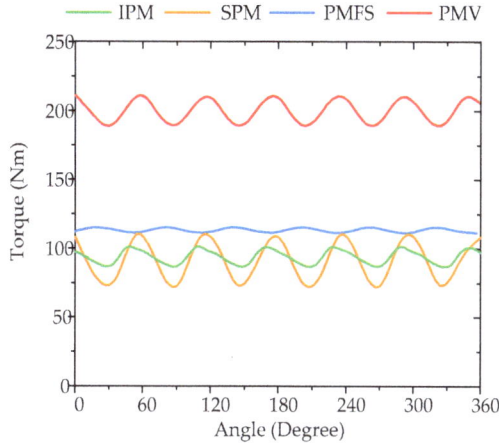

Figure 8. Steady torque of proposed six-phase outer-rotor PM machines.

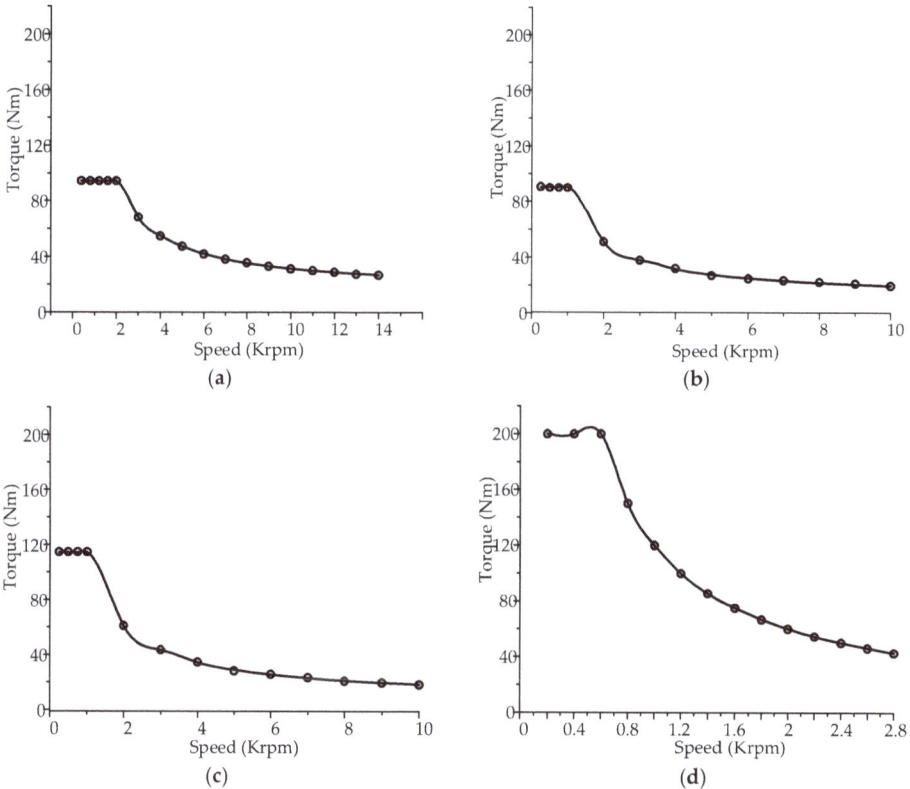

Figure 9. Torque-speed characteristic of proposed six-phase outer-rotor machines: (**a**) IPM type; (**b**) SPM type; (**c**) PMFS type; (**d**) PMV type.

In addition to this, the torque ripple is given as:

$$T_{rip} = \frac{T_{max} - T_{in}}{T_{avg}} \times 100\% \tag{13}$$

so, the torque ripples of the four proposed machines are 15.52%, 40.25%, 3.83% and 11.88%, respectively.

In addition, Figure 10 shows the cogging torque of the four proposed machines. Cogging torque is a distinctive problem of PM machines which results from the interaction between the PMs and the stator iron core when the armature winding is not electrified [29]. Cogging torque causes mechanical vibration and noise [30,31], and it can be reduced by certain ways such as changing the magnet shape [32]. The period of cogging torque is:

$$T_{cog} = \frac{360}{LCM(z, 2p)} \tag{14}$$

where *LMC* is the least common multiple of the stator slots number *z* and number of poles *2p*. It can be seen that the SPM type suffers from a large cogging torque because of its slot number and pole-pair number combination [33,34]. Compared to the SPM type, the IPM type has a remarkably reduced cogging torque with a similar steady torque. The percentages of cogging torque to steady torque of the four machines are 0.58%, 17.56%, 1.59% and 1.00%, respectively. Hence, it can be concluded that except for the SPM type, the other three types are desirable in terms of cogging since they are well below 3%.

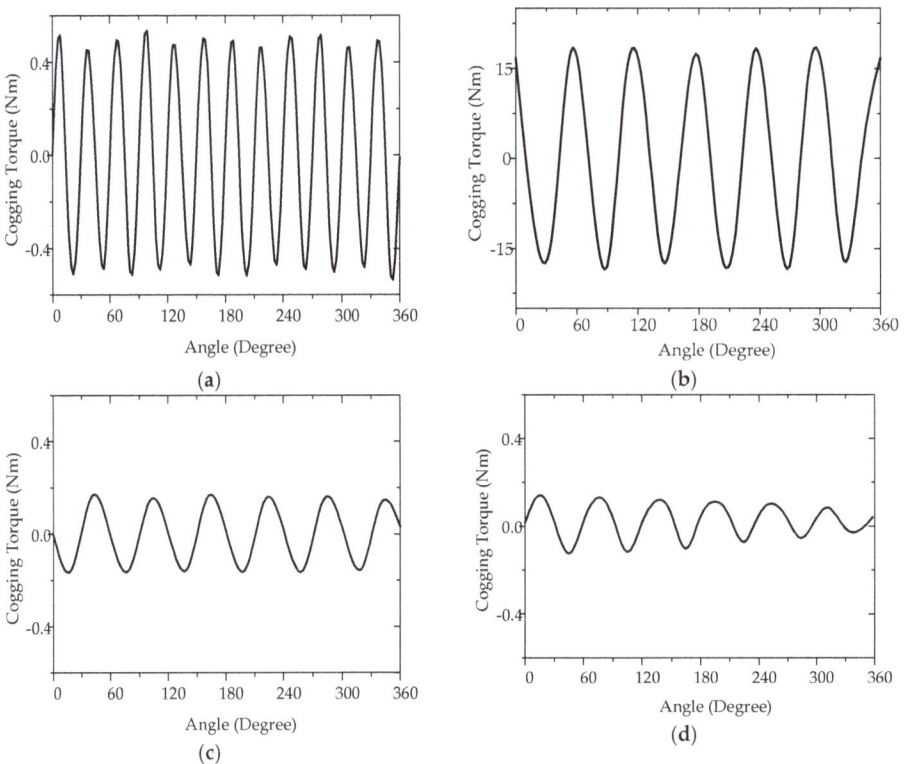

Figure 10. Cogging torque of proposed six-phase outer-rotor PM machines: (**a**) IPM type; (**b**) SPM type; (**c**) PMFS type; (**d**) PMV type.

Moreover, the power losses and efficiency of the proposed machines are calculated by FEM. The iron losses of the IPM type, SPM type, PMFS type and PMV type machine are 1014, 1914, 1438 and 649 W, respectively. The copper loss of the SPM type is the highest, since it is the only one with a distributed winding. The values of the eddy current loss, hysteresis loss, copper loss and efficiency of each machine, are shown in Table 3.

Table 3. Iron Loss of Proposed Six-phase Machines.

Item	IPM	SPM	PMFS	PMV
Eddy current loss	874 W	1763 W	1291 W	424 W
Hysteresis loss	140 W	151 W	147 W	225 W
Copper loss	163 W	283 W	178 W	197 W
Other loss	200 W	200 W	200 W	200 W
Overall power	21,047 W	11,461 W	13,686 W	13,678 W
Efficiency	93.45%	79.09%	86.73%	92.89%

3.3. Fault-Tolerant Operation Performance

Reliability is a vital characteristic of machines used for EV applications. One can improve the operation reliability by adopting the multiphase than the traditional three-phase. In order to verify the fault-tolerant operation performance for the proposed six-phase machines, the open-phase fault case and short-circuit fault case are simulated and discussed.

Figure 11 shows the steady torque of the proposed machines with A-phase open fault, and A-phase and D-phase both open faults. Also, the normal operation steady torque under motoring mode is given for comparison. It can be seen that the IPM type and SPM type have a large reduction of the average torque, when they meet the open-phase fault, while the PMV type has the smallest influence by the open-phase fault.

In addition, there may also be the short-circuit faults during the EV operation. Actually, it's considered as the most severe fault over all the winding faults in electrical machines [35], since the current value of the fault turns will become much higher than the healthy turns and there will be excessive heat generated in the fault phase [36]. In fact, this kind of fault is mostly caused by winding insulation humidity, overvoltage, overheating, etc. Figure 12 shows the electric model of an A-phase short-circuit fault. When a short-circuit fault occurs to A-phase, a resistor r_f which represents the fault connection in the model and $r_f \to 0$. Once the fault occurs, a circuit through r_f is created. The system equation is (14). This equation shows that during a short-circuit fault, the healthy windings the six-phase PM machines can remain working [37]. Rows 1 to 6 show the healthy part and the influence of the fault part. The seventh row shows the fault circuit current model which is influenced by the winding turns included in the fault.

$$
\begin{pmatrix} v_a \\ v_b \\ v_c \\ v_d \\ v_e \\ v_f \\ 0 \end{pmatrix} =
\begin{pmatrix}
r_s & 0 & 0 & 0 & 0 & 0 & -r_{af} \\
0 & r_s & 0 & 0 & 0 & 0 & 0 \\
0 & 0 & r_s & 0 & 0 & 0 & 0 \\
0 & 0 & 0 & r_s & 0 & 0 & 0 \\
0 & 0 & 0 & 0 & r_s & 0 & 0 \\
0 & 0 & 0 & 0 & 0 & r_s & 0 \\
-r_{af} & 0 & 0 & 0 & 0 & 0 & r_{af}+r_f
\end{pmatrix}
\begin{pmatrix} i_a \\ i_b \\ i_c \\ i_d \\ i_e \\ i_f \\ i_{f'} \end{pmatrix} +
$$

$$
\begin{pmatrix}
L_{ah} & M_{ah-b} & M_{ah-c} & M_{ah-d} & M_{ah-e} & M_{ah-f} & M_{ah-af} \\
M_{ah-b} & L_b & M_{b-c} & M_{b-d} & M_{b-e} & M_{b-f} & M_{b-af} \\
M_{ah-c} & M_{b-c} & L_c & M_{c-d} & M_{c-e} & M_{c-f} & M_{c-af} \\
M_{ah-d} & M_{b-d} & M_{c-d} & L_d & M_{d-e} & M_{d-f} & M_{d-af} \\
M_{ah-e} & M_{b-e} & M_{c-e} & M_{d-e} & L_e & M_{e-f} & M_{e-af} \\
M_{ah-f} & M_{b-f} & M_{c-f} & M_{d-f} & M_{e-f} & L_f & M_{f-af} \\
M_{ah-af} & M_{b-af} & M_{c-af} & M_{d-af} & M_{e-af} & M_{f-af} & L_{af}
\end{pmatrix}
\frac{d}{dt}
\begin{pmatrix} i_a \\ i_b \\ i_c \\ i_d \\ i_e \\ i_f \\ i_{f'} \end{pmatrix} +
\begin{pmatrix} e_a \\ e_b \\ e_c \\ e_d \\ e_e \\ e_f \\ -e_{af} \end{pmatrix}
\tag{15}
$$

Figure 13 shows that the steady torque of 50% short-circuit of A-phase occurs to the proposed machines for both the six-phase winding and three-phase winding configurations. It can be seen that the torque ripple of the three-phase winding is larger than six-phase winding configuration in normal operation. When a short-circuit fault occurs, the torque ripple of a three-phase winding is much larger than that of a six-phase as well. Moreover, under six-phase winding conditions, the torque of all the proposed machines are still periodical and the average torque only shows a little decrease. This indicates that the four proposed machines have good fault tolerance ability when short-circuit faults occur.

It should be noted that for practical normal conditions, the results generally won't be changed. This is well proven by most researchers with the help of FEA design [38–40]. However, for practical harsh conditions, the results may be different to some degree. As we know, machines suffer from many unexpected problems in harsh conditions, such as high temperature, large overcurrent, iron breakage issues, PM fixing problems, etc. In general, the salient machine should have a high robustness for intermittent operation [41–43], and the PMFS type meets this condition by having a relatively high robustness for harsh conditions.

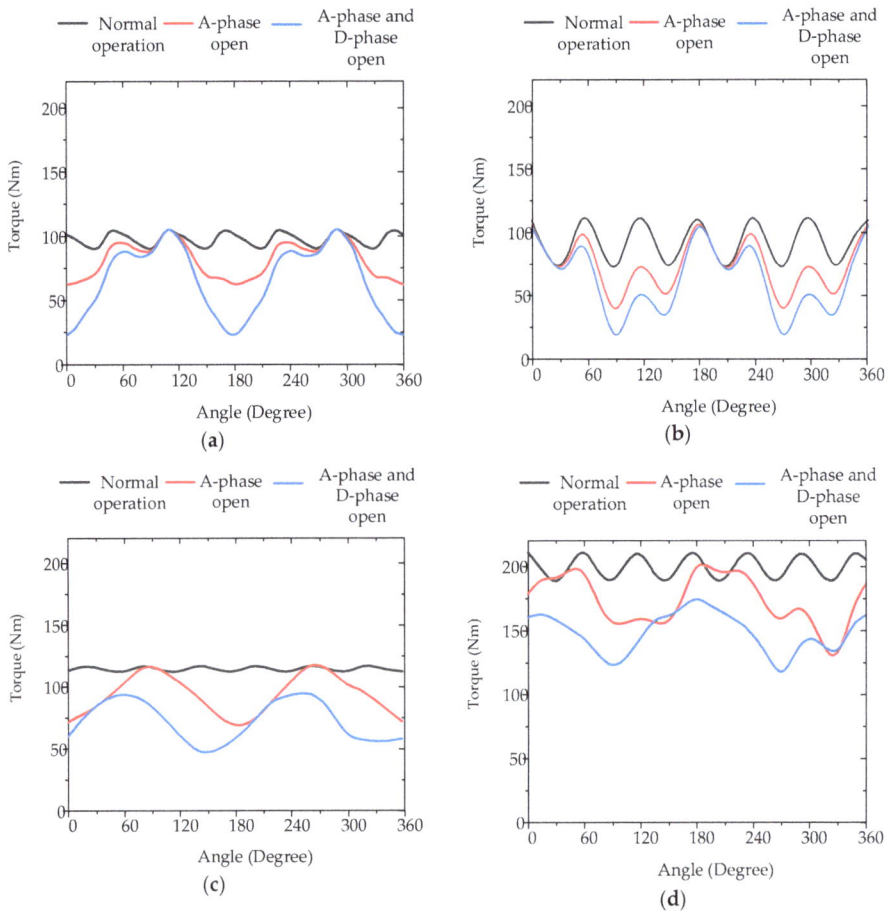

Figure 11. Torque at open-phase fault of proposed six-phase outer-rotor PM machines: (**a**) IPM type; (**b**) SPM type; (**c**) PMFS type; (**d**) PMV type.

Energies **2018**, *11*, 2141

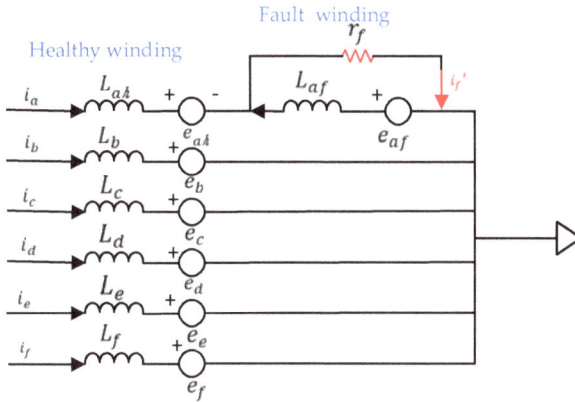

Figure 12. Six-phase star connection electrical model of A-phase short circuit fault.

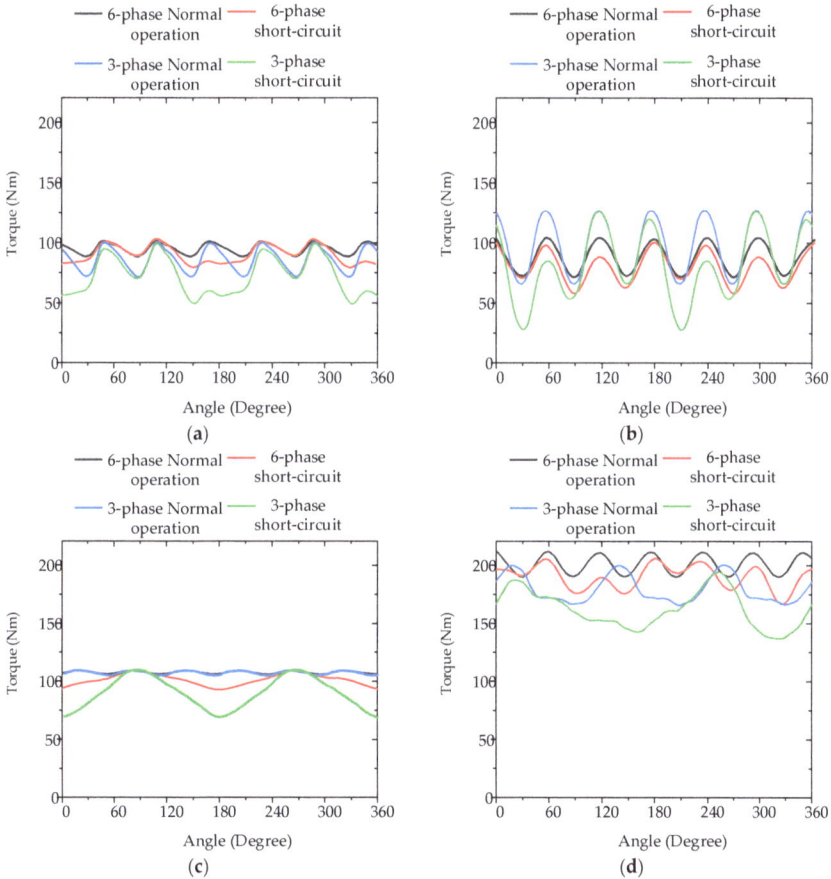

Figure 13. Steady torque of proposed outer-rotor PM machines of both six-phase and three-phase occur to short-circuit condition: (**a**) IPM type; (**b**) SPM type; (**c**) PMFS type; (**d**) PMV type.

4. Conclusions

This paper presents four six-phase outer-rotor PM machines with different working principles, which include the IPM type, SPM type, PMFS type and PMV type. The working principle of each machine has been elaborated and discussed. By analyzing the operating performance of the four machines, it is shown that the PMV type has the best operating performance and fault tolerance ability among the four machines, so it is more suitable for electric vehicle applications. The key parameters of working performance are summarized in Table 4. Moreover, the evaluation of the four types of machines proposed is summarized in Table 5.

Table 4. Working Performance of Proposed Six-phase Outer-rotor Machines.

Items	IPM	SPM	PMFS	PMV
Torque ripple	15.52%	40.25%	3.83%	11.88%
Torque density	24.71 kN·m/m^3	23.98 kN·m/m^3	29.83 kN·m/m^3	52.58 kN·m/m^3
Efficiency	93.45%	79.09%	86.73%	92.89%
Power	21,047 W	11,461 W	13,686 W	13,678 W
Base speed	2000 r/min	1000 r/min	1000 r/min	600 r/min

Table 5. Evaluation of Proposed Machines.

Items	IPM	SPM	PMFS	PMV
Efficency	high	moderate	moderate	high
Torque density	moderate	moderate	moderate	high
Thermal dissipaition	moderate	moderate	good	moderate
Cost effectivelyness	moderate	low	moderate	high

Thus, based on the above analysis and discussion, we can conclude the following from the comparison:

- The PMV type can produce the largest steady torque under the low rotation speed.
- Based on the operation principle, the PMV type can be used for in-wheel direct-drive EV applications.
- The arrangement of PMs for the PMFS type can protect the PMs from rotational stresses, which is also suitable for EV applications.
- The SPM type should be carefully considered for EV applications, since it has the high cogging torque and the lowest related efficiency.
- The outer-rotor topology can be directly connected with the tire rim of the EV.
- The multiphase machine has a good fault tolerance ability, which is suitable for EV applications.

Author Contributions: The work presented in this paper is the output of the research projects undertaken by C.L. In specific, Y.Y. and C.L. developed the topic, designed the system, analyzed the results, and wrote the paper. C.H.T.L. helped to provide the guidance and review the paper.

Acknowledgments: This work was supported by a grant (Project No.: ECF Project 92/2016) from Energy and Conservation Fund of HKSAR, China. Also, it was supported a grant (Project No.: ITS/353/16) from ITF Tier3 of Innovation and Technology Commission (ITC) of HKSAR, China.

Conflicts of Interest: The authors declare no conflict of interest.

References

1. Un-Noor, F.; Padmanaban, S.; Mihet-Popa, L.; Mollah, M.N.; Hossain, E. A comprehensive study of key electric vehicle (EV) components, technologies, challenges, impacts, and future direction of development. *Energies* **2017**, *10*, 1217. [CrossRef]

2. Chau, K.T.; Chan, C.C.; Liu, C. Overview of Permanent-Magnet Brushless Drives for Electric and Hybrid Electric Vehicles. *IEEE Trans. Ind. Electron.* **2008**, *55*, 2246–2257. [CrossRef]

3. De Santiago, J.; Bernhoff, H.; Ekergård, B.; Eriksson, S.; Ferhatovic, S.; Waters, R.; Leijon, M. Electrical Motor Drivelines in Commercial All-Electric Vehicles: A Review. *IEEE Trans. Veh. Technol.* **2012**, *61*, 475–484. [CrossRef]

4. Lulhe, A.M.; Date, T.N. A technology review paper for drives used in electrical vehicle (EV) & hybrid electrical vehicles (HEV). In Proceedings of the 2015 International Conference on Control, Instrumentation, Communication and Computational Technologies (ICCICCT), Kumaracoil, Tamilnadu, 18–19 December 2015; pp. 632–636.

5. Lee, C.H.T.; Chau, K.T.; Liu, C. Design and Analysis of an Electronic-Geared Magnetless Machine for Electric Vehicles. *IEEE Trans. Ind. Electron.* **2016**, *63*, 6705–6714. [CrossRef]

6. Jenal, M.; Sulaiman, E.; Kumar, R. Effects of rotor pole number in outer rotor permanent magnet flux switching machine for light weight electric vehicle. In Proceedings of the 4th IET Clean Energy and Technology Conference (CEAT 2016), Kuala Lumpur, Malaysia, 14–15 November 2016; pp. 1–6.

7. Rahim, N.A.; Ping, H.W.; Tadjuddin, M. Design of an In-Wheel Axial Flux Brushless DC Motor for Electric Vehicle. In Proceedings of the 2006 International Forum on Strategic Technology, Ulsan, Korea, 18–20 October 2006; pp. 16–19.

8. Reddy, G.V.V.; Reddy, B.P.; Sivakumar, K. Design constraints of multiphase induction motor drives for electric vehicles. In Proceedings of the 2017 National Power Electronics Conference (NPEC), Pune, India, 18–20 December 2017; pp. 234–239.

9. Ulu, C.; Korman, O.; Kömürgöz, G. Electromagnetic and thermal analysis/design of an induction motor for electric vehicles. In Proceedings of the 2017 8th International Conference on Mechanical and Aerospace Engineering (ICMAE), Jeju, Korea, 21–24 November 2017; pp. 6–10.

10. Gu, W.; Zhu, X.; Quan, L.; Du, Y. Design and optimization of permanent magnet brushless machines for electric vehicle applications. *Energies* **2015**, *8*, 13996–14008. [CrossRef]

11. Zhang, L.; Fan, Y.; Li, C.; Liu, C. Design and Analysis of a New Six-Phase Fault-Tolerant Hybrid-Excitation Motor for Electric Vehicles. *IEEE Trans. Magn.* **2015**, *51*, 1–4. [PubMed]

12. Mbadiwe, E.I.; Sulaiman, E.; Khan, F. Consideration of permanent magnet flux switching motor in segmented rotor for in-wheel vehicle propulsion. In Proceedings of the 2018 International Conference on Computing, Mathematics and Engineering Technologies (iCoMET), Sukkur, Pakistan, 3–4 March 2018; pp. 1–6.

13. Tang, Y.; Ilhan, E.; Paulides, J.J.H.; Lomonova, E.A. Design considerations of flux-switching machines with permanent magnet or DC excitation. In Proceedings of the 2013 15th European Conference on Power Electronics and Applications (EPE), Lille, France, 3–5 September 2013; pp. 1–10.

14. Oner, Y.; Zhu, Z.Q.; Chu, W. Comparative Study of Vernier and Interior PM Machines for Automotive Application. In Proceedings of the 2016 IEEE Vehicle Power and Propulsion Conference (VPPC), Hangzhou, China, 17–20 October 2016; pp. 1–6.

15. Gudivada, R.; Bodnapu, K.K.; Vavillapalli, K.R. Virtual characterization of Interior Permanent Magnet (IPM) motor for EV traction applications. In Proceedings of the 2017 IEEE Transportation Electrification Conference (ITEC-India), Pune, India, 13–15 December 2017; pp. 1–4.

16. Shu, Z.; Zhu, X.; Quan, L.; Du, Y.; Liu, C. Electromagnetic performance evaluation of an outer-rotor flux-switching permanent magnet motor based on electrical-thermal two-way coupling method. *Energies* **2017**, *10*, 677. [CrossRef]

17. Abdel-Khalik, A.S.; Ahmed, S.; Massoud, A. A new permanent-magnet vernier machine using a single layer winding layout for electric vehicles. In Proceedings of the 2014 IEEE 23rd International Symposium on Industrial Electronics (ISIE), Istanbul, Turkey, 1–4 June 2014; pp. 703–708.

18. Tong, C.; Wu, F.; Zheng, P.; Sui, Y.; Cheng, L. Analysis and design of a fault-tolerant six-phase permanent-magnet synchronous machine for electric vehicles. In Proceedings of the 2014 17th International Conference on Electrical Machines and Systems (ICEMS), Hangzhou, China, 22–25 October 2014; pp. 1629–1632.

19. Huang, J.; Kang, M.; Yang, J.Q.; Jiang, H.B.; Liu, D. Multiphase machine theory and its applications. In Proceedings of the 2008 International Conference on Electrical Machines and Systems, Wuhan, China, 17–20 October 2008; pp. 1–7.

20. Zhao, J.; Zheng, Y.; Zhu, C.; Liu, X.; Li, B. A Novel Modular-Stator Outer-Rotor Flux-Switching Permanent-Magnet Motor. *Energies* **2017**, *10*, 937. [CrossRef]
21. Wang, A.; Jia, Y.; Soong, W.L. Comparison of Five Topologies for an Interior Permanent-Magnet Machine for a Hybrid Electric Vehicle. *IEEE Trans. Magn.* **2011**, *47*, 3606–3609. [CrossRef]
22. Zhang, X.; Ji, J.; Zheng, J.; Zhu, X. Improvement of Reluctance Torque in Fault-Tolerant Permanent-Magnet Machines with Fractional-Slot Concentrated-Windings. *IEEE Trans. Appl. Superconduct.* **2018**, *28*, 1–5. [CrossRef]
23. Fan, Y.; Chen, S.; Tan, C.; Cheng, M. Design and investigation of a new outer-rotor IPM motor for EV and HEV in-wheel propulsion. In Proceedings of the 2016 19th International Conference on Electrical Machines and Systems (ICEMS), Chiba, Japan, 17 March 2016; pp. 1–4.
24. Yu, J.; Liu, C. Design of a double-stator hybrid flux switching permanent magnet machine for direct-drive robotics. In Proceedings of the 2017 20th International Conference on Electrical Machines and Systems (ICEMS), Sydney, Australia, 11–14 August 2017; pp. 1–6.
25. Chen, J.T.; Zhu, Z.Q.; Thomas, A.S.; Howe, D. Optimal combination of stator and rotor pole numbers in flux-switching PM brushless AC machines. In Proceedings of the 2008 International Conference on Electrical Machines and Systems, Wuhan, China, 17–20 October 2008; pp. 2905–2910.
26. Atallah, K.; Howe, D. A novel high-performance magnetic gear. *IEEE Trans. Magn.* **2001**, *37*, 2844–2846. [CrossRef]
27. Shan, Z.; Liu, C.; Lee, C.H.T.; Chen, W.; Yu, F. Design and Comparison of Direct-Drive Stator-PM Machines for Electric Power Generation. In Proceedings of the 2016 IEEE Vehicle Power and Propulsion Conference (VPPC), Hangzhou, China, 17–20 October 2016; pp. 1–6.
28. Saeed, M.S.R.; Mohamed, E.E.M.; Ali, A.I.M. Parallel Partitioned-Rotor Switched flux PM machine for light hybrid/electric vehicles. In Proceedings of the 2018 International Conference on Innovative Trends in Computer Engineering (ITCE), Aswan, Egypt, 19–21 February 2018; pp. 380–385.
29. Zhu, Z.Q.; Howe, D. Influence of design parameters on cogging torque in permanent magnet machines. *IEEE Trans. Energy Convers.* **2000**, *15*, 407–412. [CrossRef]
30. Li, Z.; Chen, J.H.; Zhang, C.; Liu, L.; Wang, X. Cogging torque reduction in external-rotor permanent magnet torque motor based on different shape of magnet. In Proceedings of the 2017 IEEE International Conference on Cybernetics and Intelligent Systems (CIS) and IEEE Conference on Robotics, Automation and Mechatronics (RAM), Ningbo, China, 19–21 November 2017; pp. 304–309.
31. Liu, Y.; Yin, J.; Gong, B.; Yang, G. Comparative analysis of cogging torque reduction methods of variable flux reluctance machines for electric vehicles. In Proceedings of the 2017 20th International Conference on Electrical Machines and Systems (ICEMS), Sydney, Australia, 11–14 August 2017; pp. 1–6.
32. Setiabudy, R.; Rahardjo, A. Cogging torque reduction by modifying stator teeth and permanent magnet shape on a surface mounted PMSG. In Proceedings of the 2017 International Seminar on Intelligent Technology and Its Applications (ISITIA), Surabaya, Indonesia, 28–29 August 2017; pp. 227–232.
33. Lee, D.H.; Jeong, C.L.; Hur, J. Analysis of cogging torque and torque ripple according to unevenly magnetized permanent magnets pattern in PMSM. In Proceedings of the 2017 IEEE Energy Conversion Congress and Exposition (ECCE), Cincinnati, OH, USA, 1–5 October 2017; pp. 2433–2438.
34. Ma, G.; Li, G.; Zhou, R.; Guo, X.; Ju, L.; Xie, F. Effect of stator and rotor notches on cogging torque of permanent magnet synchronous motor. In Proceedings of the 2017 20th International Conference on Electrical Machines and Systems (ICEMS), Sydney, Australia, 11–14 August 2017; pp. 1–5.
35. Wu, F.; Zheng, P.; Jahns, T.M. Analytical Modeling of Interturn Short Circuit for Multiphase Fault-Tolerant PM Machines with Fractional Slot Concentrated Windings. *IEEE Trans. Ind. Appl.* **2017**, *53*, 1994–2006. [CrossRef]
36. Zhang, S.; Habetler, T.G. A transient model of interior permanent magnet machines under stator winding inter-turn short circuit faults. In Proceedings of the IECON 2017—43rd Annual Conference of the IEEE Industrial Electronics Society, Beijing, China, 5–8 November 2017; pp. 1765–1770.
37. Kim, K.-H.; Choi, D.-U.; Gu, B.-G.; Jung, I.-S. Fault model and performance evaluation of an inverter-fed permanent magnet synchronous motor under winding shorted turn and inverter switch open. *IET Electr. Power Appl.* **2010**, *4*, 214–225. [CrossRef]
38. Zhang, Y.; Chau, K.T.; Jiang, J.Z.; Zhang, D.; Liu, C. A finite element–analytical method for electromagnetic field analysis of electric machines with free rotation. *IEEE Trans. Magn.* **2006**, *42*, 3392–3394. [CrossRef]

39.   Liu, C.; Chau, K.T.; Zhong, J.; Li, W.; Li, F. Quantitative Comparison of Double-Stator Permanent Magnet Vernier Machines With and Without HTS Bulks. *IEEE Trans. Appl. Supercond.* **2012**, *22*, 5202405.

40.   Liu, C.; Chau, K.T.; Li, W. Loss analysis of permanent magnet hybrid brushless machines with and without HTS field windings. *IEEE Trans. Appl. Supercond.* **2012**, *22*, 1077–1080.

41.   Liu, C. Emerging electric machines and drives—An overview. *IEEE Trans. Energy Convers.* **2018**. [CrossRef]

42.   Chen, M.; Chau, K.T.; Li, W.; Liu, C. Development of Non-rare-earth Magnetic Gears for Electric Vehicles. *J. Asian Electr. Veh.* **2012**, *10*, 1607–1613. [CrossRef]

43.   Liu, C.; Luo, Y. Overview of advanced control strategies for electric machines. *Chin. J. Electr. Eng.* **2017**, *3*, 53–61.

energies

MDPI

Article

Stability Analysis of Deadbeat-Direct Torque and Flux Control for Permanent Magnet Synchronous Motor Drives with Respect to Parameter Variations [†]

Jae Suk Lee

Department of electrical engineering, Chonbuk National University, Jeollabuk-do 54896, Korea;
jaesuk@jbnu.ac.kr; Tel.: +82-270-2398
† 2013 15th European Conference on Power Electronics and Applications (EPE), Lille, France,
2–6 September 2013.

Received: 11 July 2018; Accepted: 30 July 2018; Published: 4 August 2018

Abstract: This paper presents a stability analysis and dynamic characteristics investigation of deadbeat-direct torque and flux control (DB-DTFC) of interior permanent magnet synchronous motor (IPMSM) drives with respect to machine parameter variations. Since a DB-DTFC algorithm is developed based on a machine model and parameters, stability with respect to machine parameter variations should be evaluated. Among stability evaluation methods, an eigenvalue (EV) migration is used in this paper because both the stability and dynamic characteristics of a system can be investigated through EV migration. Since an IPMSM drive system is nonlinear, EV migration cannot be directly applied. Therefore, operating point models of DB-DTFC and CVC (current vector control) IPMSM drives are derived to obtain linearized models and to implement EV migration in this paper. Along with DB-DTFC, current vector control (CVC), one of the widely used control algorithms for motor drives, is applied and evaluated at the same operating conditions for performance comparison. For practical analysis, the US06 supplemental federal test procedure (SFTP), one of the dynamic automotive driving cycles, is transformed into torque and speed trajectories and the trajectories are used to investigate the EV migration of DB-DTFC and CVC IPMSM drives. In this paper, the stability and dynamic characteristics of DB-DTFC and CVC IPMSM drives are compared and evaluated through EV migrations with respect to machine parameter variations in simulation and experiment.

Keywords: PMSM (permanent magnet synchronous motor); DB-DTFC (deadbeat-direct torque and flux control); torque control; stability

1. Introduction

Deadbeat direct torque and flux control (DB-DTFC) was developed by combining the features of deadbeat control and direct torque control (DTC) for induction motor drive systems [1,2]. DB-DTFC has been implemented for various types of electrical machine drives such as IPMSM (interior permanent magnet synchronous motor) drives [3], wound field synchronous machine drives [4], and synchronous reluctance machine drives [5]. Recently, DB-DTFC has been applied for implementation of self-sensing control [6,7] and fault-tolerant control [8]. In [2–5], it has been presented that DB-DTFC shows advantages over other motor control algorithms, for example, fast dynamic performance and less torque ripple. However, parameter sensitivity is one of the critical issues to be investigated because DB-DTFC algorithm is developed based on an inverse electrical machine model. The parameter sensitivity issue can be reduced by using online parameter identification methods but the methods are typically complicated to implement [9–11]. In [12,13], robustness evaluation of DB-DTFC is presented with respect to parameter variations for induction machine (IM) drives and IPMSM drives, respectively. While the torque error, command tracking performance of DB-DTFC for IM, and IPMSM drives have been analyzed, the stability of DB-DTFC with respect to machine parameter variations has not been investigated. In [14], disturbance to a system is estimated using a hybrid Kalman estimator and improvement of robustness is achieved. Not only robustness but also stable operation of motor drives is important, especially for transportation applications, for example, aircraft and electric vehicles.

Though DB-DTFC has shown higher dynamic performance and robustness from previous research, parameter sensitivity is still an important issue to investigate because DB-DTFC is an algorithm developed based on machine model and estimated parameters. In particular, the stability of DB-DTFC with respect to parameter variations should be investigated for practical applications and has not been investigated in other research. Therefore, investigating the stability of a DB-DTFC motor drive with respect to parameter variations is necessary. A few stability evaluation methods for motor drive systems have been presented. The Lyapunov stability theory is applied for stability evaluation of an induction machine drive in [15–17]. Using the Lyapunov stability theory, it can be determined if a nonlinear system is (1) Lyapunov stable, (2) asymptotically stable, or (3) exponentially stable. However, the dynamic characteristics of the system are disregarded. In [18], a load angle limiting method is applied to determine a stable region of motor drives. However, the load angle limiting method is torque limitation methods for stable operation of motor drives. A pole-zero migration method is used to investigate stability of motor drives in [19–21]. From the pole-zero migration, not only system stability can be investigated but also system dynamic properties, such as frequency of oscillation and rate of decay. Also, the pole-zero migration can be clearly shown in the s-domain and z-domain for analog and digital systems, respectively. Since DB-DTFC is developed based on discrete time model of electrical motors, eigenvalue migration at z-domain is presented in this paper. The eigenvalue migration can be implemented in a linear system while CVC (current vector control) and DB-DTFC IPMSM drives are nonlinear systems. Therefore, deriving linear system models of both IPMSM drives is required. A small signal model or an operating point model is a modeling method used to approximate the dynamics of systems, including nonlinear components, with a linear equation. Using a small signal model of a nonlinear system, the dynamic response and characteristics of the system can be investigated when a small perturbation is applied. Therefore, small signal modeling has been applied for stability analysis of motor control systems and power converters [22–24]. In this paper, small signal models of CVC and DB-DTFC IPMSM drives are derived for implementation of EV (eigenvalue) migration.

This paper begins with a brief introduction of a DB-DTFC algorithm and state observers for IPMSM drives. Using the DB-DTFC equation, an operating point model is derived. Then, the eigenvalue migration of DB-DTFC and CVC at z-domain is compared by applying the US06 supplemental federal test procedure (SFTP) driving cycle for practical verification in simulations and experiments.

2. DB-DTFC and State Observers for IPMSM Drives

A DB-DTFC algorithm for IPMSM drives is initially presented in [3]. As stated in Section 1, DB-DTFC shows faster transient dynamics and less torque ripple for various types of electrical machines comparing to other control algorithms but parameter sensitivity is one of the critical issues to be investigated. The DB-DTFC algorithm is briefly introduced in this section because an operating point model is derived based on the DB-DTFC algorithm. Also, state observers used for a DB-DTFC IPMSM drive are reviewed because state observers play an important role in a DB-DTFC system.

2.1. DB-DTFC for IPMSM Drives

Three-phase electrical motors such as PMSMs are typically modeled and analyzed at a rotor reference frame. By utilizing a rotor reference frame, the system complexity becomes simpler than a three-phase model and the direct current (DC)signal can be manipulated for analysis and control instead of alternating current (AC) signals. As shown in Equations (1) and (2), Clarke transformation is applied to transform a three-phase model (*a-b-c* model) to a stationary reference frame model and

Park transformation is applied to transform a stationary reference frame model into a rotor reference frame model.

$$
\begin{bmatrix} f_d^s \\ f_q^s \end{bmatrix} = \begin{bmatrix} \frac{2}{3} & -\frac{1}{3} & -\frac{1}{3} \\ 0 & \frac{2}{\sqrt{3}} & -\frac{2}{\sqrt{3}} \end{bmatrix} \times \begin{bmatrix} f_a \\ f_b \\ f_c \end{bmatrix} = \begin{bmatrix} 1 & 0 & 0 \\ \frac{1}{\sqrt{3}} & \frac{2}{\sqrt{3}} & 0 \end{bmatrix} \times \begin{bmatrix} f_a \\ f_b \\ f_c \end{bmatrix} \tag{1}
$$

$$
\begin{bmatrix} f_d^r \\ f_q^r \end{bmatrix} = \begin{bmatrix} \cos\theta & \sin\theta \\ -\sin\theta & \cos\theta \end{bmatrix} \begin{bmatrix} f_d^s \\ f_q^s \end{bmatrix} \tag{2}
$$

A graphical representation of the reference frame of a PMSM is shown in Figure 1.

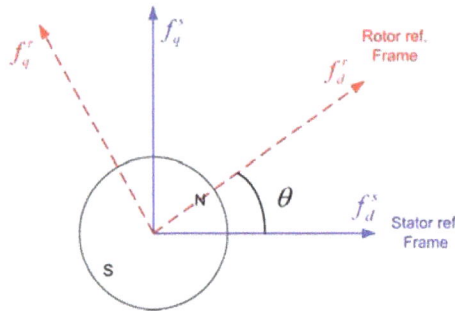

Figure 1. Stationary and rotor reference frame notation of a PMSM (permanent magnet synchronous motor).

In Equations (1) and (2), subscripts d and q represent the d and q axes and superscripts s and r represent a stationary reference frame and a rotor reference frame, respectively. In (2), θ is the rotor position of a PMSM, as shown in Figure 1. Applying the Clarke and Park transformation, an IPMSM model at a rotor reference frame can be derived. In Equations (3) and (4), differential equations of stator flux linkage, stator current and air-gap torque for IPMSM drives at a rotor reference frame are shown, where p indicates the time differential.

$$
p\lambda_{ds}^r = v_{ds}^r - R_s\frac{\lambda_{ds}^r}{L_d} + R_s\frac{\lambda_{pm}}{L_d} + \omega_r\lambda_{qs}^r \tag{3}
$$

$$
p\lambda_{qs}^r = v_{qs}^r - R_s\frac{\lambda_{qs}^r}{L_q} - \omega_r\lambda_{ds}^r \tag{4}
$$

$$
pi_{ds}^r = \frac{1}{L_d}[v_{ds}^r - R_s\frac{\lambda_{ds}^r}{L_d} + R_s\frac{\lambda_{pm}}{L_d} + \omega_r\lambda_{qs}^r] \tag{5}
$$

$$
pi_{qs}^r = \frac{1}{L_q}[v_{qs}^r - R_s\frac{\lambda_{qs}^r}{L_d} - \omega_r\lambda_{ds}^r] \tag{6}
$$

$$
pT_{em} = \frac{3}{4}P[p\lambda_{ds}^r i_{qs}^r + \lambda_{ds}^r pi_{qs}^r - p\lambda_{qs}^r i_{ds}^r - \lambda_{qs}^r pi_{ds}^r] \tag{7}
$$

Assuming that the rate of change of torque is constant during the one pulse width modulation (PWM) period (that is, when high switching frequency is applied), the approximate torque difference equation in a discrete time domain is derived as Equation (8) by substituting Equations (3)–(6) into Equation (7). Equations (3)–(6) are presented as a function of stator flux linkages so that the torque equation in Equation (7) does not include stator current vectors when Equations (3)–(6) are substituted into Equation (7).

$$\frac{\Delta T_{em}(k)}{T_s} = \frac{3}{4}P \left[\begin{array}{l} v^r_{ds}(k)\lambda^r_{qs}(k)\left(\frac{L_d-L_q}{L_dL_q}\right) + v^r_{qs}(k)\frac{(L_d-L_p)\lambda^r_{ds}(k)+\lambda_{pm}L_q}{L_dL_q} \\ + \frac{\omega_r(k)}{L_dL_q}\left((L_q-L_d)\left(\lambda^r_{ds}(k)^2 - \lambda^r_{qs}(k)^2\right) - L_q\lambda^r_{ds}(k)\lambda_{pm}\right) + \frac{R_s\lambda^r_{qs}(k)}{L_d^2L_q^2}\left(\left(L_q^2-L_d^2\right)\lambda^r_{ds}(k) - L_q^2\lambda_{pm}\right) \end{array} \right] \quad (8)$$

The Volt-sec solution that results in deadbeat control of IPMSM drives can be obtained as Equation (9) by rearranging Equation (8). It is called the "torque line."

$$v^r_{qs}(k)T_s = m v^r_{ds}(k)T_s + b \quad (9)$$

where, $\quad m \quad = \quad \left(\frac{(L_q-L_d)\lambda^r_{qs}(k)}{(L_d-L_q)\lambda^r_{ds}(k)+L_q\lambda_{pm}}\right) \quad$ and $\quad b \quad = \quad \left(\frac{L_dL_q}{(L_d-L_q)\lambda^r_{ds}(k)+L_q\lambda_{pm}}\right) \times$

$$\left[\begin{array}{l} \frac{\omega_r(k)}{L_dL_q}\left((L_q-L_d)\left(\lambda^r_{ds}(k)^2 - \lambda^r_{qs}(k)^2\right) - L_q\lambda^r_{ds}(k)\lambda_{pm}\right) \\ + \frac{R_s\lambda^r_{qs}(k)\left[\left(L_q^2-L_d^2\right)\lambda^r_{ds}(k) - L_q^2\lambda_{pm}\right]}{L_d^2L_q^2} + \frac{4\Delta T_{em}(k)}{3P} \end{array} \right]$$

This torque line is used in every switching period to solve for the desired inverter volt-seconds that achieve the commanded torque and stator flux linkage [3].

2.2. State Observers for a DB-DTFC IPMSM Drive

For the development and implementation of a DB-DTFC algorithm for IPMSM drives, a stator current observer and a stator flux linkage observer are required. As a part of the review, a stator current observer and a stator flux linkage observer used for DB-DTFC implementation are briefly introduced in this section. To implement deadbeat direct torque control algorithm, the next sample time instant stator current should be estimated. The estimated stator current is used to calculate the estimated torque and estimated stator flux linkage for the prediction of dynamics one sample time instant ahead [3]. The current at the next sample time instant can be estimated using a rotor reference frame-based stator current observer. The stator current observer is developed based on IPMSM state equations, and its block diagram in a discrete time domain is shown in Figure 2.

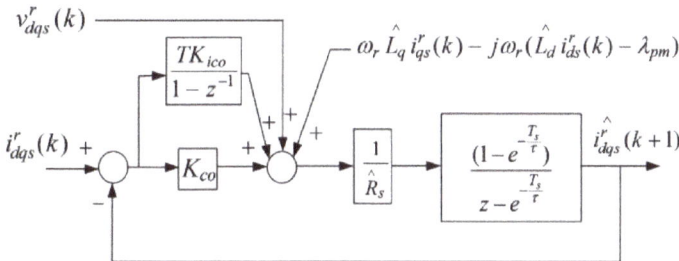

Figure 2. A block diagram of a stator current observer in a discrete time domain.

The stator current observer is implemented experimentally and its estimation accuracy characteristic with respect to q-axis inductance variation at a frequency domain is presented in [3], as shown in Figure 3.

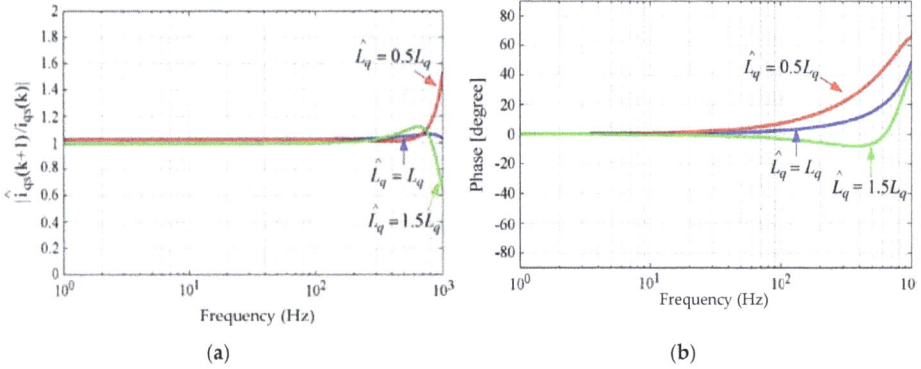

Figure 3. Estimation accuracy characteristic of a stator current observer with respect to q-axis inductance variation at a frequency domain [3]. (**a**) Magnitude of estimation accuracy; (**b**) Phase of estimation accuracy.

During the experiment for estimation accuracy, estimated q-axis inductance is detuned ±50% from its correct estimation value to investigate the parameter sensitivity characteristic of a stator current observer of which the bandwidth is 300 Hz. From Figure 3, it is seen that zero steady state error can be achieved within a bandwidth of the stator current observer regardless of parameter variations. Also, the frequency response function shows leading property because the stator current observer estimates the next sample time stator current. A stator flux linkage observer used for DB-DTFC implementation is shown in Figure 4.

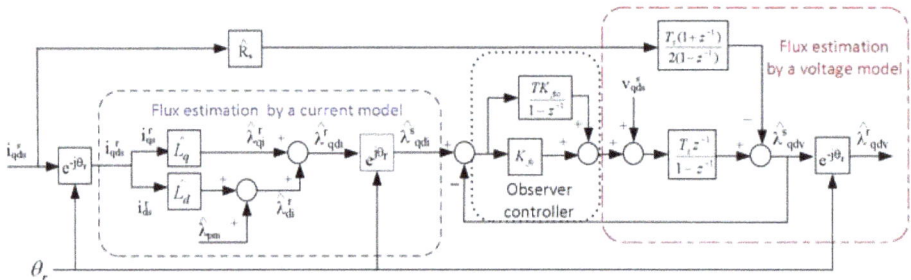

Figure 4. A block diagram of a stator flux linkage observer used for DB-DTFC (deadbeat-direct torque and flux control) implementation [25].

As shown in Figure 4, a current model and a voltage model are used for stator flux linkage estimation of a PMSM. Estimation of stator flux linkage using a current model is not affected by noise signals and dead time but is sensitive to parameter variation. Estimation of stator flux linkage using a voltage model is robust to parameter variations but is affected by voltage distortion such as noise and dead time, especially at low speeds. Therefore, a stator flux linkage is estimated from a current model at low speeds and is estimated from a voltage model at high speed. A cross-over frequency between the current model-based estimation and voltage model estimation is determined by a bandwidth of an observer controller located between a current model and a voltage model. Figure 5 shows the simulation results of a frequency response of the stator flux linkage observer with respect to parameter variations.

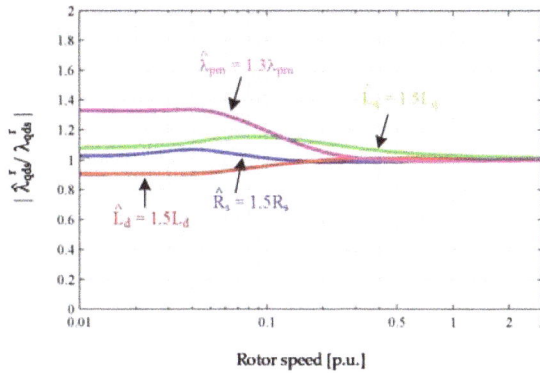

Figure 5. Simulation results of a frequency response of the stator flux linkage observer with respect to motor parameter variations [25].

As shown in Figure 5, it is verified that accuracy of stator flux linkage estimation is affected by the parameter variations when the operating speed of a PMSM is below a bandwidth of a stator flux linkage observer. Then, stator flux linkage estimation becomes accurate and robust to parameter variation when the operating speed of a PMSM is beyond a crossover frequency of the stator flux linkage observer controller. Characteristics of state observers used in DB-DTFC are reviewed in this section because observers play an important role in a DB-DTFC system and understanding its features and limitation is necessary.

3. Derivation of Operating Point Model of CVC and DB-DTFC IPMSM Drives

3.1. Operating Point Model of IPMSMs

It is known that a nonlinear system behaves similarly to its linearized approximation around an equilibrium point [26]. Therefore, the nonlinear model can be linearized using an operating point model (small signal model). In addition, the stability of a system can be investigated using the operating point model. The operating point models of IPMSM drives can be described by the following equations from [20]:

$$p\Delta i_{ds}^r = -\frac{R_s}{L_d}\Delta i_{ds}^r + \frac{L_q\omega_{ro}}{L_d}\Delta i_{qs}^r + \frac{L_q i_{qso}^r}{L_d}\Delta\omega_r + \frac{1}{L_d}\Delta v_{ds}^r \tag{10}$$

$$p\Delta i_{qs}^r = -\frac{L_d\omega_{ro}}{L_q}\Delta i_{ds}^r - \frac{R_s}{L_q}\Delta i_{qs}^r - \frac{L_d i_{dso}^r + \lambda_{pm}}{L_q}\Delta\omega_r + \frac{1}{L_q}\Delta v_{qs}^r \tag{11}$$

$$\Delta T_{em} = \frac{3P}{4J_p}\left((L_d - L_q)i_{qso}^r\Delta i_{ds}^r + (\lambda_{pm} + (L_d - L_q)i_{dso}^r)\Delta i_{qs}^r \right) \tag{12}$$

$$\Delta\lambda_s = \frac{L_d i_{dso}^r + L_d\lambda_{pm}}{\lambda_{so}}\Delta i_{ds}^r + \frac{L_d i_{qso}^r}{\lambda_{so}}\Delta i_{qs}^r \tag{13}$$

$$p\Delta\omega_r = \left(\frac{3P}{4J_p}(L_d - L_q)i_{qso}^r\right)\Delta i_{ds}^r + \left(\frac{3P}{4J_p}(\lambda_{pm} + (L_d - L_q)i_{dso}^r)\Delta i_{qs}^r - \frac{\Delta T_L}{J_p}\right) \tag{14}$$

In Equations (10) to (14), Δ indicates the perturbation of each state, p represents the derivatives of corresponding variables, and o denotes steady state values. The operating point model of IPMSM drives can be formed in a state-space representation. Based on the operating point models in continuous time, Equations (10) through (14), the operating point models in discrete time can be derived as in Equations (15) through (19).

$$\Delta i_{ds}^r z - \Delta i_{ds}^r + \frac{R_s}{L_d}\Delta i_{ds}^r - \frac{L_q \omega_{ro}}{L_d}\Delta i_{qs}^r - \frac{L_q i_{qso}^r}{L_d}\Delta \omega_r = \frac{1}{L_d}\Delta v_{ds}^r \tag{15}$$

$$\Delta i_{qs}^r z - \Delta i_{qs}^r + \frac{R_s}{L_q}\Delta i_{qs}^r + \frac{L_d \omega_{ro}}{L_q}\Delta i_{ds}^r + \frac{L_d i_{dso}^r + \lambda_{pm}}{L_q}\Delta \omega_r = \frac{1}{L_q}\Delta v_{qs}^r \tag{16}$$

$$\Delta T_{em} = \frac{3P}{4J_p}\left((L_d - L_q)i_{qso}^r \Delta i_{ds}^r + \left(\lambda_{pm} + (L_d - L_q)i_{dso}^r\right)\Delta i_{qs}^r \right) \tag{17}$$

$$\Delta \lambda_s = \frac{L_d i_{dso}^r + L_d \lambda_{pm}}{\lambda_{so}}\Delta i_{ds}^r + \frac{L_d i_{qso}^r}{\lambda_{so}}\Delta i_{qs}^r \tag{18}$$

$$\begin{aligned}\Delta \omega_r &= \left(\frac{3P}{4J_p(z-1)}(L_d - L_q)i_{qso}^r\right)\Delta i_{ds}^r + \left(\frac{3P}{4J_p(z-1)}\left(\lambda_{pm} + (L_d - L_q)i_{dso}^r\right)\right)\Delta i_{qs}^r - \frac{\Delta T_L}{J_p}\\ &= K_{\omega d}\Delta i_{ds}^r + K_{\omega q}\Delta i_{qs}^r - \frac{\Delta T_L}{J_p}\end{aligned} \tag{19}$$

The voltage and current operating point (small signal) model equations, Equations (15) and (16), can be formed in a matrix as in Equation (20):

$$\begin{bmatrix} a & b \\ c & d \end{bmatrix}\begin{bmatrix} \Delta i_{ds}^r \\ \Delta i_{qs}^r \end{bmatrix} = \begin{bmatrix} \frac{T_s}{L_d} & 0 \\ 0 & \frac{T_s}{L_q} \end{bmatrix}\begin{bmatrix} \Delta v_{ds}^r \\ \Delta v_{qs}^r \end{bmatrix} \tag{20}$$

where

$$\begin{bmatrix} a & b \\ c & d \end{bmatrix} = \begin{bmatrix} z - 1 + \frac{R_s T_s}{L_d} - \frac{L_q i_{qso}^r}{L_d}T_s K_{\omega d} & -\frac{L_q}{L_d}T_s \omega_{ro} - \frac{L_q i_{qso}^r}{L_d}T_s K_{\omega q} \\ \frac{L_d}{L_q}T_s \omega_{ro} + \frac{L_d i_{dso}^r}{L_q}T_s K_{\omega d} + \frac{T_s \lambda_{pm}}{L_q}K_{\omega d} & z - 1 + \frac{R_s T_s}{L_q} + \frac{L_d i_{dso}^r}{L_d}T_s K_{\omega q} + \frac{T_s \lambda_{pm}}{L_q}K_{\omega q} \end{bmatrix}$$

The torque and stator flux linkage operating point (small signal) model equations, Equations (17) and (18), can be formed in a matrix as in Equation (21):

$$\begin{bmatrix} \Delta T_{em} \\ \Delta \lambda_s \end{bmatrix} = \begin{bmatrix} e & f \\ g & h \end{bmatrix}\begin{bmatrix} \Delta i_{ds}^r \\ \Delta i_{qs}^r \end{bmatrix} \tag{21}$$

where $\begin{bmatrix} e & f \\ g & h \end{bmatrix} = \begin{bmatrix} \frac{3P}{4J_p}(L_d - L_q)i_{qso}^r & \frac{3P}{4J_p}\left(\lambda_{pm} + (L_d - L_q)i_{dso}^r\right) \\ \frac{L_d^2 i_{dso}^r + \lambda_{pm}L_d}{\Delta\lambda_{so}} & \frac{L_q^2 i_{qso}^r}{\Delta\lambda_{so}} \end{bmatrix}.$

Then, the operating point model between torque and voltage can be written as in Equation (22):

$$\begin{bmatrix} \Delta T_{em} \\ \Delta \lambda_s \end{bmatrix} = \begin{bmatrix} e & f \\ g & h \end{bmatrix}\frac{1}{ad-bc}\begin{bmatrix} d & -b \\ -c & a \end{bmatrix}\begin{bmatrix} \frac{T_s}{L_d} & 0 \\ 0 & \frac{T_s}{L_q} \end{bmatrix}\begin{bmatrix} \Delta v_{ds}^r \\ \Delta v_{qs}^r \end{bmatrix} = \begin{bmatrix} A & B \\ C & D \end{bmatrix}\begin{bmatrix} \Delta v_{ds}^r \\ \Delta v_{qs}^r \end{bmatrix} \tag{22}$$

where $\begin{bmatrix} A & B \\ C & D \end{bmatrix} = \frac{T_s}{ad-bc}\begin{bmatrix} \frac{de-cf}{L_d} & \frac{af-be}{L_q} \\ \frac{dg-ch}{L_d} & \frac{ah-bg}{L_q} \end{bmatrix}.$

The operating point model, Equation (22), only covers a physical system, i.e., an IPMSM, and its dynamics. Therefore, a controller of an IPMSM is not included in Equation (22). It should be noted that the derived operating point model is sensitive to machine parameter variations such as q-axis inductance saturation with respect to stator current magnitude and permanent magnet flux linkage with respect to temperature. For more accurate modeling, either of the following can be applied as parameters in the operating point model: (1) Look-up table-based parameters varying as a function of operating conditions, or (2) calculated parameters by online estimation method. In this paper, constant parameters are used for analysis. In the following section, operating point models including DB-DTFC and CVC are derived.

3.2. Operating Point Model of DB-DTFC IPMSM Drives

In Figure 6, a block diagram of a DB-DTFC IPMSM drive is shown.

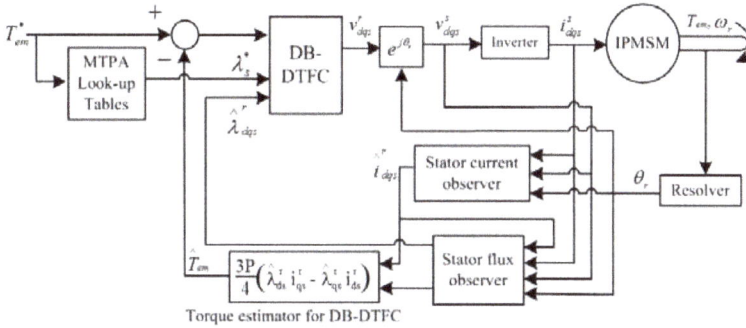

Figure 6. A block diagram of a DB-DTFC IPMSM (interior permanent magnet synchronous motor) drive.

As shown in Figure 6, torque and stator flux linkage commands, estimated torque and estimated stator flux linkage are input signals of the DB-DTFC algorithm block. For efficient operation of a DB-DTFC IPMSM drive, pre-calculated copper loss minimizing stator flux linkage is selected from a maximum torque per ampere (MTPA) look-up table as stator flux linkage command. For estimation of torque and stator flux linkage, a stator current observer and a stator flux linkage observer are used in a DB-DTFC IPMSM drive. Block diagrams of observers can be found in Figures 2 and 4.

In this section, an operating point model (or a small signal model) of a DB-DTFC IPMSM drive is derived. Since DB-DTFC is a model inverse solution, a closed-loop system including DB-DTFC can be written as in Equation (23):

$$
\begin{bmatrix} \Delta T_{em} \\ \Delta \lambda_s \end{bmatrix} = \begin{bmatrix} A & B \\ C & D \end{bmatrix} \begin{bmatrix} \Delta v_{ds}^r \\ \Delta v_{qs}^r \end{bmatrix} = \begin{bmatrix} A & B \\ C & D \end{bmatrix} \begin{bmatrix} \hat{A} & \hat{B} \\ \hat{C} & \hat{D} \end{bmatrix}^{-1} \begin{bmatrix} \frac{1}{z-1} & 0 \\ 0 & \frac{1}{z-1} \end{bmatrix} \begin{bmatrix} \Delta T_{em}^* - \Delta T_{em} \\ \Delta \lambda_s^* - \Delta \lambda_s \end{bmatrix} \tag{23}
$$

A transfer function of a closed loop DB-DTFC system can be derived as in Equation (24) by expanding Equation (23):

$$
\frac{\Delta T_{em}}{\Delta T_{em}^*} = \frac{A\hat{D} - B\hat{C}}{(z-1)(\hat{A}\hat{D} - \hat{B}\hat{C}) + A\hat{D} - B\hat{C}} \tag{24}
$$

If the estimated machine parameters match the actual machine parameters, that is, $A = \hat{A}$, $B = \hat{B}$, $C = \hat{C}$, and $D = \hat{D}$, the transfer function becomes deadbeat as in Equation (25):

$$
\frac{\Delta T_{em}}{\Delta T_{em}^*} = \frac{1}{z} \tag{25}
$$

If estimated parameters do not match the actual machine parameters, the characteristic equation, the denominator of the transfer function of the closed loop DB-DTFC system becomes Equation (26):

$$
\begin{aligned}
z(ad - bc)\frac{eh - fg}{L_d L_q} ad + \frac{eh}{L_d L_q} ad + \frac{eh}{L_d L_q} bc - \frac{eh}{L_d L_q} bc + \frac{fg}{L_d L_q} ad - \frac{fg}{L_d L_q} ad \\
- \frac{fg}{L_d L_q} bc + \frac{fg}{L_d L_q} bc - \frac{fh}{L_d L_q} ac + \frac{fh}{L_d L_q} ac - \frac{eg}{L_d L_q} \hat{b}\hat{d} + \frac{eg}{L_d L_q} b\hat{d}
\end{aligned} \tag{26}
$$

The roots of the characteristic equations are the eigenvalues (EVs) of the closed loop DB-DTFC system. The EV migration on a z-plane shows properties of IPMSM dynamics, such as stability, response time, and oscillation.

3.3. Operating Point Model of CVC IPMSM Drives

A block diagram of a current vector controlled (CVC) IPMSM drive is seen in Figure 7.

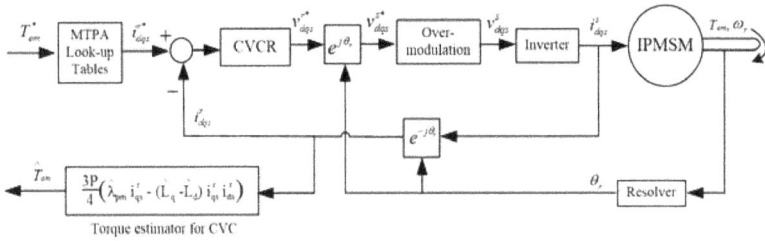

Figure 7. A block diagram of an MTPA (maximum torque per ampere) based CVC IPMSM drive.

For CVC, a closed loop system including PI controllers can be derived as in Equations (27) and (28):

$$v_{ds}^* = \left(i_{ds}^{r*} - i_{ds}^r\right)\left(K_{pd} + \frac{K_{id}T_s z}{z-1}\right) = R_s i_{ds}^r + L_d \frac{i_{ds}^r z - i_{ds}^r}{T_s} - \omega_r L_q i_{qs}^r \tag{27}$$

$$v_{qs}^* = \left(i_{qs}^{r*} - i_{qs}^r\right)\left(K_{pq} + \frac{K_{iq}T_s z}{z-1}\right) = R_s i_{qs}^r + L_d \frac{i_{qs}^r z - i_{qs}^r}{T_s} + \omega_r L_d i_{ds}^r - \omega_r \lambda_{pm} \tag{28}$$

The operating point model equations of the closed loop CVC system can be written as in Equations (29) and (30):

$$\left(K_{pd}(z-1) + K_{id}T_s z\right)\Delta i_{ds}^{r*} = \left(K_{pd}(z-1) + K_{id}T_s z\right)\Delta i_{ds}^r + R_s(z-1)\Delta i_{ds}^r + \frac{L_d}{T_s}(z-1)^2 \Delta i_{ds}^r$$
$$- \Delta \omega_r L_q i_{qso}^r(z-1) - \omega_{ro} L_q \Delta i_{qs}^r(z-1) \tag{29}$$

$$\left(K_{pq}(z-1) + K_{iq}T_s z\right)\Delta i_{qs}^{r*} = \left(K_{pq}(z-1) + K_{iq}T_s z\right)\Delta i_{qs}^r + R_s(z-1)\Delta i_{qs}^r + \frac{L_q}{T_s}(z-1)^2 \Delta i_{qs}^r$$
$$+ \Delta \omega_r \lambda_{pm}(z-1) + \Delta \omega_r L_d \Delta i_{dso}^r(z-1) + \omega_{ro} L_d \Delta i_{ds}^r(z-1) \tag{30}$$

By substituting Equation (19) into Equations (29) and (30), Equations (29) and (30) become functions of $\Delta irds$ and $\Delta irqs$, as shown in Equations (31) and (32):

$$\left(K_{pd}(z-1) + K_{id}T_s z\right)\Delta i_{ds}^{r*} = \left(K_{pd}(z-1) + K_{id}T_s z + R_s(z-1) + \frac{L_d}{T_s}(z-1)^2 - L_q i_{qso}^r K_{\omega d}(z-1)\right)\Delta i_{ds}^r$$
$$+ \left(L_q i_{qso}^r(z-1)K_{\omega d} + \omega_{ro} L_q(z-1)\right)\Delta i_{qs}^r \tag{31}$$

$$\left(K_{pq}(z-1) + K_{iq}T_s z\right)\Delta i_{qs}^{r*} = \left((\lambda_{pm} + L_d i_{dso}^r)(z-1)K_{\omega d} + \omega_{ro} L_d(z-1)\right)\Delta i_{ds}^r$$
$$+ \left(K_{pq}(z-1) + K_{iq}T_s z + R_s(z-1) + \frac{L_q}{T_s}(z-1)^2 + (\lambda_{pm} + L_d i_{dso}^r)K_{\omega q}(z-1)\right)\Delta i_{qs}^r \tag{32}$$

Equations (31) and (32) can be written in a matrix form as in Equation (33):

$$\begin{bmatrix} \Delta i_{ds}^r \\ \Delta i_{qs}^r \end{bmatrix} = \frac{1}{uy - wx}\begin{bmatrix} y & w \\ x & u \end{bmatrix}\begin{bmatrix} C_d & 0 \\ 0 & C_q \end{bmatrix}\begin{bmatrix} \Delta i_{ds}^{r*} \\ \Delta i_{qs}^{r*} \end{bmatrix} \tag{33}$$

where

$$u = K_{pd}(z-1) + K_{id}T_s z + R_s(z-1) + \frac{L_d}{T_s}(z-1)^2 - L_q i_{qso}^r K_{wd}(z-1)$$
$$w = L_q i_{qso}^r(z-1)K_{wq} + \omega_{ro}L_q(z-1)$$
$$x = \left(\lambda_{pm} + L_d i_{dso}^r\right)(z-1)K_{wq} + \omega_{ro}L_d(z-1)$$
$$y = \left(K_{pq}(z-1) + K_{iq}T_s z + R_s(z-1) + \frac{L_q}{T_s}(z-1)^2 + \left(\lambda_{pm} + L_d i_{dso}^r\right)K_{wq}(z-1)\right)$$
$$C_d = K_{pd}(z-1) + K_{id}T_s z$$
$$C_q = K_{pq}(z-1) + K_{iq}T_s z$$

The operating point (small signal) models for the DB-DTFC and CVC IPMSM drives are derived in Equations (26) and (33), respectively. Using the operating point (small signal) models, the EV migration of each motor drive is analyzed in simulations and experiments.

4. Simulation Results

Since most driving cycles cover a wide operating space of a motor drive, a driving cycle is chosen as the load and command trajectories for a stability evaluation of DB-DTFC and CVC IPMSM drives. Among automotive driving cycles, the US06 SFTP automotive driving cycle is selected as a stability test trajectory in this paper because it contains high acceleration conditions, which are more suitable for dynamic operation testing of a motor drive. Torque command trajectory for US06 SFTP is developed by multiplying traction force and wheel radius of a test vehicle as in Equation (34), where T_{tr} is the traction torque, F_{tr} is the fraction force, and R_{wh} is the wheel radius of a test vehicle:

$$T_{tr} = F_{tr} \times R_{wh} \tag{34}$$

$$F_{tr} = K_m \times m_v \times a_v + F_{drag} \tag{35}$$

Equation (35) is an equation to calculate traction force; K_m is the rotational inertia, m_v is the mass of a test vehicle in kg, a_v is the acceleration of a test vehicle, and F_{drag} is the drag force. The drag force in Equation (35) can be calculated using Equation (36):

$$F_{drag} = D_{air} \times C_d \times A_f \times (v_x)^2 \tag{36}$$

In Equation (36), D_{air} is air density and C_d is the aerodynamic parameter, which is typically 0.2~0.4. A_f is the front area of a test vehicle and v_x is the velocity of a test vehicle in meters per second. For more accurate torque trajectory development, additional forces such as rolling resistance force can be applied. Since the actual speed and torque trajectories of the US06 driving cycle cannot be applied directly, the trajectories are adjusted to fit to the rated capacity of the test IPMSM. Time range and scale are also changed so that the data size of the simulation results does not exceed the memory capacity of a computer.

The simulation results for the US06 SFTP automotive driving cycle using DB-DTFC and CVC are shown in Figure 8. Since simulation results using both control algorithms are the same, the results are overlaid. As shown in the simulation results, the operating speed range is increased over the rated speed (Figure 8b), including flux weakening operation (Figure 8e,f), and the torque trajectory of the driving cycle covers up to the rated torque of the test IPMSM (Figure 8a). In addition to operation within a rated operating condition, operation at voltage and current limits (Figure 8c,d) is also investigated when the test IPMSM is driven along the US06 automotive driving cycle. Since the MTPA look-up tables applied for DB-DTFC and CVC shown in Figures 6 and 7 are developed using identical parameters and conditions, the d and q axis current vectors are the same for both control methods. For implementation of DB-DTFC, stator flux linkage and stator current observers are required. Therefore, the complexity and computational effort are higher than in CVC. From the simulation results in Figure 8, the advantages of DB-DTFC over CVC are not directly shown. Though the US06 SFTP is one representative dynamic driving cycle, the corresponding speed and torque trajectories are relatively smooth to present the performance difference between the CVC and

DB-DTFC algorithms. A comparison of the CVC and DB-DTFC algorithms based on the simulation results is given in Table 1.

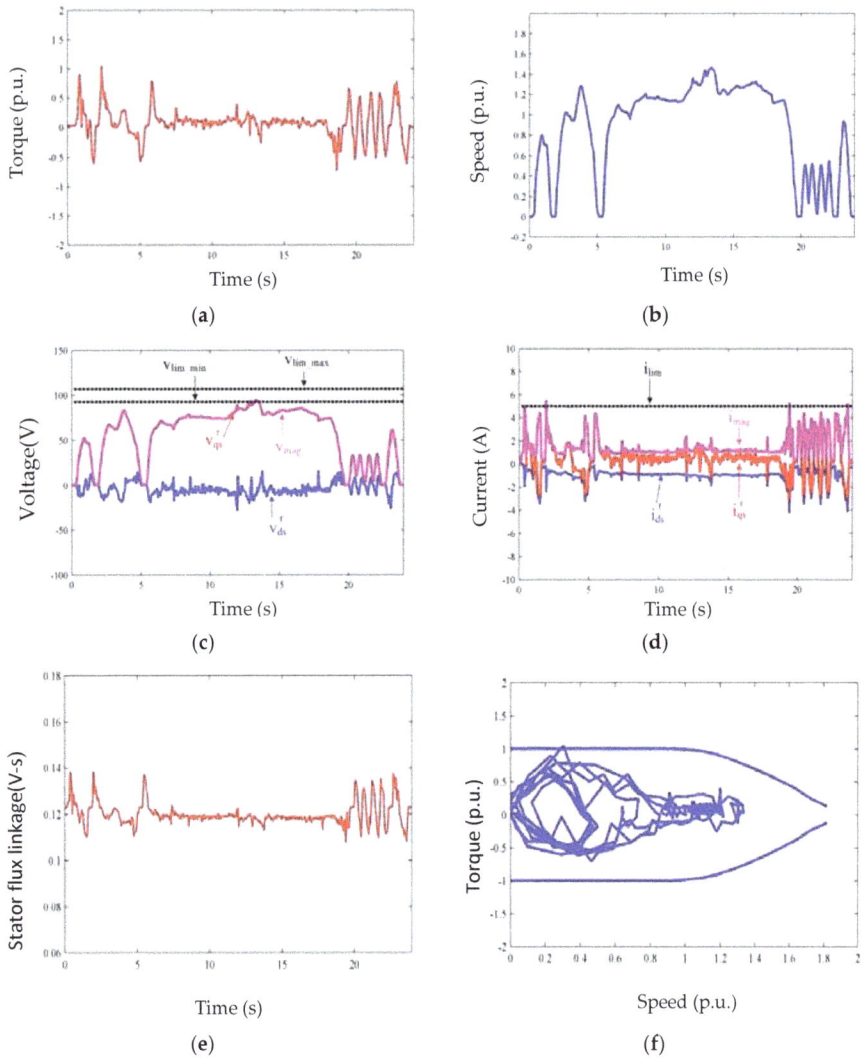

Figure 8. Simulation results of motor dynamics along the US06 SFTP (supplemental federal test procedure) driving cycle. (**a**) Torque; (**b**) Speed; (**c**) Voltage; (**d**) Current; (**e**) Stator flux linkage; (**f**) Torque-speed.

Table 1. Comparison of CVC (current vector control) and DB-DTFC (deadbeat-direct torque and flux control) based on results in Figure 8.

Characteristic	CVC	DB-DTFC
Complexity	Medium	High
Computation time using a diginal signal processor (DSP) [27]	31.4 (µs)	34 (µs)
Command tracking	Satisfactory	Satisfactory
Efficient operation	Satisfactory	Satisfactory

In addition to analyzing CVC and DB-DTFC using time domain simulation results, the EV migration of CVC and DB-DTFC is compared at different points. Utilizing data from the simulation results in Figure 8 and the derived operating point (small signal) models of DB-DTFC and CVC IPMSM drives, the EV migration of each system can be investigated as shown in Figure 9. To investigate the EV migration characteristics of the CVC and DB-DTFC IPMSM drive systems with respect to machine parameter variations, q-axis inductance is detuned by 50% of its actual value intentionally, as shown in Figure 9a,b, and permanent magnetic flux linkage is also detuned by 30% of its actual value, as shown in Figure 9c,d. In the case of a DB-DTFC IPMSM drive, poles are always located at the center of a z-plane, as seen in Figure 9a,c, if the estimated parameters used for the DB-DTFC algorithm exactly match the actual electrical machine parameters. This means that the deadbeat response is achieved along the torque and speed trajectories. If the parameters in the DB-DTFC and the machine parameters are different, the poles move away from a center of a z-plane, as seen in the red trajectories in Figure 9a,c. As a pole moves away from the center of a z-plane, the dynamic response becomes slower, which means that deadbeat response is not achieved any more. Though deadbeat response is not achieved when machine parameters do not match, faster dynamic response can still be achieved using DB-DTFC than CVC. In the case of a CVC IPMSM drive, poles are located and move around near the bandwidth of the current vector controller, as shown in Figure 9a,c. The location of poles is changed as shown in Figure 9b when detuned q-axis inductance is used because inductance is used for calculation of controller gains of a current vector controller.

Since 50% higher q-axis inductance is used for controller gain calculation, the bandwidth of the controller becomes higher and a faster dynamic response is achieved than the desired dynamic response shown in Figure 9b. Unlike q-axis inductance variation, a change in the permanent magnet flux linkage does not affect the dynamic characteristics, as shown in Figure 9d, because permanent magnet flux linkage is not used for controller gain tuning. As seen in the simulation results in Figure 9, eigenvalues are located within the unit circle in the z-plane for both the CVC and the DB-DTFC IPMSM drives. This means that the operation of both motor drive systems is stable along the US06 automotive driving cycle regardless of the machine parameter variations. Figure 10 shows corresponding impulse responses with respect to EV locations of the DB-DTFC and CVC IPMSM drives specified in Figure 9.

As shown in Figure 10a, deadbeat response can be achieved when an accurately estimated parameter is used for the implementation of DB-DTFC. However, DB-DTFC sometimes results in forced oscillations, as seen in Figure 10b, or a deadbeat response is not achieved, as seen in Figure 10c, when an inaccurate machine parameter is used for the implementation of DB-DTFC. Though forced oscillation occurs, the oscillation does not continue for a long time. This means that the dynamic performance of DB-DTFC is still faster than that of CVC, regardless of the forced oscillation.

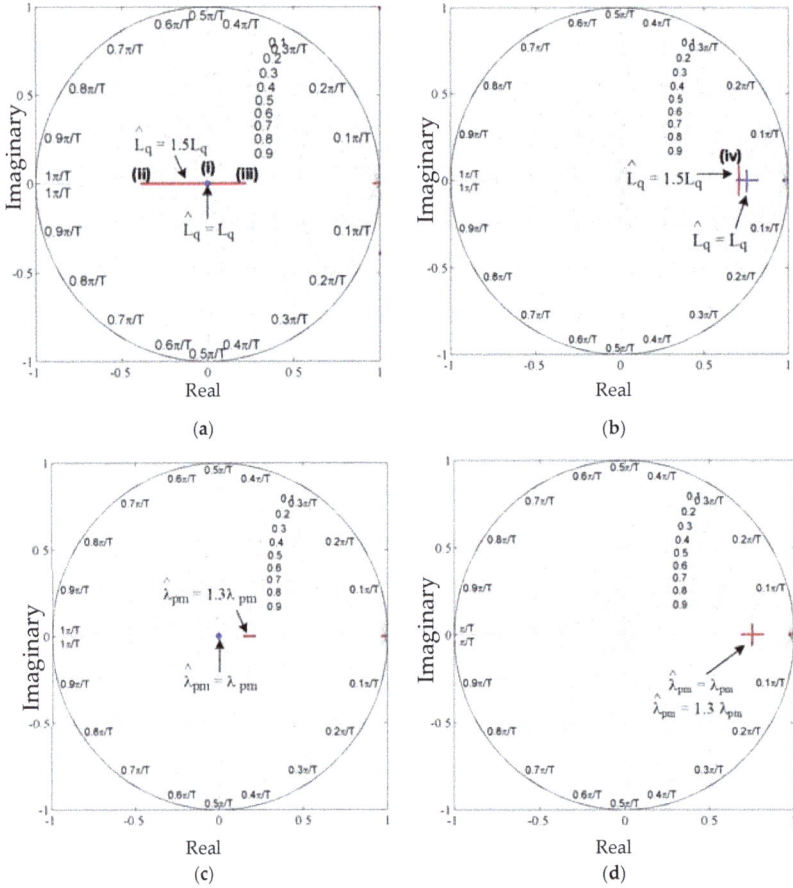

Figure 9. Eigenvalue (EV) migration of DB-DTFC and CVC along the U.S.06 driving cycle when $\hat{L}_q$ and $\hat{\lambda}_{pm}$ are detuned by 50% and 30%, respectively. (**a**) When L_q is detuned in DB-DTFC; (**b**) When L_q is detuned in CVC; (**c**) When λ_{pm} is detuned in DB-DTFC; (**d**) When λ_{pm} is detuned in CVC.

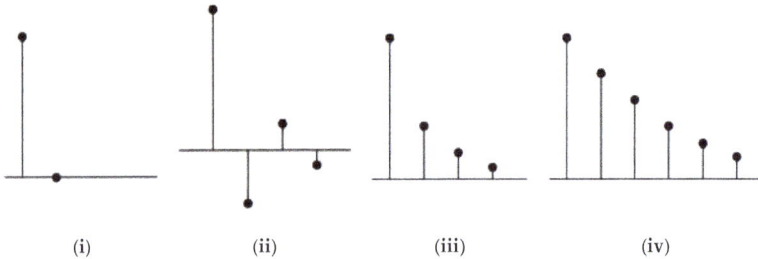

Figure 10. Corresponding impulse responses with respect to EV locations specified in Figure 9a,b. (**i**) deadbeat response; (**ii**) forced oscillation; (**iii**) underdamped response of DB-DTFC; (**iv**) underdamped response of CVC.

5. Experimental Results

Following simulation, the eigenvalue migration of DB-DTFC and CVC IPMSM drives is implemented and verified experimentally. The experimental test set-up is shown in Figure 11 and the specifications of the test IPMSM are summarized in Table 2.

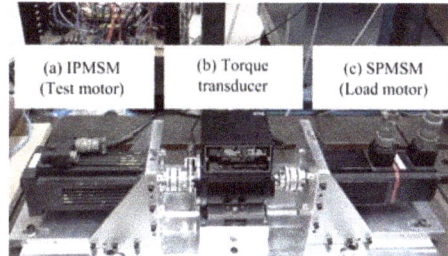

Figure 11. Experimental test set-up.

Table 2. IPMSM (interior permanent magnet synchronous motor) specifications.

Parameter	Value	Parameter	Value
R_s	1.4 (Ω)	Rated Power	1.5 (kW)
λ_{pm}	0.121 (Volt-s)	Rate Torque	2.26 (N.m.)
L_d	8.5 (mH)	Rated Speed	6200 (rpm)
L_q	20 (mH)	Rated current (Continuous)	5.5 (A)
Poles	4	Max. current (Instantaneous)	17 (A)
T_s	100 (μs)	J_p	1.0 (kg·m$^2 \times 10^{-4}$)

In Figure 11, the test IPMSM is controlled in a torque control mode and a load SPMSM is controlled in a speed control mode. Between the IPMSM and the SPMSM in Figure 11, a torque detector (SS-505, Ono Sokki, Yokohama, Japan) is mounted and a shaft torque of the IPMSM is measured; the torque measurement is displayed through a torque meter. During the experiment, an AIX DSP control development system (XCS2000, Analog Devices, Norwood, MA, United States) is used for controlling and sensing signals. Control algorithms are developed using C++ language and downloaded to a DSP. The line current of a motor is measured by a hall-type LEM LA 55-P current sensor and 10 kHz of PWM sampling frequency is applied during experiment. For an inverter drive, a Semikron power converter (Semitop Stack, Semikron, Nuremberg, Germany) rated for 350 V$_{dc}$ and 26 A is used. Figure 12 shows the Semikron power converter used during the experiment.

For stable operation of CVC and DB-DTFC motor drives, the US06 SFTP automotive driving cycle was applied. Rather than using actual torque and speed trajectories, the trajectories of the US06 driving cycle are scaled. During the experiment, a reduced DC link voltage is applied for safe operation of the test IPMSM drive, the load machine drive, and the torque sensor. The time range is also scaled down so that the saved data size does not exceed the memory capacity of a controller. The experimental results for the US06 SFTP automotive driving cycle using DB-DTFC are shown in Figure 13.

The torque trajectory increases up to the rated torque of the test IPMSM and the speed trajectory is increased beyond the rated speed of the test IPMSM as shown in Figure 13a,b,f. Also, the voltage and current limited operation can be observed in Figure 13c,d when the test motor drive is operated along the US06 automotive driving cycle. In Figure 13d, a rated current of the test IPMSM is marked as the current limit. A city driving cycle and a high driving cycle are combined in the US06 SFTP. Therefore, it is observed in Figure 13f that the test IPMSM operates at both low speeds and high torque conditions and at high speeds and low torque operating conditions, which correspond to a city driving cycle and a highway driving cycle, respectively.

Figure 12. A Semikron power converter.

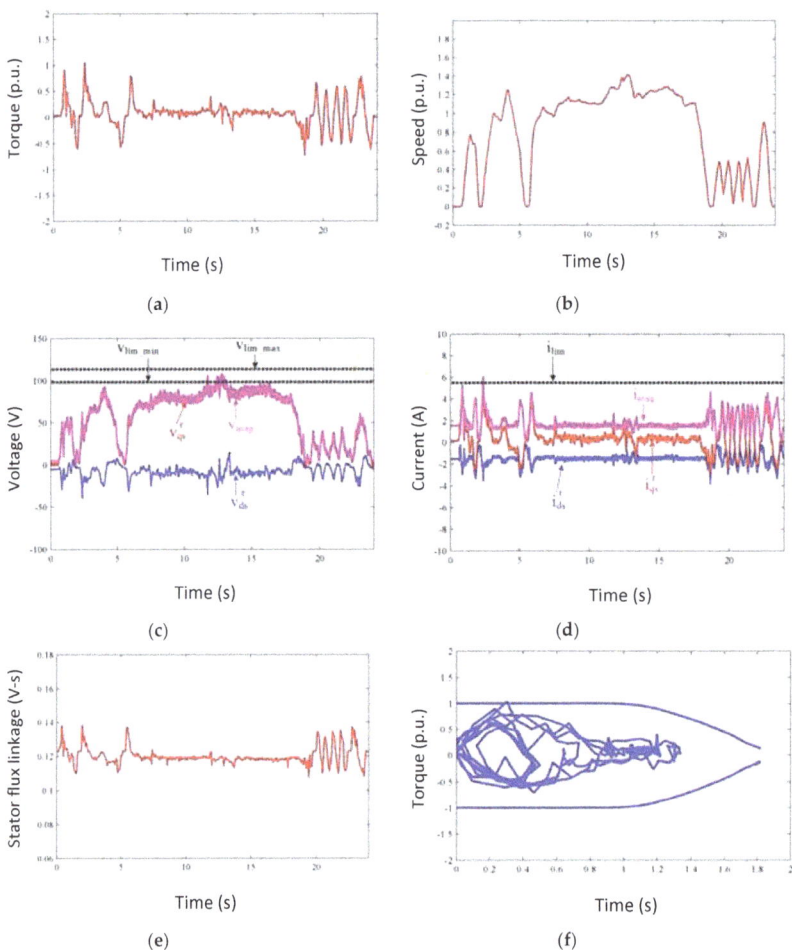

Figure 13. Experimental results of dynamics of the DB-DTFC IPMSM drive along the US06 SFTP driving cycle. Operating conditions: f_{SW} = 10 (kHz) and V_{dc} = 170 (V). (**a**) Torque; (**b**) Speed; (**c**) Voltage; (**d**) Current; (**e**) Stator flux linkage; (**f**) Torque-speed.

The operating point (small signal) models for the DB-DTFC and CVC IPMSM drives are derived in Equations (21) and (28), respectively. The EV migration of each motor drive is analyzed using the operating point (small signal) models derived for DB-DTFC and CVC IPMSM drives in Equations (21) and (28) and the experimental data shown in Figure 14. For evaluation of the parameter sensitivity of eigenvalue migration for each control algorithm, the q-axis inductance is detuned by 50% of its actual value and the permanent magnetic flux linkage is detuned by 30% of its actual value. The EV migration results of DB-DTFC and CVC IPMSM drives are shown in Figure 14.

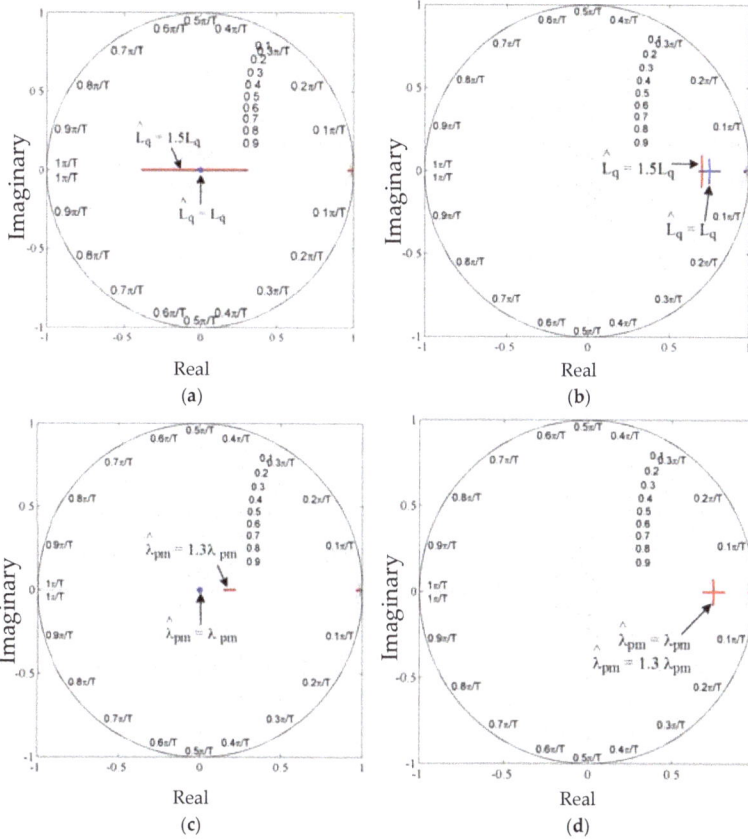

Figure 14. Eigenvalue (EV) migration of DB-DTFC and CVC along the US06 driving cycle when $\hat{L}_q$ and $\hat{\lambda}_{pm}$ are detuned by 50% and 30%, respectively. (**a**) When L_q is detuned in DB-DTFC; (**b**) When L_q is detuned in CVC; (**c**) When λ_{pm} is detuned in DB-DTFC; (**d**) When λ_{pm} is detuned in CVC.

Both the DB-DTFC and the CVC IPMSM drives are shown to be operating at a stable region when the torque and speed trajectories for the US06 automotive driving cycle are applied because the eigenvalues are located within the unit circle in the z-plain, as seen in Figure 14. As seen in Figure 14a,c, a deadbeat response can be achieved when the accurate estimated parameter is used. When the machine parameters, a permanent magnetic flux linkage and a q-axis inductance, are detuned, the DB-DTFC IPMSM drive still shows faster dynamic performance than the CVC IPMSM drive. However, a deadbeat response cannot be perfectly achieved, and a well-damped forced oscillation property is observed in the DB-DTFC IPMSM drive.

6. Conclusions

In this paper, the stability and dynamic characteristics of DB-DTFC and CVC IPMSM drives are analyzed with respect to machine parameter variations. Among the stability evaluation methods reviewed in Section 1, the EV migration method is applied to investigate both the stability and dynamic characteristics of each control algorithm along torque and speed trajectories of the US06 SFTP driving cycle. For eigenvalue migration of a nonlinear system, an operating point model is required. Since an IPMSM drive is a nonlinear system, the operating point models for IPMSM itself, DB-DTFC, and CVC IPMSM drive systems are derived in this paper. Using the operating point models and a driving cycle, eigenvalue migration of DB-DTFC and CVC IPMSM drives is investigated along the torque and speed trajectories of the US06 SFTP driving cycle in simulation and experiment. The simulation and experimental results show that both control systems operate within a stable region regardless of the parameter variation over the wide operating space of a motor drive system. From the simulation and experimental results of this paper, it can be concluded that the stability of a DB-DTFC IPMSM drive with respect to parameter variation is not a critical issue even though a DB-DTFC algorithm is developed for an IPMSM drive based on a machine model and parameters. When inaccurate estimated machine parameters are used for a DB-DTFC algorithm, EV migration in simulations and experiments shows that forced oscillation and well-damped responses are observed instead of a deadbeat response. However, the DB-DTFC IPMSM drive still shows faster dynamic performance than the CVC IPMSM drive, though the machine parameters, a q-axis inductance, and a permanent magnetic flux linkage are detuned within a stable operating region.

Funding: This research was funded by the Korea Electric Power Corporation [Grant number: R18XA04], the Base Science Research Program through the National Research Foundation of Korea (NRF) by the Ministry of Education (2017R1D1A1B03031979) and research funds for newly appointed professors of Chonbuk National University in 2017.

Acknowledgments: This research was supported by the Korea Electric Power Corporation [Grant number: R18XA04] and the Base Science Research Program through the National Research Foundation of Korea (NRF) by the Ministry of Education (2017R1D1A1B03031979). This research was also supported by research funds for newly appointed professors of Chonbuk National University in 2017.

Conflicts of Interest: The author declares no conflict of interest.

References

1. Habetler, T.G.; Profumo, F.; Pastorelli, M.; Tolbert, L.M. Direct torque control of induction machines using space vector modulation. *IEEE Trans. Ind. Appl.* **1992**, *28*, 1045–1053. [CrossRef]
2. Kenny, B.H.; Lorenz, R.D. Stator- and rotor-flux-based deadbeat direct torque control of induction machines. *IEEE Trans. Ind. Appl.* **2003**, *39*, 1093–1101. [CrossRef]
3. Xu, W.; Lorenz, R.D. Low-sampling-frequency stator flux linkage observer for interior permanent-magnet synchronous machines. *IEEE Trans. Ind. Appl.* **2015**, *51*, 3932–3942. [CrossRef]
4. Nie, Y.; Brown, I.P.; Ludois, D.C. Deadbeat-Direct Torque and Flux Control for Wound Field Synchronous Machines. *IEEE Trans Ind. Electr.* **2018**, *65*, 2069–2079. [CrossRef]
5. Saur, M.; Ramos, F.; Perez, A.; Gerling, D.; Lorenz, R.D. Implementation of deadbeat-direct torque and flux control for synchronous reluctance machines to minimize loss each switching period. In Proceedings of the 2016 IEEE Applied Power Electronics Conference and Exposition (APEC), Long Beach, CA, USA, 20–24 March 2016.
6. Wang, K.; Lorenz, R.D.; Baloch, N.A. Improvement of back-emf self-sensing for induction machines when using deadbeat-direct torque and flux control (db-dtfc). In Proceedings of the 2016 IEEE Energy Conversion Congress and Exposition (ECCE), Milwaukee, WI, USA, 18–22 September 2016.
7. Wang, K.; Lorenz, R.D.; Baloch, N.A. Enhanced methodology for injection based real time parameter estimation to improve back-emf self-sensing in induction machine deadbeat direct torque and flux control drives. In Proceedings of the 2017 IEEE Energy Conversion Congress and Exposition (ECCE), Cincinnati, OH, USA, 1–5 October 2017.

8. Pulvirenti, M.; Scarcella, G.; Scelba, G.; Lorenz, R.D. Fault tolerant capability of deadbeat—direct torque and flux control for three-phase pmsm drives. In Proceedings of the 2016 IEEE Energy Conversion Congress and Exposition (ECCE), Milwaukee, WI, USA, 18–22 September 2016.

9. Kim, H.; Lorenz, R.D. Using on-line parameter estimation to improve efficiency of IPM machine drives. In Proceedings of the 2002 IEEE 33rd Annual IEEE Power Electronics Specialists Conference, Cairns, QLD, Australia, 23–27 June 2002.

10. Alonge, F.; D'lppolito, F.; Feranta, G.; Ramondi, F.M. Parameter identification of induction motor model using genetic algorithms. *IEE Proc. Control Theor. Appl.* **1998**, *145*, 587–593. [CrossRef]

11. Mercorelli, P. Parameters identification in a permanent magnet three-phase synchronous motor of a city-bus for an intelligent drive assistant. *Int. J. Model. Identif. Control.* **2014**, *21*, 352. [CrossRef]

12. Heinbokel, B.; Lorenz, R.D. Robustness evaluation of deadbeat direct torque and flux control for induction machine driver. In Proceedings of the 2009 13th European Conference on Power Electronics and Applications, Barcelona, Spain, 8–10 September 2009.

13. Lee, J.S.; Lorenz, R.D. Robustness analysis of deadbeat-direct torque and flux control for ipmsm drives. *IEEE Trans. Ind. Electr.* **2016**, *63*, 2775–2784. [CrossRef]

14. Chen, L.; Mercorelli, P.; Liu, S. A Kalman estimator for detecting repetitive disturbance. In Proceedings of the 2005 American Control Conference, Portland, OR, USA, 8–10 June 2005.

15. Liu, Z.; Wang, Q. Robust control of electrical machines with load uncertainty. *J. Electr. Eng.* **2005**, *59*, 760–765.

16. Wai, R.J. Robust control for induction servo motor drive. *IEE Proc. Electr. Power Appl.* **2001**, *148*, 279–286. [CrossRef]

17. Vaclavek, P.; Blaha, P. Lyapunov function based design of PMSM state observer for sensorless control. In Proceedings of the 2009 IEEE Symposium on Industrial Electronics & Applications, Kuala Lumpur, Malaysia, 4–6 October 2009.

18. Qiang, S.; Xingming, Z.; Hongzhi, C. Limitation of the load angle for direct torque control of PMSM. In Proceedings of the 2014 IEEE Transportation Electrification Asis-Pacific (ITEC Asia-Pacific), Beijing, China, 31 August–3 September 2014.

19. Luukko, J.; Pyrhonen, O.; Niemela, M.; Pyrhonen, J. Limitation of the load angle in a direct-torque-controlled synchronous machine drive. *IEEE Trans. Ind. Electr.* **2004**, *51*, 793–798. [CrossRef]

20. Tsuji, M.; Kojima, K.; Mangindaan, G.M.C.; Akafuji, D.; Hamasaki, S. Stability study of a permanent magnet synchronous motor sensorless vector control system based on extended emf model. *IEEJ J. Ind. Appl.* **2012**, *1*, 148–154. [CrossRef]

21. Tsuji, M.; Mizusaki, H.; Hamasaki, S. Stability comparison of ipmsm sensorless vector control systems using extended emf. In Proceedings of the 2014 International Power Electronics Conference (IPEC-Hiroshima 2014-ECCE ASIA), Hiroshima, Japan, 18–21 May 2014. [CrossRef]

22. Guha, A.; Narayanan, G. Small-signal stability analysis of an open-loop induction motor drive including the effect of inverter deadtime. *IEEE Trans. Ind. Appl.* **2016**, *52*, 242–253. [CrossRef]

23. Grdenić, G.; Delimar, M. Small-Signal Stability Analysis of Interaction Modes in VSC MTDC Systems with Voltage Margin Control. *Energies* **2017**, *10*, 873. [CrossRef]

24. Zhang, B.; Yan, X.; Li, D.; Zhang, X.; Han, J.; Xiao, X. Stable Operation and Small-Signal Analysis of Multiple Parallel DG Inverters Based on a Virtual Synchronous Generator Scheme. *Energies* **2018**, *11*, 203. [CrossRef]

25. Park, C.; Lee, J. A torque error compensation algorithm for surface mounted permanent magnet synchronous machines with respect to magnet temperature variations. *Energies* **2017**, *10*, 1365. [CrossRef]

26. Slotine, J.; Li, W. *Applied Nonlinear Control*; Prentice Hall: Upper Saddle River, NJ, USA, 1991; p. 199.

27. Lee, J.S.; Lorenz, R.D. Deadbeat-direct torque and flux control of ipmsm drives using a minimum time ramp trajectory method at voltage and current limits. In Proceedings of the 2013 IEEE Energy Conversion Congress and Exposition, Denver, CO, USA, 15–19 September 2013. [CrossRef]

MDPI

St. Alban-Anlage 66

4052 Basel

Switzerland

Tel. +41 61 683 77 34

Fax +41 61 302 89 18

www.mdpi.com

Energies Editorial Office

E-mail: energies@mdpi.com

www.mdpi.com/journal/energies

www.ingramcontent.com/pod-product-compliance
Lightning Source LLC
Chambersburg PA
CBHW051721210326
41597CB00032B/5561